AA001150

2008
International Symposium on System-on-Chip Proceedings

Editors:

Jari Nurmi

Jarmo Takala

Olli Vainio

Tampere University of Technology
Tampere, Finland

Copyright © 2008 by the Institute of Electrical and Electronic Engineers, Inc
All Rights Reserved

Copyright and Reprint Permissions: Abstracting is permitted with credit to the source. Libraries are permitted to photocopy beyond the limit of U.S. copyright law for private use of patrons those articles in this volume that carry a code at the bottom of the first page, provided the per-copy fee indicated in the code is paid through Copyright Clearance Center, 222 Rosewood Drive, Danvers, MA 01923.

For other copying, reprint or republication permission, write to IEEE Copyrights Manager, IEEE Service Center, 445 Hoes Lane, Piscataway, NJ 08854. All rights reserved.

***This publication is a representation of what appears in the IEEE Digital Libraries. Some format issues inherent in the e-media version may also appear in this print version.**

IEEE Catalog Number: CFP08554-PRT
ISBN 13:   978-1-4244-2541-9
Library of Congress No.: 2008904133

Additional Copies of This Publication Are Available From:

Curran Associates, Inc
57 Morehouse Lane
Red Hook, NY 12571 USA
Phone: (845) 758-0400
Fax: (845) 758-2633
E-mail: curran@proceedings.com

TABLE OF CONTENTS

PER Performance Enhancement through Antenna And Transceiver Co-Design for Multi-band OFDM UWB Communication 1
Peng Wang, Hannu Tenhunen, Dian Zhou, Li-Rong Zheng

Trade-offs in Mapping High-level Dataflow Graphs onto ASIPs 6
V.Guzma, S.S. Bhattacharyya, P. Kellomaki, J. Takala

An ASIC-Design-Based Configurable SOC Architecture for Networked Media 10
Ning Ma, Zhibo Pang, Hannu Tenhunen, Li-Rong Zheng

Congestion-aware Task Mapping in Heterogeneous MPSoCs 14
Ewerson Carvalho, Fernando Moraes

Optimizing Routing Tables on Systems-on-Chip with Content–Addressable Memories 18
Stergios Stergiou, Jawahar Jain

Balancing Wrapper Chains of SoC Core Based on Best Interchange Decreasing 24
Maoxiang Yi, Huaguo Liang, Zhengfeng Huang

Area-Efficient Low-Cost Low-Dropout Regulators Using MOS Capacitors 28
Hamed Aminzadeh, Reza Lotfi, Khalil Mafinezhad

A 65nm CMOS Down-Sampling Micromixer with Enhanced DC Current Capability 32
Kurt Schweiger, Horst Zimmermann

A 1V Current-Mode Filter in 65nm CMOS Using Capacitance Multiplication 36
Heimo Uhrmann, Horst Zimmermann

FPGA Implementation of a 2G Fibre Channel Link Encryptor with Authenticated Encryption Mode GCM 40
L. Henzen, F. Carbognani, N. Felber, W. Fichtner

On the Credibility of Load-latency Measurement of Network-on-chips 44
Erno Salminen, Ari Kulmala, Timo D. Hamalainen

Realizing a flexible constraint length Viterbi decoder for software radio on a de Bruijn interconnection network 51
Ganesh Garga, Mythri Alle, Keshavan Varadarajan, S.K Nandy, H.S Jamadagni

Tlmco-simulation for an Open Source Mpsoc Platformunder Starsoc Environment 55
Sami Boukhechem, El-Bay Bourennane

A Flexible Modeling and Simulation Framework for Design Space Exploration 61
Camille Jalier, Didier Lattard, Gilles Sassatelli

Energy Analysis of Re-Injection Based Deadlock Recovery Routing Algorithms 65
H. Kooti, M. Mirza-Aghatabar, S. Hessabi, A. Tavakkol

UML profile for Estimating Application Worst Case Execution Time on System-On-Chip 69
Fateh Boutekkouk, Sébastien Bilavarn, Michel Auguin, Mohammed Benmohammed

Specification of GNSS Application for Multiprocessor Platform 75
Heikki Hurskainen, Jussi Raasakka, Jari Nurmi

Inherent Reliability Evaluation of Networks-on-Chip Based on Analytical Models 81
Mojtaba Valinataj, Siamak Mohammadi, Saeed Safari

Implementation of W-CDMA Slot Synchronization on a Reconfigurable System-on-Chip 85
Fabio Garzia, Claudio Brunelli, Carmelo Giliberto, Roberto Airoldi, Jari Nurmi

FlexPath NP - A Network Processor Architecture with Flexible Processing Paths...............89
Michael Meitinger, Rainer Ohlendorf, Thomas Wild, Andreas Herkersdorf

Micronmesh for Fault-Tolerant GALS Multiprocessors on FPGA...............95
Heikki Kariniemi, Jari Nurmi

Configuring Smart Objects over Cognitive Radio...............103
K. Nikunen, H. Heusala, J. Komulainen

Real-Time Execution Monitoring on Multi-Processor System-on-Chip...............107
Kalle Holma, Tero Arpinen, Erno Salminen, Marko Hannikainen, Timo D. Hamalainen

Using Soft Processors for Component Design in SOC: A Case-Study of Timers...............113
M. Ortiz, M. Brox, F. Quiles, A. Gersnoviez, C. Moreno, M. Montijano

Synthesis for Variable Pipelined Function Units...............119
Yosi Ben-Asher, Nadav Rotem

A Two-Phase Return-to-Zero (RZ) Asynchronous Transceiver Circuit for Pipe-Lined SoC Interconnects...............121
Muhammad E. S. Elrabaa

High Resolution Flash Time-to-Digital Converter with Sub-Picosecond Measurement Capabilities...............125
Nikolaos Minas, David Kinniment, Gordon Russell, Alex Yakovlev

RF Transmitter Architecture Investigation for Power Efficient Mobile WiMAX Applications...............129
Liang Rong, Fredrik Jonsson, Lirong Zheng, Mats Carlsson, Charlotta Hedenäs

Evaluation of Heterogeneous Multiprocessor Architectures by Energy and Performance Optimization...............133
Heikki Orsila, Erno Salminen, Marko Hannikainen, Timo D. Hamalainen

Integrating High Speed Multipliers in Coarse Grain Reconfigurable Arrays...............139
Stavros Georgiopoulos, Grigoris Dimitroulakos, Costas E. Goutis

Analyzing Models of Computation for Software Defined Radio Applications...............143
Heikki Berg, Claudio Brunelli, Ulf Lücking

Multi-Objective Genetic Optimized Multiprocessor SoC Design...............147
Mohammad Arjomand, Hamid Sarbazi-Azad, S. Hamid Amiri

A 110 dB, 3-mW Fourth-order Σ-Δ Modulator for Atmospheric Pressure Sensor...............151
Taeyoon Kim, Wonki Park, Heesun Ahn, Kyongwon Min, Sangyong Lee, Jongchan Choi, Chulwoo Kim, Kynnyun Kim, Sungchul Lee

A State Based Framework for Efficient System-level Power Estimation of Of Custum Reconfigurable Cores...............155
Ali Ahmadinia, Balal Ahmad, Tughrul Arslan

Low Noise Amplifier Architecture Analysis for UWB System...............159
Peng Wang, Fredrik Jonsson, Dian Zhou, Li-Rong Zheng

Impact of Power-Management Granularity on The Energy-Quality Trade-off for Soft And Hard Real-Time Applications...............163
Aleksandar Milutinovic, Kees Goossens, Gerard J.M. Smit

Author Index

Foreword

International Symposium on System-on-Chip 2008 is the 10th annual SoC event in Tampere, it builds on a tradition started back in 1999. In addition to the invited lectures, commercial exhibit and vendor programme, the conference is open for contributions from researchers on this broad but focused field. The symposium also features a panel discussion on topics of high interest to the SoC community.This will reflect the theme of the year, Software-Defined and Cognitive Radio. The mission of SoC 2008 is to provide a forum that is fully and comprehensively dedicated to SoC. We enjoy the privilege to have IEEE Circuits and Systems Society as our technical co-sponsor.

The event was the first to use solely "SoC" as its name and focus. Later on, many counterparts have emerged worldwide, adopting this magnificent abbreviation in their names. Still, it is the major international SoC event in the Northern Europe, equally appreciated by the companies and academics in Europe but also increasingly in Americas and Far East. This is also reflected in the spectrum of countries where the papers presented in SoC 2008 originate from, they come from 19 countries all over the world (first author). We think that a very interesting thing is that the four top countries are Finland, Iran, Sweden, and France. Also Austria, Belgium, Greece, UK and USA came in strongly. The original submissions represented 21 countries. Thanks to all contributors for their submissions, whether or not exceeding the publication threshold this time.

We would like to acknowledge the sponsorship received from Nokia Corporation, Xilinx, and IEEE Finland Section. Even more than that, we are especially pleased about the presence of numerous Nokia representatives in the event, which has also become a tradition. We also thank the exhibitors for their support.

We are grateful to the technical program committee members and other reviewers of the submitted papers, with their help we could provide valuable feedback to the authors to improve the quality of the Proceedings. Last but definitely not least, we extend our thanks to the invited speakers of this year. Traditionally, the backbone of the event has been formed by the invited talks. We believe that with the selected seven distinguished people from the academy and industry all over the world, the event will be in a good shape. They all approach the theme of the year from different viewpoints.

Thanks also to our steering committee comprising Prof. Jan Rabaey, Prof. Heinrich Meyr, Prof. Hannu Tenhunen, and Dr. Mika Kuulusa, chaired by the permanent general chair Prof. Jari Nurmi.

Welcome to Tampere, the SoC City!

The General Chair   The Scientific Program Chair   The Proceedings Chair
Jari Nurmi   Jarmo Takala   Olli Vainio

Organizing Committee

General Chairman

| Jari Nurmi | Tampere University of Technology | Finland |

Scientific Program Chair

| Jarmo Takala | Tampere University of Technology | Finland |

Proceedings Chair

| Olli Vainio | Tampere University of Technology | Finland |

Exhibit and Sponsor Chair

| Timo Rintakoski | Tampere University of Technology | Finland |

Scientific Program Committee

Andrea Aquaviva	University of Verona	Italy
Brian Bailey	independent consultant	USA
Davide Bertozzi	University of Ferrara	Italy
Shuvra S. Bhattacharyya	University of Maryland	USA
Abdelhafid Bouhraoua	KFUPM	Saudi-Arabia
Fabio Campi	ST Microelectronics	Italy
Jean-Luc Dekeyser	LIFL	France
William Fornaciari	Politecnico di Torino	Italy
Kees Goossens	NXP Semiconductors	The Netherlands
Tariq Jamil	Sultan Qaboos University	Oman
Murali Jayapala	IMEC	Belgium
Kimmo Kuusilinna	Nokia	Finland
Vesa Lahtinen	Nokia	Finland
Steve Leibson	Tensilica	USA
Rainer Leupers	RWTH Aachen	Germany
Oz Levia	independent consultant	USA
Tobias Noll	RWTH Aachen	Germany
Wei Qin	Boston University	USA
Tero Rissa	Nokia	Finland
Stefan Rusu	Intel	USA
Gerard J.M. Smit	University of Twente	The Netherlands
Dirk Stroobandt	Ghent University	Belgium
Wonyong Sung	Seoul National University	Korea
Lionel Torres	LIRMM	France

External Reviewers Appointed
(in addition to Scientific Program Committee members and chairmen)

Tapani Ahonen	Tampere University of Technology	Finland
Claudio Brunelli	Nokia	Finland
Fabio Garzia	Tampere University of Technology	Finland
Hannu Heusala	University of Oulu	Finland
Kalle Holma	Tampere University of Technology	Finland
Timo D. Hämäläinen	Tampere University of Technology	Finland
Marko Hännikäinen	Tampere University of Technology	Finland
Pekka Jääskeläinen	Tampere University of Technology	Finland
Heikki Kariniemi	Tampere University of Technology	Finland
Pasi Liljeberg	University of Turku	Finland
Dragomir Milojevic	Université Libre de Bruxelles	Belgium
Heikki Orsila	Tampere University of Technology	Finland
Yang Qu	Nokia	Finland
Tero Partanen	Tampere University of Technology	Finland
Teemu Pitkänen	Tampere University of Technology	Finland
Ville Rantala	University of Turku	Finland
Perttu Salmela	Tampere University of Technology	Finland
Marco Santambrogio	Politecnico di Milano	Italy
Tero Sihvo	University of Jyväskylä	Finland
Jorma Skyttä	Helsinki University of Technology	Finland
Leandro Soares Indrusiak	TU Darmstadt	Germany
Lars Wanhammar	Linköping University	Sweden

Invited Presentations

KEYNOTE: PAST AND FUTURE OF COGNITIVE RADIO
Joseph Mitola, III, *MITRE, USA*

FLEXIBLE AIR INTERFACE BASEBAND PLATFORM FOR MOBILE SDR
Liesbet van der Perre, *IMEC, Belgium*

SDR – COMPUTERIZING RADIO MODEM
Jukka Wallinheimo, *Nokia, Finland*

FLEXIBLE TRANSCEIVERS BASED ON TIME-FREQUENCY REPRESENTATION THEORY
Harri Saarnisaari, *University of Oulu, Finland*

SELF MANAGEMENT OF COGNITIVE RADIO NETWORKS
Eleni Patouni, *University of Athens, Greece*

Sponsors of the Symposium

Technical Co-Sponsorship

Financial Sponsors

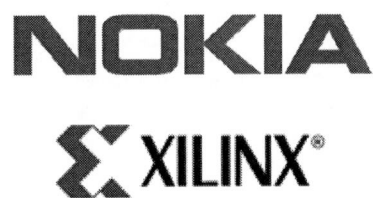

Exhibitors

Delta

Mentor Graphics

Synopsys

Target Compiler Technology

Xilinx

PER Performance Enhancement through Antenna And Transceiver Co-Design for Multi-band OFDM UWB Communication

Peng Wang[1,2], Hannu Tenhunen[1], Dian Zhou[2], Li-Rong Zheng[1]

[1]Royal Institute of Technology (KTH), ECS/ICT, ELECTRUM 229, SE-164 40 Kista-Stockholm, Sweden
Email: {pengw, hannu, and lirong}@kth.se

[2]School of Microelectronics of Fudan University, 825 Zhangheng Road, Shanghai 201203, China
Email: zhoud@fudan.edu.cn

Abstract— **This paper investigates the packet error rate (PER) performance enhancement through the antenna and the transceiver co-design for MB-OFDM UWB system. Five different UWB antennas, covering the whole UWB spectrum, are selected for study. Through the link-margin analysis and PER performance simulation, radio transceiver design specifications are optimized according to different antennas' performance at different band groups. Transmitter pre-distortion and receiver equalization are applied at the front-end of the transceiver to co-design with the antenna, which further enhance PER performance. Our study reveals that, antenna is an important part of radio transceiver which has to be considered in advance during the chip design [1], particularly in the ultra-wide band radio system. Through the antenna and radio front-end co-design, not only could PER performance be enhanced, but also could design parameters be relaxed for the power amplifier (PA) and the low noise amplifier (LNA), particularly for higher band groups.**

Index Terms— **PER, Antenna, Transceiver, Co-design, UWB.**

I. INTRODUCTION

The potential for UWB to provide solutions of wireless communications under the indoor environments lies in its use of extremely wide transmission bandwidths, high data rate and low power consumption. Federal Communication Commission (FCC) requires that all UWB devices occupy more than 500 MHz of sub-bandwidth through 3.1-10.6GHz band, while the UWB physical layer standard [2] divides the whole available ultra wideband spectrum into 14 sub-bands. There are 6 band groups, band group 1 is mandatory, other groups are optional.

Most of the conducted researches were focused on group 1(3.1-4.8GHz). Band group 2, 3, and 6 are included in emerging chipsets. When considering circuit implementation for higher band groups, higher frequency fading resulting in more path loss in link budget and lower performance of CMOS devices will push us to harder design challenges. In a traditional design approach, chip designers assume that UWB antenna is omni-directional with 0dBi gain. All the physical layer analysis, including the link budget, transceiver architecture, and circuit implementation, are based on such an assumption. This facilitates the design process by clearly dividing a boundary between chip designers and antenna designers. On the other hand, it also puts both into a difficult situation that RF chip designers do a tough work to realize the circuits with good performance using CMOS technology, while antenna designers try their whole efforts to approach an ideal 0dBi UWB antenna (isotropic antenna, the frequency independent antennas [3]).

In this paper, we consider the performance of the frequency dependent antenna for circuit design specifications. We analyze the system performance and make the link budget through the antenna and radio front-end co-design. This will not only improve UWB spectrum efficiency under FCC regulation, but also relax circuit design specifications, and thus facilitate the implementation and enhance PER performance. The method is found useful for ultra-wideband system where both CMOS device and the antenna cover a large performance variation through the whole frequency bandwidth. The indoor UWB channel model proposed by IEEE 802.15.3a working group [5] is selected in the simulation. All simulations are completed with Matlab and Simulink in the channel model CM3 (non-line of sight, 4~10m) for 110Mbps data rate with the target PER 8%/500 packets, where each packet includes 1k bytes.

The organization of this paper arranges as follows. First, we introduce UWB architecture and different antennas. Next, we analyze the link margin and PER performance through the antenna-transmitter co-design. This is followed by the analysis results of the antenna-receiver co-design. Conclusions are given in the last chapter.

II. UWB SYSTEM AND CHANNEL MODEL

A. MB-OFDM UWB System Architecture

The fundamental MB-OFDM UWB system architecture is illustrated in Fig.1. Because the baseband is normalized by UWB standard [2], the analog/radio part is mainly focused on to improve the UWB system performance. At the transmitter, a pre-distorter is applied with the tunable power gain which compensates the variations of the antenna gain for each band group to reach the maximum spectrum efficiency under EIRP limitation (-41.25dBm/MHz) ruled by FCC, while improve the power of PA. At the receiver, the antenna by co-designing with the pre-equalizer can compensate the additive path loss due to more frequency fading in higher band groups and relax the requirements of the low noise amplifier.

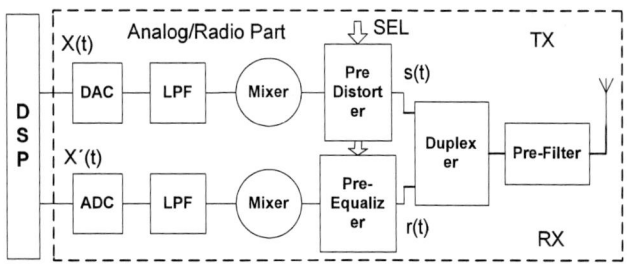

Fig. 1. UWB System Architecture

B. UWB Antennas

For the narrow band systems, the antenna gain is nearly constant through the whole band, and the isotropic antenna is available. Different from the narrow-band system, for UWB systems, the frequency dependent antennas' gain varies through the whole bandwidth, which can be expressed as follows

$$G_{ANT} = 4\pi A_e / \lambda^2 = 4\pi A_e f^2 / c^2 \qquad (1)$$

Where A_e is the effective aperture, λ is the wavelength, c is the speed of light, f is the signal frequency. From (1), it is evident that antenna's gain is proportional to the square of the signal frequency. This interesting characteristic of antenna can be utilized to co-design with the transceiver front-end. The relationship between the antenna directivity and the link performance for both omni-directional antenna and directional antenna is shown in [4]. Table I gives five kinds of UWB antennas with the varied gain in each band group (here we consider the maximum gain to meet the FCC frequency spectrum mask limitation). Basically the tendency of the antenna gain is in accordance with (1). In this context, details about the co-design will be given in chapter III and IV.

III. TRANSMITTER AND ANTENNA CO-DESIGN

A. Transmitter Architecture Analysis

The architecture of the transmitter with antenna is shown in Fig.2. Equation (2) expresses the function of the co-design through the pre-distorter and the antenna which leads to a flat frequency spectrum envelope sending from the antenna,

$$EIRP \geq P_T = P_i + G_p + G_{ANT} \approx Const \qquad (2)$$

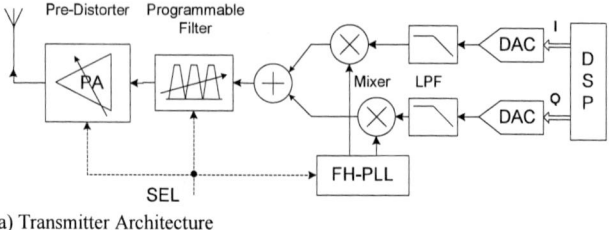

(a) Transmitter Architecture

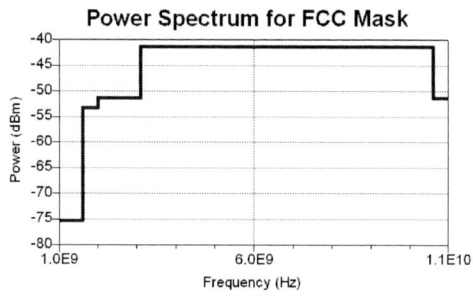

(b) Co-design the power gain between pre-distorter in PA and antenna

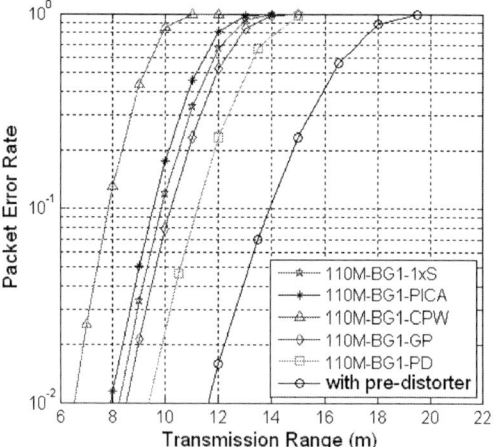

(c) FCC Mask for UWB application which the UWB transmitter has to meet
Fig. 2. Architecture of UWB Transmitter with Pre-Distorter and FCC Mask

Fig. 3. Average PER comparison between different antennas without pre-distorter and the best case with pre-distorter in Band Group 1. Here the antenna at RX is assumed to be omni-direction with 0dBi gain.

TABLE I
SELECTED UWB ANTENNAS WITH GAIN IN EACH BAND GROUP

Antenna Band Group Gain (dBi)	Group 1 (3168~4752 MHz)	Group 2 (4752~6336 MHz)	Group 3 (6336~7920 MHz)	Group 4 (7920~9504 MHz)	Group 5 (9504~10560 MHz)
1x Spiral [6]	6.0	6.2	7.0	7.8	9.0
PICA [7]	5.5	6.9	8.2	8.2	8.8
CPW-Fed Monopole [8]	1.6	3.1	3.2	4.7	6.6
Groundplane [9]	3.9	5.0	4.4	4.4	5.6
Planar Dipoles[10]	2.0	1.8	3.0	4.2	4.5

978-1-4244-2541-9/08 $25.00 © 2008 IEEE

In accordance with Fig.2 (b), where P_T is the transmitting power from the antenna terminal, $P_i(P_o)$ is the input (output) power of PA and G_P is the gain of PA. In this way, it is better to improve the spectrum efficiency than the architecture in which the single PA without pre-distorter is connected to the antenna directly. The challenge for UWB PA is the tradeoff between the ultra bandwidth and the power efficiency. The larger the bandwidth of PA, the less matching between PA and the antenna, it results in lower power efficiency of PA and worse frequency spectrum efficiency. For low power UWB application, lower power efficiency means higher power consumption, worse spectrum efficiency decreases the link margin for UWB communication.

B. PER Performance Enhancement

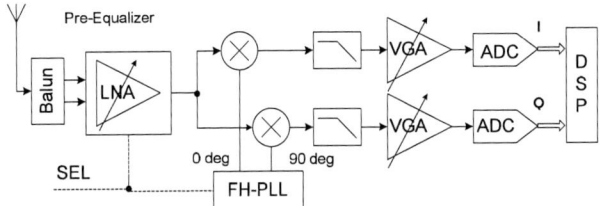

Fig. 4. UWB Receiver Architecture with Pre-Equalizer

Two kinds of transmitter architectures are compared by PER performance through 5 kind UWB antennas. One is the architecture with pre-distorter, and the other is single PA without pre-distorter. Here the single PA model is assumed to have a constant power gain through the whole UWB bandwidth. For the transmitter architecture without the pre-distorter, since (2) is unavailable due to the antenna gain variation, it has to degrade the transmitting spectrum efficiency and the link margin, and cannot follow the EIRP limitation well. Fig.3 compares the PER performance between the transmitter which does not adopt the pre-distorter to co-design with different antennas, and the transmitter with the pre-distorter. From Fig.3, it is obvious that the PER performance is degraded much due to the transmitter architecture without pre-distorter in contrast to the transmitter architecture with pre-distorter. Using pre-distorter, the perfect output spectrum matching with FCC EIRP limitation can be realized.

IV. ANTENNA AND RECEIVER CO-DESIGN

A. Receiver Architecture Analysis

Fig.4 is the receiver architecture with pre-equalizer recommended in this paper. The co-design between antenna and receiver have three advantages: 1. utilize the antenna gain to compensate the extra path loss due to the frequency fading in higher band groups comparing with band group 1, which enables the transmission range to capture the requirement in the standard even for the high end UWB application; 2. relax the requirements of LNA; and 3. provide a flat gain through the antenna and LNA co-design to decrease the required dynamic range for the following circuits.

B. PER Performance Enhancement

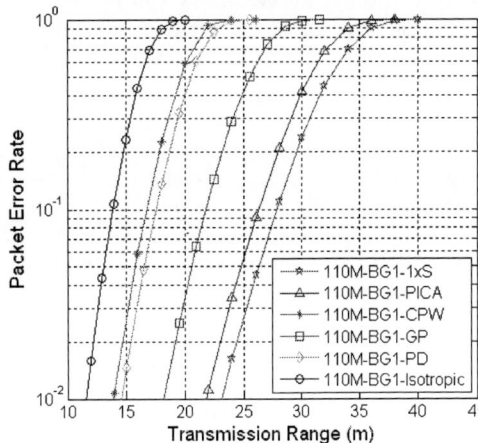

Fig. 5. Enhancement of Average PER in Band Group 1 due to the gain of real Antennas at Receiver comparing with Isotropic Antenna. Here it is assumed the pre-distorter is used at TX, the perfect output frequency spectrum from TX matching with FCC EIRP limitation is realized.

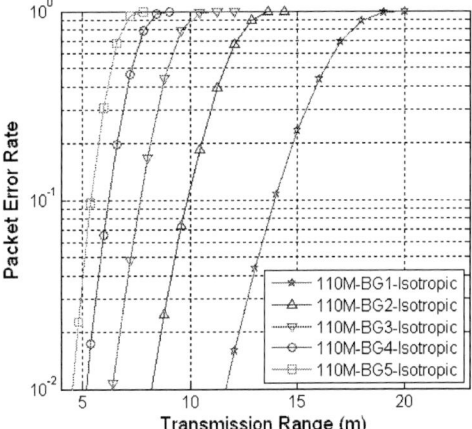

(a). Istropic antenna at both TX and RX for communication

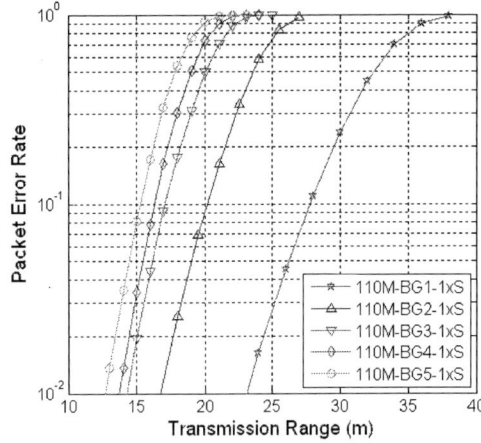

(b). 1x Spiral at RX and with pre-distorter at TX for communication
Fig. 6. (a) Average PER vs. Transmission Range through whole five Band Groups for Isotropic Antenna. (b) Average PER vs. Transmission Range through whole five Band Groups for 1x Spiral Antenna with pre-distorter.

Equation (3) indicates that the antenna gain at the receiver can compensate the path loss due to the frequency fading. For an

978-1-4244-2541-9/08 $25.00 © 2008 IEEE

ideal frequency dependent antenna described in (1), the channel fading caused by the frequency can be compensated completely. Fortunately, the actual antennas have the similar characteristics shown in Table I.

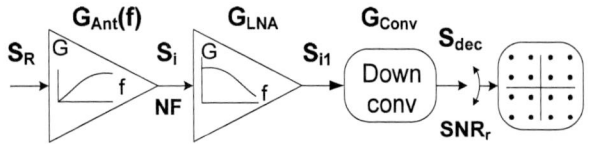

Fig. 7. Co-design through Low Noise Amplifier and Antenna

$$10\log_{10}(G_{ANT}) - L_1 = 10\log10(A_e / 4\pi) \quad (3)$$

In equation (3) $L_1 = 20\log_{10}(4\pi f / c)$ is the path loss at 1m. Comparing with the isotropic antenna, Fig.5 illustrates the PER performance enhancement because the real antennas shown in Table I provide more gain than the isotropic antenna.

Fig.6 illustrates the PER performance of the isotropic antenna and 1xSpiral antenna with the pre-distorter at both transmitter and receiver through the whole five band groups. For other kinds of antennas, the PER performance can be estimated according to Table I and Fig.6.

C. Impact on Low Noise Amplifier Design Parameters

The OFDM UWB system demands the LNA have a gain around 15 dB with the noise figure (NF) less than 4 dB. The couple of parameters are very strict, particularly for noise figure, while it is a big challenge to realize an UWB LNA with flat gain and constant NF through the whole bandwidth. There are two limitations for demodulating the received signal correctly. One is the minimum signal to noise ratio (SNR), and the other is the minimum energy of the signal [11] for detection. Trading off the antenna gain and low noise amplifier parameters together will be significant to make the RF circuits design more comfortable. Fig.7 illustrates the principle of the co-design at the receiver. The expression of signal to noise
ratio in Fig.7 is shown by (4),

$$SNRr = S_R + (G_{ANT} - NF)\big|_{Const} - KT - 10\log_{10} B \quad (4)$$

Where, $SNRr$ is the signal to noise ratio at the receiver before demodulation, S_R is the received signal at the antenna, KT is the thermal noise, and B is the channel bandwidth. The term in brackets indicates the function of co-design through the antenna gain and LNA noise figure. For most LNAs, the NF will increase as the operation frequency is higher. Antenna gain can compensate the extra noise figure at higher band groups to relax the NF for LNA design, and keep the equivalent NF of LNA as a constant.

The co-design between the antenna gain and LNA gain is shown in equation (5).

$$S_{dec} = S_R + (G_{ANT} + G_{LNA})\big|_{Const} + G_{Conv} \quad (5)$$

As shown in Fig.7, S_{dec} is the signal energy before the demodulation, G_{Conv} is the gain after the down-conversion. Due to the parasitic capacitor at the output, LNA gain will be lower at higher frequency band. The term in brackets of (5) shows the compensation for LNA gain if a positive antenna gain is supported. The total gain of antenna and LNA is flat through the co-design, like a pre-equalizer for received signals in different operation band groups. The dynamic range of the latter circuits is relaxed too. Fig.8 illustrates the relation between the link margin, gain, noise figure of LNA, output power of PA, and the 1x spiral antenna gain in five band groups for UWB system. At the transmitter, the pre-distorter can tune the output power of the power amplifier. At the receiver, with the extra link margin due to the antenna gain, the designers can trade off the transmission range, noise figure, and the gain of LNA. With different antennas in the UWB system, the specification of the radio circuits can be changed correspondingly.

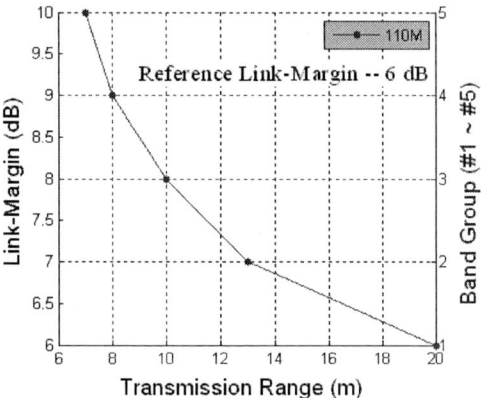

(a) Link Margin vs. Transmission Range with 1x Spiral Antenna at RX and TX with pre-distorter

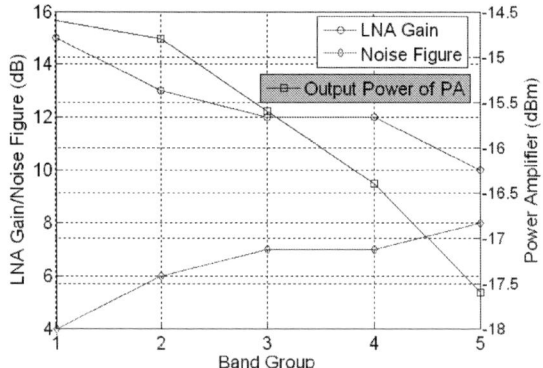

(b) Parameters relation among PA, LNA and 1x Spiral Antenna

Fig. 8.(a) Link Margin vs. Transmission Range with 1x Spiral Antenna in five Band Groups for 110Mbps data rate. (b) Power amplifier and low noise amplifier Parameters Requirements of Co-design with 1x Spiral Antenna in five Band Groups for 110Mbps.

V. CONCLUSION

This paper investigates the PER performance enhancement through the antenna and the transceiver co-design for MB-OFDM UWB system. Five different UWB antennas, covering the whole UWB spectrum, are selected to analyze the link-margin and PER performance. Radio part of the transceiver parameters and the design specifications are optimized according to different antenna performance at different band groups. Transmitter pre-distortion and receiver equalization are applied at the front-end according to the new design specification, which further enhance PER performance. Through the study of co-design between the antenna and the transceiver, it reveals that the antenna is an important part of the transceiver circuits which has to be considered in advance during the chip design. In this paper, we find out that through the antenna and the radio front-end co-design, not only PER performance can be enhanced, but also the design parameters for LNA can be relaxed, particularly for higher band groups.

VI. ACKNOWLEDGMENT

One of the authors (Peng Wang) wishes to acknowledge China scholarship council (CSC) for the study grant and iPack Center at KTH for co-financing. Valuable advices and discussions from members of iPack Center and colleagues at KTH-ECS are highly appreciated. Expecially appreciate the help from Jingfeng Du and Jinliang Huang.

REFERENCES

[1] M. Pelissier, F. Demeestere, F. Hameau, D. Morche, C. Delaveaud, "LNA-Antenna codesign for UWB systems", *Circuits and Systems, 2006. ISCAS 2006. Proceedings. 2006 IEEE International Symposium on*, 21-24 May 2006.

[2] "High Rate Ultra Wideband PHY and MAC Standard," Standard ECMA 368, 2nd Edition, Dec. 2007.

[3] V. Rumsey, "Frequency Independent Antennas," *IRE National Convention Record*, 1957.

[4] H. G. Schantz, "A brief history of UWB antennas", *IEEE Aerospace and Electronic Systems Magazine, Volume 19*, Issue 4, April 2004, pp. 22-26.

[5] J. R. Foerster, M. Pendergrass, and A. F. Molisch, "A channel model for ultrawideband indoor communication," IEEE 802.15.3a standardization group, 2003

[6] M. Karlsson, Gong Shaofang, "An integrated spiral antenna system for UWB", *European Radar Conference*, Oct. 2005, pp. 283-286.

[7] S.-Y. Suh, W. L. Stutzman, and W. A. Davis, "A New Ultra-wideband Printed Monopole Antenna: The Planar Inverted Cone Antenna (PICA)", *IEEE Transactions on Antennas and Propagation, Volume 52*, Issue 5, May 2004. pp. 1361-1364.

[8] Kuan-Jung Hung, Yi-Cheng Lin, "Open-Slot Loaded Monopole Antennas for WLAN and UWB Application", *IEEE Antennas and Propagation Society International Symposium*, 2006, pp. 4653-4656.

[9] Sergio Curto, Matthias John, and Max J. Ammann, "Groundplane Dependent Performance of Printed Antenna for MB-OFDM-UWB", *IEEE 65th Vehicular Technology Conference*, April 2007, pp. 352-356.

[10] Xuanhui Wu, Zhining Chen, "Comparison of Planar Dipoles in UWB Applications", *IEEE Transactions on Antennas and Propagation, Volume 53*, Issue. 6, June 2005, pp. 1973-1983.

[11] Bernard Sklar, "Digital Communications – Fundamentals and Applications", Second Edition, Prentice Hall PTR, January 2001.

978-1-4244-2541-9/08 $25.00 © 2008 IEEE

Trade-offs in Mapping High-level Dataflow Graphs onto ASIPs

Vladimír Guzma*, Shuvra S. Bhattacharyya†, Pertti Kellomäki*, and Jarmo Takala*

*Department of Computer Systems
Tampere University of Technology, Tampere, FI-33720, Finland
{vladimir.guzma, pertti.kellomaki, jarmo.takala}@tut.fi
†Department of Electrical and Computer Engineering
University of Maryland, College Park, MD 20742, USA
ssb@umd.edu

Abstract—**Data-flow based design environments bring advantages of specification, validation and synthesis to embedded systems design by decoupling computation from transfer of data. The former is performed by actors, and data transfer between actors and an execution order of actors is determined by scheduling and buffering strategies. In this work, we examine code sizes and cycle counts resulting from combinations of scheduling and buffering techniques. The experiments were carried out by designing an application specific instruction-set processor streamlined for each of the benchmarks, using a codesign environment called TCE. We also show what additional overhead is introduced when an architecture implemented using our approach is employed for an application outside its targeted domain.**

I. INTRODUCTION

The dataflow programming model represents a program as a set of tasks (actors), and data dependencies (FIFO queues) between actors. Individual actors consume data from their inputs and produce data on their outputs when they are executed. The functionality of whole program is defined by the functionality of the individual actors together with the semantics of their interconnections. In area of digital signal processing (DSP), the applications often work on a streams of data. Therefore, the scheduled dataflow graph needs to be executed in an iterative manner, running within a loop (often an infinite loop) without deadlocks, and using only a finite amount of physical memory.

The synchronous dataflow (SDF) model [1] supports these requirements well for an important class of signal processing applications. With the application written as an SDF graph, the actual work is performed by the actors, while a *schedule* for the graph defines the order in which the actors actors are executed, and also defines requirements for buffer management between actors. Schedules and their associated buffer management requirements in general add some overhead to the execution time, code size and consequently instruction memory, and data memory requirements of an SDF-based application.

In this work, we model benchmarks as SDF graphs using the dataflow interchange format [2] (DIF), which is a tool for developing and experimenting with DSP-oriented dataflow models of computation. We use the DIF-to-C tool [3] to synthesize C code for each of the benchmarks using different combinations of SDF-based scheduling and buffering strategies.

Our generation of ASIPs is done using the TTA Codesign Environment [4] (TCE). We use the TCE compiler to compile synthesized C code for different benchmarks onto different ASIP instances, and we use the TCE simulator [5] to obtain the count of executed instructions.

We also explore the relative quality of critical and non-critical applications in this framework. Specifically, we select one benchmark as being *critical* (highest priority for optimized implementation), and tune the processor architecture to minimize execute cycle count for the critical application. We then recompile other (*non-critical*) benchmarks for the derived architecture, and measure the overhead observed when executing the non-critical benchmarks on a processor that was not tuned specifically for those applications.

Our experiments demonstrate important trade-offs and interactions among SDF-based applications; SDF techniques for scheduling and buffer management; and critical and non-critical application support in ASIPs. Our work also demonstrates a novel design flow that integrates SDF techniques from the DIF environment with ASIP technique from the TCE environment.

II. RELATED WORK

Other approach to introduce dataflow programming to DSP area is presented in SPEX language extension [6]. SPEX adds constructs to programming language (C++ in presented work) to allow programmers to describe inherent parallelism within a DSP system, including describing streaming computation and communication patterns of DSP systems.

Focusing on distributed control and memory, Kahn process networks are used in [7], as a method for programming high-throughput multimedia on a platforms consisting of multiple microprocessors and reconfigurable components. From application written in subset of Matlab, Kahn process network is automatically derived.

In [8] authors present a high-level heterogeneous functional specification and verification part of system for modeling and simulation of embedded systems, *El Greco*. Their system allows specifications in forms of cyclo-static dataflow (CSDF)

978-1-4244-2541-9/08 $25.00 © 2008 IEEE

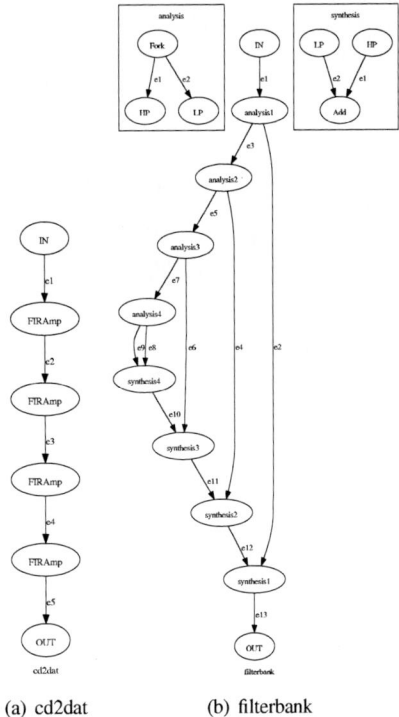

(a) cd2dat   (b) filterbank

Fig. 1. Graphical representation of SDF benchmarks

and hierarchical finite state machines, with ability to nest models at any level.

III. EXPERIMENTAL SETUP

In our experiments, we use four benchmarks from the DSP domain: multi-stage CD-DAT (cd2dat) and DAT-CD (dat2cd) sample rate conversion, a four-level tree-structured filter bank performing the bi-orthogonal wavelet decomposition (fb4bwd), and a JPEG encoder subsystem with RGB-YCbCr, 2d-DCT, Quantization and ZigZag sequencing (jpeg). Filter bank and CD-DAT are shown in Fig. 1 as a SDF applications. We used the DIF-to-C software synthesis framework [3] to produce C code for the applications for all compatible combinations of buffering and scheduling strategies. Our choice of buffering strategies includes:

- buffer sharing [9] (abbreviated BS)
- modified circular buffering [10] (abbreviated C)
- static read-write pointer resetting [1] (abbreviated SRW)
- in-place buffer merging for in-place actors [11] in JPEG (abbreviated IPBM)

Our choice of scheduling strategies includes:

- dynamic programming post optimization SDPPO [9] (abbreviated SD)
- acyclic pairwise grouping of adjacent nodes [12] and DPPO post optimization (abbreviated AD)
- flat scheduling, using topological sorting (abbreviated F)
- recursive procedure call based multiple appearance schedule [13] (abbreviated RPC)

The filter bank benchmark is modeled using a hierarchical SDF graph (with the actors for analysis and synthesis defined as SDF subgraphs), so we apply scheduling on the hierarchical as well as flattened graphs. This gives us two alternative schedules — one that respects the hierarchy of the original application specification, with each SDF subgraph scheduled independently, and another that is not necessarily constrained to follow this hierarchy. We use the TTA Codesign Environment (TCE) [4] to generate target architectures and compile and simulate each benchmark. TCE allows the designer to select the number of function units and what operations each unit performs; the number of register files, including the number of their read and write ports and the number of registers in each of them; the number of transport buses; and how function units and register files are connected to each bus. We use TCE with disabled inlining of procedures to get a clear view of how many cycles are spent in synthesized functions in the code generated by DIF-to-C.

We start by finding an architecture for each benchmark group, determining the minimal cycle count required for each of the benchmarks to process its sample input, and compiling each of the benchmarks for such an architecture. Since the architecture does not change when combinations of scheduling and buffering strategies change for a particular benchmark group, the intra-actor code scheduling is performed in the same way and the number of clock cycles spent inside each actor is unchanged. Therefore in section IV, we only show cycle counts spent in synthesized functions of each benchmark, and leave out the cycle counts for the actors themselves. The synthesized functions are where the actual differences due to choices of scheduling and buffering strategies shows. We also present the code size for the whole application (total) and the code size for the synthesized functions only (synthesized) for each of the benchmarks and strategies combination. We also estimate the area required for each architecture [14]. This estimate does not include the I/O function unit, since I/O is implementation specific. Then we perform experiments related to application criticality, as described in Section IV-B. Specifically, we take the JPEG benchmark (image processing) to be critical, and the cd2dat, dat2cd, fb4bwd benchmarks (audio processing) to be non-critical, and evaluate the overhead — due to lack of dedicated support — incurred when executing the non-critical benchmarks.

IV. EXPERIMENTAL RESULTS

In this section, we first show counts of executed instructions as well as static code sizes for the applications and their synthesized *main* functions. The synthesized *main* functions embody the parts of code that are generated automatically by DIF-to-C; the remaining code is taken from the actor library components associated with the applications. Afterwards, we show results for cd2dat, dat2cd and the filter bank application compiled for an architecture that has been optimized for jpeg, and we show the increases in cycle count and code size resulting from executing these applications with such a non-critical status.

978-1-4244-2541-9/08 $25.00 © 2008 IEEE

TABLE I
CYCLE COUNTS AND NUMBERS OF INSTRUCTIONS

(a) cd2dat

	cycles	synthesized	total
BS-AD, BS-SD	18684	892 (23%)	3933
BS-RPC	16884	5698 (65%)	8739
C-AD	20918	742 (19%)	3772
C-F	34757	738 (19%)	3777
C-RPC	25412	2315 (43%)	5350
SRW-AD, SRW-SD	16912	2130 (41%)	5166
SRW-RPC	22535	1897 (38%)	4938

(b) dat2cd

	cycles	synthesized	total
BS-AD, BS-SD	9836	726 (19%)	3763
BS-RPC	10105	1489 (33%)	4530
C-AD	15134	497 (14%)	3534
C-F	20900	590 (16%)	3628
C-RPC	17445	1421(31%)	4458
SRW-AD, SRW-SD	9816	726 (19%)	3757
SRW-RPC	10700	1456 (32%)	4489

(c) fb4bwd

Hierarchical	cycles	synthesized	total
BS-AD, BS-RPC	5712	2079 (40%)	5127
BS-SD	6042	2258 (42%)	5316
C-AD, C-RPC	6851	2248 (42%)	5288
C-F	8330	3286 (60%)	5388
SRW-AD, SRW-RPC	6075	2077 (40%)	5117
SRW-SD	6087	2273 (43%)	5325
Flat	cycles	synthesized	total
BS-AD, BS-RPC	4208	2351 (43%)	5403
BS-SD	4164	2346 (43%)	5392
C-AD, C-RPC	5957	1978 (39%)	5023
C-F	9475	2803 (48%)	5851
SRW-AD, SRW-RPC	4160	2314 (43%)	5369
SRW-SD	4216	2390 (44%)	5441

(d) jpeg

	cycles	synthesized	total
BS-AD, BS-RPC	8124	603 (27%)	2214
BS-SD	8124	605 (27%)	2216
C-AD, C-RPC	10508	734 (31%)	2341
C-F	19783	1387 (46%)	2996
SRW-AD, SRW-RPC	7926	591 (27%)	2192
SRW-SD	7926	600 (27%)	2201
IPBM-AD	7714	565 (26%)	2148

TABLE II
CYCLE COUNTS AND NUMBERS OF INSTRUCTIONS FOR JPEG ARCHITECTURE TARGET

(a) cd2dat

	cycles	synthesized	total
BS-AD, BS-SD	19615	952 (22%)	4327
BS-RPC	17547	5943 (64%)	9315
C-AD	22028	794 (19%)	4168
C-F	36756	803 (19%)	4180
C-RPC	27765	2469 (42%)	5848
SRW-AD, SRW-SD	17677	2223 (39%)	5610
SRW-RPC	23427	2008 (37%)	5388

(b) dat2cd

	cycles	synthesized	total
BS-AD, BS-SD	10452	777 (19%)	4156
BS-RPC	10773	1585 (32%)	4959
C-AD	16916	539 (13%)	3922
C-F	22637	643 (16%)	4027
C-RPC	19623	1529 (31%)	4901
SRW-AD, SRW-SD	10429	779 (18%)	4167
SRW-RPC	10898	1568 (32%)	4947

(c) fb4bwd

Hierarchical	cycles	synthesized	total
BS-AD, BS-RPC	6334	2202 (39%)	5586
BS-SD	6311	2395 (41%)	5789
C-AD, C-RPC	7225	2366 (41%)	5758
C-F	8863	3355 (57%)	5876
SRW-AD, SRW-RPC	6311	2181 (39%)	5567
SRW-SD	6339	2412 (42%)	5809
Flat	cycles	synthesized	total
BS-AD, BS-RPC	4377	2454 (42%)	5835
BS-SD	4345	2457 (42%)	5849
C-AD, C-RPC	6488	2144 (38%)	5538
C-F	9475	2966 (46%)	6354
SRW-AD, SRW-RPC	4336	2420 (41%)	5805
SRW-SD	4398	2501 (43%)	5884

A. Optimized architecture for each benchmark group

Table I(a) and Table I(b) show cycle counts and code sizes for the cd2dat and dat2cd benchmarks. Table I(c) shows the same kind of data for fb4bwd, but with two sets of data corresponding to hierarchical and flattened schedules (see Section III). Table I(d) shows results for jpeg.

In some cases, the structure of the SDF graph causes different scheduling strategies to produce same schedule. In particular, for cd2dat and dat2cd, dynamic programming post optimization (SD) and pairwise grouping of adjacent nodes with DPPO post optimization (AD) produce the same schedule. It can be seen from the results that the synthesized code contributes significantly to the code size of whole application. For the jpeg and filter bank cases, the AD scheduling and recursive procedure call based multiple appearance scheduling produce the same schedule. For the cd2dat, dat2cd and jpeg

benchmarks written as flat SDF graphs, this contribution is between 14% to 65%. Buffer sharing with recursive procedure call (BS-RPC) caused the largest increase in code size for the chain-structured SDF graphs of the cd2dat and dat2cd benchmarks. The more complex topology of the jpeg benchmark caused the largest overhead for the circular buffering with flat scheduling (C-F) strategy. For all three of the benchmarks, C-F added the largest overhead to the cycle count.

From the results, we see that the smallest overhead in cycle count does not correspond to the smallest overhead in code size. Buffer sharing with dynamic programming post optimization (BS-SD) appears to be a strategic choice in regards to this trade-off. For fb4bwd-hierarchical (the hierarchical version of fb4bwd), the synthesized code contributes between 40% and 60% of the total code size. When the flattening strategy is applied, the hierarchy of SDF graphs is flattened before scheduling starts, and therefore, only one SDF graph is scheduled. The relative contribution of synthesized code is generally smaller, in this case — between 39 and 48% for fb4bwd. In both cases, the C-F strategy combination causes the largest code size increase, due to the complex SDF topology that is associated with the application. This strategy also adds the largest overhead to the executed cycle count.

Most effective, in terms of minimal increase in code size and cycle count, seem to be the BS-SD and SRW strategy combinations. Furthermore, the results show that the circular buffering strategy, with any of the scheduling techniques, consistently causes larger increase in cycle count than buffer sharing or static read-write pointer resetting.

The architecture derived for jpeg has an area estimate of 48 kgates, and the architectures for cd2dat, dat2cd and fb4bwd all have area estimates of about 90 kgates. Significant similarity results in the architectures for cd2dat, dat2cd and fb2bwd. This is caused by the use of floating point computation in the benchmarks, and consequent use of software floating point emulation.

B. Recompiled benchmarks for the critical 'jpeg' architecture

Table II(a) shows cycle counts and code sizes for the cd2dat benchmark compiled for an architecture that has been designed specifically for jpeg. Table II(b) and Table II(c) show analogous results for dat2cd and fb4bwd, respectively. Effects of different combinations of scheduling and buffering techniques on code size and cycle count are similar to the results from Tables I(a), I(b), and I(c). However, recompiling benchmarks as non-critical applications causes up to 12% increases in code size and cycle count compared to the corresponding *native* architectures for these applications. Figures 2(a), 2(b), and 2(c) show graphical comparisons of cycle count and synthesized code size overhead for cd2dat, dat2cd, and fb4bwd, respectively.

V. CONCLUSION

In this paper, we have demonstrated relationships and trade-offs involving SDF scheduling techniques, SDF buffer management techniques, ASIP implementation, and application criticality. Our work has integrated the SDF techniques of the DIF design tool with the ASIP techniques of the TCE

environment, and we have carried out extensive experiments on practical DSP applications using this integrated approach. Our results indicate various relevant trends — for example, that use of circular buffering with flat scheduling adds the highest overhead in terms of cycle count, and for complex topologies, this combination also appears to add relatively high code size overhead. More broadly, we have demonstrated a new methodology for exploring trade-offs involving various key design considerations when mapping SDF graphs onto ASIP platforms. Interesting directions for future work include developing more integrated tool support and automation for this methodology, and applying it to other kinds of high level dataflow transformations and architectural platforms.

REFERENCES

[1] E. A. Lee and D. G. Messerschmitt, "Static scheduling of synchronous data flow programs for digital signal processing," *IEEE Trans. Comput.*, vol. 36, no. 1, pp. 24–35, 1987.

[2] C. Hsu, F. Keceli, M. Ko, S. Shahpamia, and S. S. Bhattacharyya, "DIF: An interchange format for dataflow-based design tools," in *Proceedings of the International Workshop on Systems, Architectures, Modeling, and Simulation*, Samos, Greece, July 2004, pp. 423–432.

[3] C.-J. Hsu, M.-Y. Ko, and S. S. Bhattacharyya, "Software synthesis from the dataflow interchange format," in *SCOPES '05: Proceedings of the 2005 workshop on Software and compilers for embedded systems*. New York, NY, USA: ACM, 2005, pp. 37–49.

[4] P. Jääskeläinen, V. Guzma, A. Cilio, and J. Takala, "Codesign toolset for application-specific instruction-set processors," in *Proc. Multimedia on Mobile Devices 2007*, 2007, pp. 65 070X–1 – 65 070X–11, http://tce.cs.tut.fi/.

[5] P. Jääskeläinen, "Instruction Set Simulator for Transport Triggered Architectures," Master's thesis, Department of Information Technology, Tampere University of Technology, Tampere, Finland, P.O.Box 553, FIN-33101 Tampere, Finland, Sep 2005, \tthttp://tce.cs.tut.fi/.

[6] Y. Lin, Y. Choi, S. Mahlke, and T. Mudge, "A parametrized dataflow language extension for embedded streaming systems," in *Int. Conf. on Embedded Computer Systems: Architectures, Modeling, and Simulation (IC-SAMOS VIII)*, July 2008, pp. 10–17.

[7] T. Stefanov, C. Zissulescu, A. Turjan, B. Kienhuis, and E. Deprettere, "System design using Kahn process networks: The Compaan/Laura approach," in *DATE '04: Proceedings of the conference on Design, automation and test in Europe*. Washington, DC, USA: IEEE Computer Society, 2004, p. 10340.

[8] J. Buck and R. Vaidyanathan, "Heterogeneous modeling and simulation of embedded systems in El Greco," in *CODES '00: Proceedings of the eighth international workshop on Hardware/software codesign*. New York, NY, USA: ACM, 2000, pp. 142–146.

[9] P. K. Murthy and S. S. Bhattacharyya, "Shared buffer implementations of signal processing systems using lifetime analysis techniques," *IEEE Transactions on Computer-Aided Design of Integrated Circuits and Systems*, vol. 20, no. 2, pp. 177–198, February 2001.

[10] S. S. Bhattacharyya and E. A. Lee, "Memory management for dataflow programming of multirate signal processing algorithms," *IEEE Transactions on Signal Processing*, vol. 42, no. 5, pp. 1190–1201, May 1994.

[11] S. S. Bhattacharyya and P. K. Murthy, "The CBP parameter — a module characterization approach for DSP software optimization," *Journal of VLSI Signal Processing Systems for Signal, Image, and Video Technology*, vol. 38, no. 2, pp. 131–146, September 2004.

[12] S. S. Bhattacharyya, P. K. Murthy, and E. A. Lee, *Software Synthesis from Dataflow Graphs*. Kluwer Academic Publishers, 1996.

[13] M. Ko, P. K. Murthy, and S. S. Bhattacharyya, "Compact procedural implementation in DSP software synthesis through recursive graph decomposition," in *Proceedings of the International Workshop on Software and Compilers for Embedded Systems*, Amsterdam, The Netherlands, September 2004, pp. 47–61.

[14] T. Pitkänen, T. Rantanen, A. G. M. Cilio, and J. Takala, "Hardware cost estimation for application-specific processor design," in *SAMOS*, ser. Lecture Notes in Computer Science, T. D. Hämäläinen, A. D. Pimentel, J. Takala, and S. Vassiliadis, Eds., vol. 3553. Springer, 2005, pp. 212–221.

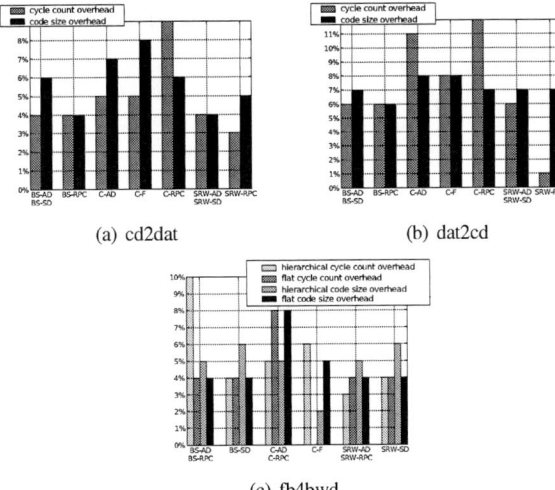

(a) cd2dat   (b) dat2cd

(c) fb4bwd

Fig. 2. Cycle count and code size increase when compiled for 'jpeg' target architecture

An ASIC-Design-Based Configurable SOC Architecture for Networked Media

Ning Ma, Zhibo Pang, Hannu Tenhunen, and Li-Rong Zheng
School of Information and Communication Technology, Royal Institute of Technology (KTH),
Forum 120, SE 164 40 Kista-Stockholm, Sweden
(E-mails: mning@kth.se | zhibo@kth.se | hannu@kth.se | lirong@kth.se)

Abstract—An ASIC-design-based configurable SOC architecture, which is high performance, flexible, programmable, and compiler-independent, is designed for networked media applications. A coarse-grained parallel computing mechanism is employed in this architecture. Mapping this architecture to a specific application is demonstrated through an example in multimedia application. The design is validated in a powerful FPGA, consisting of two CPUs, working at 81MHz and five function units, working at 40.5MHz.

I. INTRODUCTION

Networked media in portable devices is driven by technological development in area of network communication as well as embedded processors. Nowadays advanced network connects people nearly anywhere and anytime. Data rate of communication, especially wireless networks, is increased to satisfy many new applications and services. On the other hand, more and more powerful embedded processors are employed in media platform, which allows high performance processing of media content at lower power. There are two folds in context of networked media: one is network which carries out control-intensive tasks, and another is media which is compute-intensive. It is well known that general purpose processor (GPP) is good for control-intensive tasks, and digital signal processor (DSP) is more suitable for computation. Whereas, ASIC/ASSP are excellent for both control and computation. They consume very low power and have better efficiency in hardware utilization due to dedicated hardware. The advantage of GPP/DSP lies in their excellent configurability and programmability, which is often important, because today's technique develops so quick for both network and media that the period of developing ASIC/ASSP is too long to be cost effective for short product life cycle.

In order to achieve both good computability and flexibility, multi-core system-on-chip (SoC) architectures are proposed. One method is clustering a number of GPPs to increase the computability [1] [2]. It consumes huge power, and the achieved performance is still limited for compute-intensive applications. DSPs in parallelism in combination with GPP can enhance data processing ability as well as exhibit good system controllability [3]. This method acquires

high performance but suffers from high power consumption. References [4] and [5] propose a fine-grained dedicated hardware combined with DSP/GPP method, which can evidently increase performance and keep flexibility as well. However, exploiting parallelism on such a fine-grained level is very complicated and a challenging task. The reconfigurable computing platform in [6] uses embedded FPGAs to realize hardware configurability, and an embedded CPU to fulfill software flexibility. The resulting silicon area is inevitably large because of the low efficiency of hardware utilization.

In this paper, we propose an ASIC-design-based configurable architecture, which makes a tradeoff between performance and cost including power, silicon area and development period. The platform is implemented for networked media. The rest of paper is organized as follows: In section II, the overall architecture is introduced. Section III gives details of the networked media as an application example. In section IV, we give the performance and experimental results of this implementation in FPGA. Conclusions are stated in section V.

II. OVERALL ARCHITECTURE

Fig 1 shows the overview of proposed architecture. There are three main parts: CPU group, enhanced data processing unit (EDPU) and DMA control unit.

Frame I in Fig 1 is the CPU group. The number of CPUs is determined by specific applications. But at least one is needed and named dominating CPU. The dominating CPU is engaged in system tasks and manages peripheral interfaces. The rest of CPUs can be designed for multi-thread processing, and are under the control of the dominating CPU. The members of CPU group internally communicate with each other using instructions and state information through command bus (CBUS1). Local memories (LM) are deployed for each CPU to minimize the global transmitted data. The data are exchanged through data bus (DBUS1) using DMA to accelerate data transmission, which is controlled by DMA controller (DMAC).

Frame II in Fig 1 is the EDPU part, which is concentrated on compute-intensive tasks. Each function unit (FU) executes one macro operation made up of a series of

978-1-4244-2541-9/08 $25.00 © 2008 IEEE

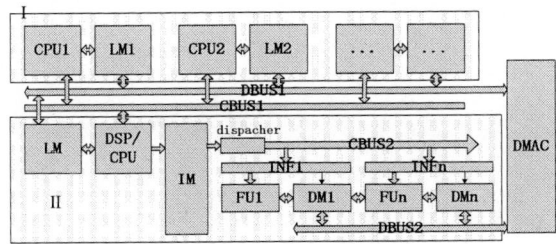

Figure 1: Proposed architecture

computation operations. FUs are specified at the design phase, but they are still configurable by abstracting the functionality of applications. Pre-designed instructions are used to configure these FUs. To decrease software development complexity, coarse-grained function partitioning is adopted to assign diverse tasks to different FUs. ASIC design method means FUs need to be redesigned if the application is changed. But it's easy to redesign because of the regular structure of compute-intensive tasks. Meanwhile, the advantage of ASIC, better efficiency of hardware utilization and low power, is brought in. In section III, an example will be exhibited to interpret detailedly.

Data memory (DM) can be configured as private memory possessed by only one FU (Fig 2 (a)) or common memory shared by two neighbor FUs (Fig 2 (b)) on the purpose of localizing data transmitting. DMs can also exchange data with off-chip DRAM using DMA operation through data bus (DBUS2).

Local memory (LM) provides the same functionality as those in CPU group. DSP/CPU is used for generating the pre-designed instructions for FUs and performing basic computations. Selecting DSP or CPU as processing unit is determined by the proportions between control and computation tasks. But the major function is to produce diverse instructions and put them into instruction memory (IM). The dispatcher surveils the state of each FU, and broadcasts instructions fetched from IM through command bus (CBUS2). The interface (INF) is very important. It

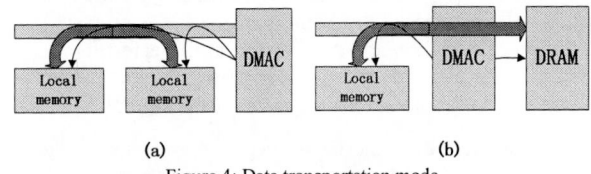

Figure 4: Data transportation mode

parses instructions received from dispatcher and supplies specified FU with needed parameters. It's the bridge between software and hardware. If the instruction is changed for some reasons, all need to do is modifying the interface only.

Fig 3 shows the data flow of instructions. The corresponding CPU writes instructions to IM, and dispatcher fetches appropriate ones and broadcasts them through CBUS2. Then the corresponding INF (e.g. INF2 in Fig 3) receives them and interprets them for related FU. What should be noticed is: the dispatcher fetches the instruction according to FUs' requirement, and it may not sequentially access the IM. On the other hand, CPU writes all instructions sequentially. A mechanism is needed to avoid the possible collision.

The last part is DMA controller (DMAC). In compute-intensive task such as image/video processing, data transmission is the bottleneck. DMA is adopted for block data transmission. DMAC controls the data exchange between local memories (Fig 4 (a)) or local memory and off-chip DRAM (Fig 4 (b)).

III. APPLICATION FOR NETWORKED MEDIA

In this section, we will give an application of our proposed architecture for networked media.

First, the design flow for mapping the architecture to a detailed application is shown in Fig 5. The first two phases

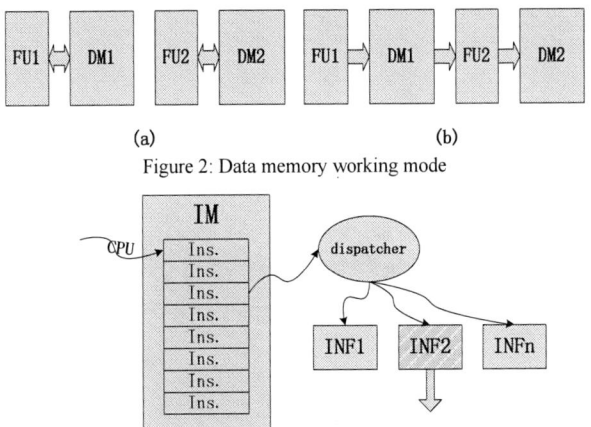

Figure 2: Data memory working mode

Figure 3: Instructions dispatch

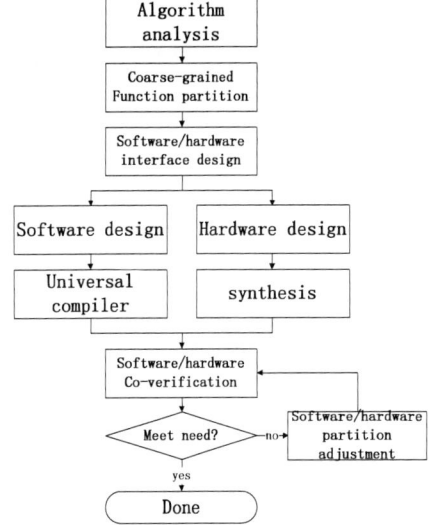

Figure 5: Design flow for mapping

978-1-4244-2541-9/08 $25.00 © 2008 IEEE

are very crucial. They make a trade-off between software and hardware, between flexibility and performance, and will determine the development period and overall performance. If improper partitioning is made, redesign will be inevitable, and it will cost vast amount of time to cover the shortage.

After defining the interfaces between software and hardware which is also known as designing instructions on coarse-grained level, software and hardware can be designed concurrently. In software design branch, universal compiler is needed only, not special designed compiler or de-compiler. The latter is usually used in fine-grained parallel architecture.

Software/hardware co-verification is to measure the validity and performance. If it meets the need, the mapping is finished. Otherwise, recursive adjustment for software/hardware partition is needed.

For simple verification, we only implement wired networked audio/video decoding. The functions are summarized in TABLE I.

In CPU group, one CPU is enough for processing system task and network protocols. An embedded operating system is run in this CPU. Based on analyse of audio/video encoding algorithms, another CPU is adopted in EDPU part for audio decoding and video decoding instruction generation. For audio decoding, general purpose processor is adequate according to state-of-the-art performance. But for video decoding, it's far more inadequate. So hardware acceleration is necessary.

By analyzing current popular video coding standards, the universal decoder should be consisted of five major parts: variable length decoding (VLD), inverse quantization (IQ), inverse transformation (IT), motion compensation (MC) and

TABLE I. SUPPORTED FUNCTIONS

Category	Supported
OS	Linux
Network	Wired Ethernet
Audio	MP3/RealAudio
Video	MPEG4/RealVideo

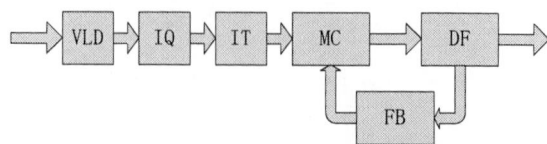

Figure 6: The basic data flow of video decoding

deblocking filter (DF), Fig 6. FB stands for frame buffer which stores frames for later frame decoding. For these parts mentioned above, a basic decoding unit comprised of 256 pixels (16 pixels width by 16 pixels height) named macro-block (MB) is applied.

Every FU of proposed architecture can be mapped accordingly. Five specified FUs which will implement the function of five major parts are designed. For different video coding standards on MB level, differences are coefficients or coefficient matrixes. In order to realize multiple standard video decoding, there are two choices: one is configuring the coefficients using pre-designed instructions through CBUS2; the other is selecting the proper results according to the specific video standard being decoded. Since only two standards are supported, the latter method is suitable, and hardware reusing is needed; otherwise, if more standards are involved and expandability is considered, the former one is preferred.

The data used by fore-and-aft FUs are highly correlated, so DMs are configured as common memory shared by neighbor FUs. Every FU can process one MB once. Operations are pipelined on MB level to speed up video decoding. FUs are driven by correlated data in shared DMs. If the data of frontal SRAM (storing data of next MB) are not ready for use or the data of back SRAM (storing data of previous MB) are not consumed yet, current FU is stalled automatically. The CPU of EDPU deals with parameters analysis, and generates instructions accordingly. It controls every FUs by instructions. Reference frames are stored in off-chip DRAM considering the huge amount of data. The data needed for reference and decoded already are transported using DMA. The final mapped architecture is shown in Fig 7.

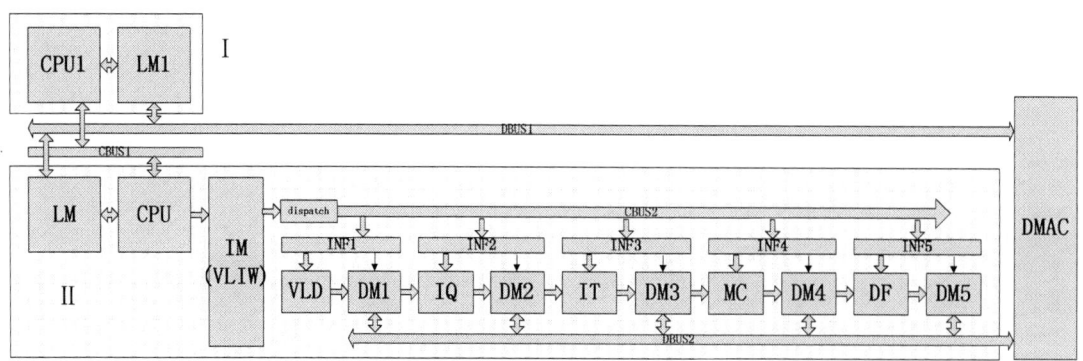

Figure 7: The mapped architecture

978-1-4244-2541-9/08 $25.00 © 2008 IEEE   12

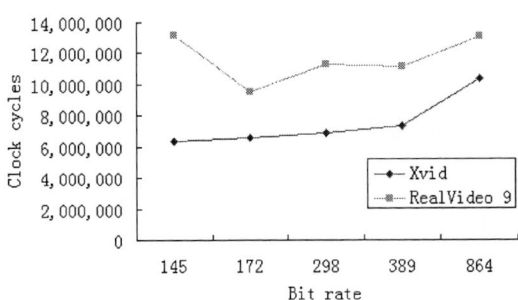

Figure 8: Simulation result for QCIF

IV. SIMULATION AND VERIFICATION IN FPGA

For simulation, VCS of Synopsys is used. Full hardware simulation platform, realized by using Verilog HDL only, is designed, so the simulation is accurate, but it will cost a lot of time. Considering this, QCIF is selected as the resolution of test sequences. The frame rate is 25fps. Different benchmark files (salesman, news, mobile, foreman, carphone) coded by Xvid [7], packed as AVI files and coded by RealVideo 9 [8], packed as RMVB files are inputted to the simulation platform. The clock frequency of CPU is set two times of that of FUs and SRAMs. Fig 8 shows the relation between bit rate and system frequency needed. This instantiation are capable of decoding QCIF video of XVID at less than 10MHz of system clock frequency (10MHz for CPU, and 5MHz for function units) and 13MHz for RealVideo 9.

For verification, the FPGA of Altera StratixII EP2S1801508C3 is adopted and Quartus II 7.0 is used for development. The development kit integrates network interface, audio input/output interface, video display interface, a touchable LCD screen, DRAM interface, some universal interfaces (such as COM/I2C), JTAG interface for debugging, and other extent interfaces. So it's suitable for verification of complex networked media systems.

System clock frequency is set to 81MHz, and CPUs work at 81MHz and FUs work at 40.5MHz. Real-time standard definition (SD) quality (720x576@25fps) video decoding is achieved. To acquire the same resolution, the system clock frequency of [9] is 166MHz, and that of [10] is 400MHz.

The result of fitting in the FPGA is shown in TABLE II

TABLE II. RESOURCE UTILIZATION

Resource	Total	Used	Percentage
Combinational ALUTs	143,520	102,049	71%
Dedicated logic registers	143,520	42,558	30%
Block memory bits	9,383,040	724,528	8 %
DSP block 9-bit elements	768	98	13%
PLLs	12	2	17%

V. CONCLUSION

An ASIC-design-based configurable architecture is proposed making trade-off among performance, flexibility, and realization complexity. Utilizing this architecture, general compiler for high level language is needed only. High performance is obtained by exploiting and utilizing the parallelism of the algorithm for specific application. It is easy to realize on coarse-grained level. Flexibility and configurability are acquired by assigning control-intensive tasks to general purpose processor or digital signal processor, and only assigning compute-intensive and regular processing tasks to specified function units, which also achieves better hardware utilization efficiency. A specific networked media platform is exhibited to verify the proposed architecture. It's implemented in FPGA development kit for verification and evaluation. Wired network with multiple standard audio and video decoding is realized in this platform.

ACKNOWLEDGMENT

Altera provided the FPGA chips and development board for this work through university donation program. The authors want to acknowledge the donation.

REFERENCES

[1] Leon, A. S.; Tam, K. W.; Shin, J. L.; Weisner, D.; Schumacher, F.; "A Power-Efficient High-Throughput 32-Thread SPARC Processor", Solid-State Circuits, IEEE Journal of Volume 42, Issue 1, Jan. 2007 Page(s):7 – 16

[2] Pham, D.; Asano, S.; Bolliger, M.; Day, M.N; "The design and implementation of a first-generation CELL processor", Solid-State Circuits Conference, 2005. Digest of Technical Papers. ISSCC. 2005 IEEE International 6-10 Feb. 2005 Page(s):184 - 592 Vol. 1

[3] Ramacher, U.; "Software-Defined Radio Prospects for Multistandard Mobile Phones", Computer Volume 40, Issue 10, Oct. 2007 Page(s):62 - 69

[4] Khailany, B.K.; Williams, T.; Lin, J.; Long, E.P.; Rygh, M.; "A Programmable 512 GOPS Stream Processor for Signal, Image, and Video Processing", Solid-State Circuits, IEEE Journal of Volume 43, Issue 1, Jan. 2008 Page(s):202 – 213

[5] Noda, H.; Nakajima, M.; Dosaka, K.; Nakata, K.; Higashida, M.; "The Design and Implementation of the Massively Parallel Processor Based on the Matrix Architecture", Solid-State Circuits, IEEE Journal of Volume 42, Issue 1, Jan. 2007 Page(s):183 – 192

[6] Bondalapati, K.; Prasanna, V.K.; "Reconfigurable computing systems", Proceedings of the IEEE Volume 90, Issue 7, July 2002 Page(s):1201 – 1217

[7] http://www.xvid.org/

[8] http://www.realnetworks.com/

[9] Yi-Shin Tung; Sung-Wen Wang; Chien-Wu Tsai; Ya-Ting Yang; "DSP-based multi-format video decoding engine for media adapter applications", Consumer Electronics, IEEE Transactions on Volume 51, Issue 1, Feb. 2005 Page(s):273 - 280

[10] Liu Feng; Guo Rui; Shi Shu; Cheng Xu; "HW/SW Co-Design and Implementation of Multi-Standard Video Decoding", Embedded Systems for Real Time Multimedia, Proceedings of the 2006 IEEE/ACM/IFIP Workshop on. Oct. 2006 Page(s):87 - 92

978-1-4244-2541-9/08 $25.00 © 2008 IEEE

Congestion-aware Task Mapping in Heterogeneous MPSoCs

Ewerson Carvalho and Fernando Moraes

Pontifícia Universidade Católica do Rio Grande do Sul (FACIN-PUCRS)
Av. Ipiranga, 6681 – P32 – 90619-900 – Porto Alegre – RS – Brasil
{ecarvalho, moraes}@inf.pucrs.br

Abstract—**Multiprocessors Systems-on-Chip (MPSoCs) are a trend in VLSI design, since they minimize the design crisis represented by the gap between the silicon technology and the actual SoC design capacity. MPSoCs may employ NoCs to integrate several programmable processors, specialized memories, and other specific IPs in a scalable way. Besides communication infrastructure, another important issue in MPSoCs is task mapping. Dynamic task mapping is needed, since the number of tasks running in the MPSoC may exceed the available resources. Most works in literature present static mapping solutions, not appropriate for this scenario. This paper investigates the performance of mapping algorithms in NoC-based heterogeneous MPSoCs, targeting NoC congestion minimization. The use of the proposed congestion-aware heuristics reduces the NoC channel load, congestion, and packet latency.**

I. INTRODUCTION

The evolution of deep-submicron technology allows increasing circuits density significantly, giving support to the development of Multiprocessors System-on-Chip. MPSoC merges System-on-Chip (SoC) concept and the multiprocessing approach. It can integrate several programmable processors, specialized memories and IP-cores. Concerning its communication infrastructure, generally the use of NoC is mandatory in MPSoC design, since traditional communication infrastructures (e.g. busses) provide low scalability and small parallelism.

In general, industrial [1] [2] and academic [3] [4] devices employs homogeneous MPSoCs, composed by a set of identical processing elements. Homogeneous MPSoCs favors task migration and load balancing. Meanwhile, heterogeneous MPSoC, composed of distinct processor elements, tends to support a wide variety of applications. Tasks can be loaded on-demand at runtime and executed as software, at different processing elements, or as hardware at embedded reconfigurable logic.

Applications running in heterogeneous MPSoCs (e.g. multimedia, networking) may have their tasks dynamically loaded. Therefore, the number of tasks simultaneously running may exceed the available MPSoC resources. Thus, it is necessary to control task management and resource occupation. This management includes *task mapping*, a NP-hard problem that consists on finding a placement for a required task in the system, according to some specific criteria, as

energy, channel load, and system fragmentation. Task mapping decisions reflect directly on the overall system performance. Most works in literature propose static mapping solutions [5] [6] [7], which are not appropriate for dynamic workloads scenarios, while few works focus on dynamic approaches [8] [9].

This work investigates the performance of a set of mapping algorithms for NoC-based MPSoCs, with dynamic workload. The main cost function is to optimize the NoC channel occupation. The mapping algorithms evaluation includes overall execution time, channel load, NoC congestion and packet latency.

The paper is organized as follows. Section II presents related works on task mapping. Section III discusses the target MPSoC architecture. Section IV introduces the task mapping algorithms. Section V discusses the experimental setup and results. Section VI presents conclusions and directions for future work.

II. RELATED WORKS

Academic works often propose homogeneous NoC-based MPSoCs [3] [4]. On the industrial side, IBM, Sony and Toshiba proposed a heterogeneous MPSoC composed of one manager processor and 8 floating-point units [10]. Intel [1] and Tilera [2] present homogeneous NoC-based MPSoCs, composed by 80 and 64 identical processors respectively.

Most works presenting static mapping algorithms employ NoC-based homogeneous MPSoCs [5] [6] [11]. Different algorithms are used: genetic approaches [7] [12], Tabu Search [11] [13], and Simulated Annealing [5] [6] [14]. Hu and Marculescu [5] and Marcon et al. [6] explore energy-aware mapping algorithms, with the purpose of reducing the overall power consumption by decreasing the consumed energy on communication. Results achieved in such works effectively improve the overall system performance, but this static mapping defines task placement at design time, for a system that will execute a fixed set of applications, with a well-known computation and communication behaviour. Therefore, static mapping approaches are not adequate for systems that support applications with dynamic workload.

Dynamic mapping approaches also target NoC-based homogeneous MPSoCs [8] [9]. Ngouanga et al. [8] present a mapping solution based on attraction forces between com-

978-1-4244-2541-9/08 $25.00 © 2008 IEEE

municating tasks, which tend to place them near to each other. A drawback of this work is that the Authors do not consider the overhead of mapping on the execution time. Wronski et al. [9] evaluate heuristics for the bin-packing problem as Best Fit and Worst Fit, mapping all tasks belonging to a given application simultaneously. Differently from [9], the strategy proposed here maps only the initial task of each application. The advantage of dynamic mapping is to improve system use, since only resources executing tasks are effectively mapped. On the other hand, dynamic mapping can increase execution time, if a required resource is not available for mapping.

III. MPSoC TARGET ORGANIZATION

A heterogeneous MPSoC is a set of different PEs interacting through a communication network [15]. PEs may support either hardware or software tasks execution. Software tasks execute in Instruction Set Processors (ISPs), while hardware tasks execute in reconfigurable logic (RL) or dedicated IPs. Figure 1 illustrates the proposed heterogeneous MPSoC organization. Among the available resources, one processor, named Manager Processor (MP), is responsible for resource control, task scheduling, task binding, task mapping, task migration, and configuration control. The MP starts the initial task of each application. New tasks are loaded into the MPSoC from the task memory when a communication to them is required, if they are not already mapped.

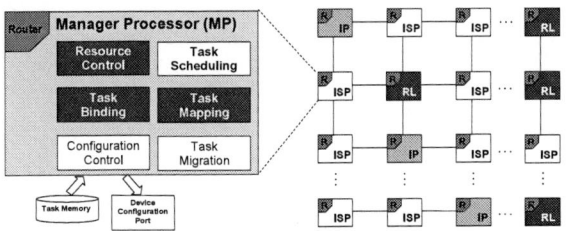

Figure 1. MPSoC target organization.

Actually, the present work emphasises resource control, task binding, and task mapping. Task scheduling is based on a queue strategy. There are three queues, one for each task type (i.e. hardware, software or initial). A task enters on a queue if there are no free resources, and it waits until this condition changes. The configuration control process is simulated based on configuration overhead results [16].

Resources are represented by R_{ij}, where i and j refer to the resource column and row respectively. The information stored into the resource matrix regards its type (hardware or software) and its status (free or used). Four matrices define the available channels: East channels (EC_{ij}); West channels (WC_{ij}); North channels (NC_{ij}); and South channels (SC_{ij}). Each element of the channel matrices contains the percentage of channel bandwidth use.

Inter-task communications use messages transmitted through the network. Four message types compose the adopted communication protocol: REQUEST, RELEASE, NOTIFY and GENERAL. A REQUEST message to the MP con-

tains the identification of a new task to be inserted into the system and the respective communication volumes and rates. The RELEASE message notifies the MP that a processing PE has finished its current task, being possible to reuse the PE for a new task. The NOTIFY message is sent by the MP, after task mapping, to both tasks involved in the communication, containing task addresses to make possible a correct packet transmission. Inter-task communications employ the GENERAL messages

Usually, application models employ tasks graphs, which define tasks dependency [5] [6], volume [5] [6] of communication between tasks, and estimated computation time [17]. Here, an acyclic directed graph models an application, where vertices represent software or hardware tasks and edges define its dependencies. Volumes and rates from/to the node define communication values between tasks. Nodes connection defines a master-slave pair. Each graph edge has four parameters that represent the *volume* and *rate* of data sent from master to slave, and vice-versa. A source task of directed edges (master) needs to request the configuration of the target task (slave) before starting its communication.

IV. DYNAMIC MAPPING HEURISTICS

The method used to define the mapping of each application initial task has a significant impact in the performance of the dynamic mapping. The method adopted in the present work divides the NoC into regions, named clusters. Each initial task is mapped to a unique cluster, reducing the probability of tasks belonging to different applications to share the same NoC region.

Two reference mapping methods are employed here: *First Free* (FF) and *Nearest Neighbor* (NN). FF is a method that starts at resource R_{00}, walking the network column by column, bottom to top. FF selects the first free resource according to binding definitions, without taking into account other metrics. FF may generate the worst results when compared to the other heuristics presented here. The NN method does not take into account other metrics except the proximity of an available resource able to execute the required task. NN starts searching for a free node able to execute the task near the requesting task. The search tests all *n*-hop neighbors, *n* varying between 1 and the NoC limits, stopping when a first resource able to execute the task is found.

A. Congestion-aware Mapping Heuristics

The *Minimum Maximum Channel Load* (MM) congestion-aware mapping heuristic evaluates all possible mappings for each new task inserted into the system. The goal of this heuristic is to globally minimize the channel occupancy peaks, reducing the occurrence of hotspots.

The *Minimum Average Channel Load* (MA) aims reducing the average occupancy of the NoC channels. This heuristic is similar to the MMC, replacing the *maximum* with the *average* function. While the MMC heuristic tries to minimize the channel occupancy peak, the MAC heuristic tries to homogenously distribute the communication load in the

NoC. The selected mapping is that resulting in the lower average channel occupancy.

The MMC and MAC heuristics consider all NoC channels while mapping a new task. Since this evaluation can take long, the *Path Load* algorithm are proposed to considers only the channels used by the task being mapped (its *communication path*). However, all possible mapping possibilities are still evaluated.

The *Best Neighbor* (BN) heuristic combines NN search strategy and the PL computation approach. The search space of BN is similar to NN, i. e., circular searches from the source node. This avoids computing all feasible mapping solutions, as in the PL heuristic, reducing the execution time for the heuristic. BN selects the best neighbor, according to PL equations, instead of the first free neighbor.

More details concerning such mapping algorithms are presented in [18]. In [18] TLM modeling was employed, while in the present work it is adopted a RTL level modeling, ensuring accurate results.

V. EXPERIMENTS AND RESULTS

This Section presents the experimental setup and results. All results are obtained from *ModelSim* co-simulation (RTL-VHDL for the NoC and *System-C* for the applications). The number of applications varies from 1 to 20, each one containing from 3 to 10 tasks. The processing time of each task is fixed to 25 microseconds. Rates are fixed from 5 to 30% of the available channels bandwidth, using a Pareto On-Off distribution. Each task transmits 200 to 500 packets, with size varying from 100 to 450 16-bit flits.

An 8x8 2D-mesh topology, described in VHDL [15], is responsible to transfer data between tasks. The maximum communication rate between routers is 50% of the available bandwidth, since the handshake control flow is used (two clock cycles to transmit one flit). The 64 processing elements (PEs) are distributed as follows: one node is used for the MP, 16 nodes are hardware resources and 47 nodes are software resources.

Three scenarios are evaluated: (*i*) 20 identical pipeline-like applications; (*ii*) 20 identical tree-like applications; and (*iii*) 20 different generic applications generated using TGFF [19]. Table I summarizes the results, comparing the relative gain of each mapping algorithms to the reference FF method.

The **average channel load** (2nd line of Table I) represents the NoC usage, primary cost function of MM, MA, PL and BN heuristics. As expected, all algorithms reduce the *average channel load* when compared to FF. Two congestion aware mapping algorithms, MM and MA, reduces 20% the channel load in average, result inferior to obtained with NN heuristic, where no traffic status is considered during the mapping. The BN heuristic have similar gains to NN. The average channel load is reduced to 38.04% when adopting PL, demonstrating its efficiency.

The **channel load standard deviation** (3rd line of Table I) measures the traffic distribution inside the network. Lower values correspond to homogeneous traffic distribution, while higher values suggest some channels with higher loads, and others not used at all. This performance figure follows the average channel load. The PL and BN heuristics reduces the channel load standard deviation, with values similar to the NN heuristic.

TABLE I. ALGORITHM PERCENT GAIN IN RELATION TO FF METHOD.

Metrics	Mapping Algorithms (% *Gain w. r. t. FF*)				
	NN	MM	MA	PL	BN
Avg. Channel Load	34.49%	21.59%	21.56%	38.04%	35.81%
Channel Load s.d.	24.67%	12.71%	12.70%	25.63%	25.44%
Avg. Packet Latency	13.90%	5.38%	6.85%	13.79%	13.79%
Packet Latency s.d.	43.84%	5.87%	27.23%	46.42%	48.42%
Congestion n.c.	56.81%	32.85%	25.88%	63.11%	61.51%
Congestion w.t.	77.35%	41.07%	44.72%	78.34%	75.46%
Execution Time	1.69%	-15.63%	-8.47%	-4.06%	0.51%

The **average packet latency** (4th line of Table I) is proportional to the distance between source and target PEs, as well the congestion in the communication path. All heuristics leads to positive gains, with similar results between the simple NN and the congestion-aware algorithms PL and BN.

The **latency standard deviation** (5th line of Table I) is an important metric for systems with QoS constraints. Packets being transmitted from a given source must keep a regular interval between then to respect the injection rate. Variation in the latency incurs in jitter, which may lead to packet loss for applications with strict temporal deadline (e.g. real-time). The PL and BN congestion aware heuristics reduces the latency standard deviation to 46.42% and 48.42% respectively. The gain obtained with NN is 43.84%, smaller when compared to the other congestion aware heuristics.

These results may suggest smaller gains with the congestion aware heuristics when compared to NN algorithm. Note that the NoC is not heavily used. Only 9 simultaneous applications may run simultaneously, and each application may have up to 10 tasks, Therefore, the maximum number of simultaneous tasks (90) is very close the number of PEs (64), favoring the locality explored by the NN mapping heuristic. Even in such situation, the congestion aware heuristics surpass the NN heuristic.

The evaluation of **congestion** uses two metrics (6th and 7th lines of Table I). The first one, *n.c.* – number of saturated channels, counts the number of channels with an occupancy equal to the maximum allowed (50%). The second one, *w.t.* – wasted time, measures the number of clock cycles packets stalled in routers buffers. Reductions are observed with PL and BN heurists when compared to NN. PL is the most effective heuristic to reduce congestion, in both metrics.

Finally, the **execution time overhead** must be considered (8th line of Table I). The cost to perform the congestion aware heuristics, using FF as reference, is 4% for PL and 0.5% to the BN. This is an acceptable cost, since the reduction on channel load and congestion implies in energy saving. In addition, longer execution times applications may compensate this mapping cost

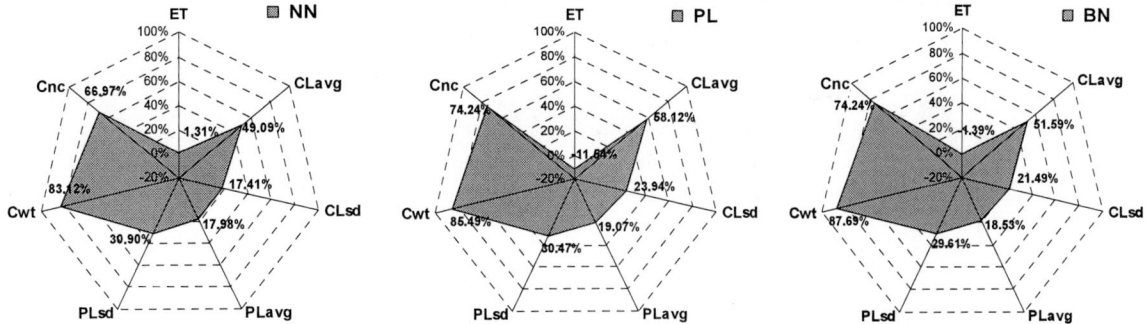

Figure 2. Results for NN, PL and BN mapping algorithms in comparison to FF, conceining: the average channel load (*CLavg*) and standard deviation (*CLsd*); the total system execution time (*ET*); average packet latency average (*PLavg*) and deviation (*PLsd*); the accumulated number of NoC congested channels (*Cnc*) and respective wasted time (*Cwt*). This results are for the third scenario (i.e. 20 different generic applications generated using TGFF).

Figure 2 summarizes the improvements of **NN** (not congestion aware), **PL** (congestion aware) and **BN** (congestion aware) algorithms when compared to FF. It takes into account the third scenario (TGFF), where a more realistic scenario is modeled, since different application graphs co-exist simultaneously. The gains obtained for each performance figure can be inferred based on figures surfaces. From one side, PL and BN improve channel load, latency and congestion when compared to FF and NN. By the other hand, such congestion aware heuristics increase the total execution time.

VI. Conclusion

Some commercial MPSoCs were recently presented, exposing the necessity for research in MPSoC area, including mechanisms for the dynamic control of the system functionalities. Communication infrastructure and task mapping are two important research topics. Concerning communication schemas, NoC seems to be a trend. Several works on task mapping were found in literature, most of them targeting static mapping, not suitable for systems with dynamic task load.

Congestion-aware dynamic mapping heuristics are proposed and evaluated here. According to obtained results, PL heuristic is the best solution when compared to the other algorithms. MA and MC are not effective since these algorithms take local decisions: when a new mapping does not reduce the average link load (or maximum load), the algorithm selects the first mapping option available. NN and BN have similar improvements, however with gains inferior to PL. The heuristics execution time is an important parameter, which increases the total execution time. While NN and FF execute from 15 to 20 clock cycles, MMC and MAC present the worse performances (1000 and 1600 clock cycles). PL and BN execute in approximately 500 and 100 clock cycles, respectively. As mentioned before, in application with larger execution time, the mapping execution time overhead will be negligible.

Currently, two related topics are being investigated by the Authors: benchmarks evaluation with higher NoC loads and energy consumption measurements. Other works deemed as essential are the evaluation of heuristics complexity and the modeling of the use of multitasking processors.

Acknowledgments

This research was supported partially by CNPq (Brazilian research agency), projects 141225/2005-0 and 300774/2006-0.

References

[1] Vangal, S.; et al. An 80-Tile 1.28TFLOPS Network-on-Chip in 65nm CMOS. In: ISSCC.2007.

[2] Tilera Corporation. TILE64™ Processor. Product Brief. 2007.

[3] Lin, L.; et al. Communication-driven task binding for multiprocessor with latency insensitive network-on-chip. In: ASPDAC, 2005.

[4] Saint-Jean, N.; et al. HS-Scale: a Hardware-Software Scalable MPSOC Architecture for embedded Systems. In: ISVLSI. 2007.

[5] Hu, J.; Marculescu, R. Energy- and Performance-Aware Mapping for Regular NoC Architectures. IEEE Transaction on Computer-Aided Design of Integrated Circuits and Systems, Vol 24-4. 2005.

[6] Marcon, C.; et al. Evaluation of Algorithms for Low Energy Mapping onto NoCs. In: ISCAS. 2007.

[7] Lei, T.; and Kumar, S. Algorithms and Tools for Networks on Chip based System Design. In: SBCCI. 2003.

[8] Ngouanga, A.; et al. A contextual resources use: a proof of concept through the APACHES platform. In: DDECS. 2006.

[9] Wronski, F.; et al. Evaluating Energy-aware Task Allocation Strategies for MPSoCS. In: DIPES. 2006.

[10] Kistler, M.; et al. Cell Multiprocessor Communication Network: Built for Speed. IEEE Micro, Vol 26-3. 2006.

[11] Murali, S.; et al. A methodology for mapping multiple use-cases onto networks on chips. In: DATE. 2006.

[12] Wu, D.; et al. Scheduling and Mapping of Conditional Task Graphs for the Synthesis of Low Power Embedded Systems. In: DATE. 2003.

[13] Manolache, S.; et al. Fault and Energy-Aware Communication Mapping with Guaranteed Latency for Applications Implemented on NoC. In: DAC. 2005.

[14] Orsila, H.; et al. Automated Memory-Aware Application Distribution for Multi-Processor System-On-Chips. Journal of Systems Architecture, Vol 53-11. 2007.

[15] Moraes, F. et al. Hermes: an Infrastructure for Low Area Overhead Packet-switching Networks on Chip. Integration, the VLSI Journal, Vol 38-1. 2004.

[16] Möller, L.; et al. A NoC-based Infrastructure to Enable Dynamic Self Reconfigurable Systems. In: ReCoSoC. 2007.

[17] Marcon, C.; et al. Exploring NoC mapping strategies: an energy and timing aware technique. In: DATE. 2005.

[18] Carvalho, E.; et al. Heuristics for Dynamic Task Mapping in NoC-based Heterogeneous MPSoCs. In: RSP. 2007.

[19] Dick, R.P.; et al. TGFF: task graphs for free. In: CODES/CASHE, 1998.

Optimizing Routing Tables on Systems-on-Chip with Content–Addressable Memories

Stergios Stergiou and Jawahar Jain
Fujitsu Laboratories of America, CA
{stergiou,jawahar}@fla.fujitsu.com

Abstract—Routing tables **are part of a critical subsystem of modern internet routers that controls the filtering and forwarding of packets. They are typically embedded in** *Content-Addressable Memories* **which in this context behave as elaborated** *Sum-Of-Products* **expression evaluators. In this work, we examine the applicability of** *Exclusive-Or Sum-Of-Products* **expressions as an alternative routing table formulation and conclude that they provide significant and practical savings in CAM utilization.**

I. INTRODUCTION

In order to implement packet forwarding and filtering, an internet router is required to perform lookup operations on its routing table based on incoming packet IP addresses. When an address matches multiple routing table entries, the entry with the longest matching prefix is selected.

Researchers have proposed various methodologies for optimizing lookups. For small-scale routers, where requirements on bandwidth are not critical, various software-based schemes have been presented [1], [2], [3].

However, modern routers are required to concurrently process packets from multiple network interfaces, each of which may have multi-gigabit physical links. Software-based solutions unfortunately do not scale at such large speeds.

Several hardware-based solutions have been proposed in the bibliography [4], [5], [6], [7], [8], [9]. Some of them require special-purpose hardware, while others make use of content-addressable memories (CAMs,) a widely used special purpose memory structure that enables single clock cycle lookups on unsorted data.

Unlike a RAM which returns the data word stored at a given address, a CAM returns the smallest address where a matching data word is found. CAMs greatly speed up the address lookup subsystem while keeping router design relatively simple. However, CAMs tend to be much more expensive than typical memories of comparable size and consume far more power [10].

Compressing routing tables helps reduce necessary CAM chips for a given design, thereby reducing both cost and power consumption. As a problem, it has been studied both in an academic [11] and industrial context [12].

In this work, we present two algorithms for compacting routing table sizes placed on TCAMs with almost no runtime lookup overhead. Both algorithms cast the problem into a boolean algebra one and subsequently leverage logic synthesis techniques in order to minimize the routing tables.

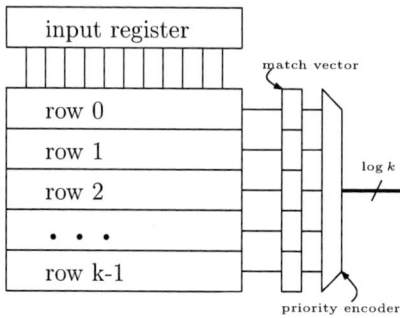

Fig. 1. Simplified CAM architecture

The first algorithm is based on manipulating Binary Decision Diagrams, while the second utilizes exclusive-or sums-of-products, a well-known two-level logic expression.

This paper is structured as follows. In subsections I-A, I-B, I-C and I-D we provide essential background information. In section II we present an extension to classic TCAM structures that enables storing ESOPs within them. The core of this work is presented in section III, followed by the experimental results in section IV. We conclude the paper in section V.

A. Border Gateway Protocol

The *Border Gateway Protocol* (BGP) is the core routing protocol of the Internet. It works by maintaining a table of IP networks or "prefixes" which designate network reachability among *autonomous systems* (AS) [13]. An AS is a collection of IP networks and routers under the control of one or more entities that presents a common routing policy to the Internet.

One of the largest problems faced by BGP, and indeed the Internet infrastructure as a whole, comes from the growth of the Internet routing table. A full BGP table can have more than 200,000 routes as can be seen, for example, on the AT&T WorldNet Common Backbone Route Monitor.

B. Ternary Content–Addressable Memories

CAMs are the popular choice used for single-cycle table lookups in internet backbone routers for the critical task of packet filtering and forwarding.

When presented with an input, a CAM compares it against all entries and provides the address of the matching entry at the output. If more than one entries match, typically the

978-1-4244-2541-9/08 $25.00 © 2008 IEEE

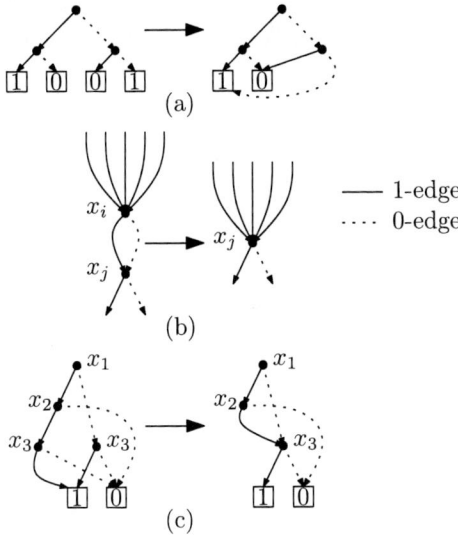

Fig. 2. Elementary BDD transformations. removing redundant (a) terminal nodes, (b) internal nodes and (c) isomorphic subgraphs

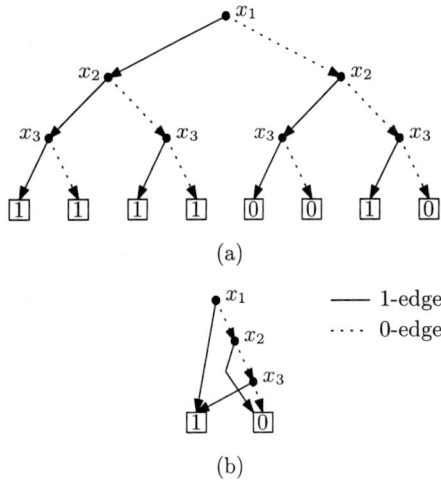

Fig. 3. An example ROBDD. (a) BDT for $f = x_1 + \bar{x}_2 x_3$ (b) respective ROBDD

smallest matching address is returned. Additionally, there exists a hit signal that denotes a matching event.

Fig. 1 shows a simplified architecture of a CAM. The contents of the input register are propagated to all rows. Each row includes a dedicated comparator that checks whether its content matches the input. The outputs of all the comparators form the match vector.

Following is a priority encoder that outputs the address of the first '1' bit in the match vector. The hit signal is simply the logical OR of all the match vector bits.

A ternary CAM (TCAM) is the most popular variation of a CAM, whereupon each cell in the memory can take 3 values: '0', '1' and 'X'. An 'X' denotes a don't care that matches any bit value on the input register. TCAMs are used almost exclusively in routers.

As a rule of thumb, a TCAM is twice as large as an SRAM of the same size [10]. The power consumption is an even more pronounced issue. Therefore, while TCAMs are a powerful asset in the arsenal of a router designer, it makes economic sense to minimize the total size of TCAMs for a given implementation.

C. Binary Decision Diagrams

A *Binary Decision Diagram* (BDD) is a fundamental data structure that is widely used in numerous fields for various practical problems associated with the analysis and representation of boolean functions. It was popularized by Randy Bryant [14].

A BDD is a *Binary Decision Tree* folded into a directed acyclic graph such that no nodes exist whose outgoing edges point to the same node, and no isomorphic subgraphs are present (see Fig. 2.)

If additionally all variables appear on the same order on each path from the root to a terminal node the BDD is called *Reduced Ordered* (ROBDD.) On Fig. 3 we show a simple Binary Decision Tree and its ROBDD representation.

For a specific variable ordering, ROBDDs are a canonical data structure. Nevertheless, variable ordering can have a dramatic impact on the size of an ROBDD and naturally, obtaining good variable orders is a heavily studied problem.

D. Exclusive-Or Sum-Of-Products

A *literal* is either a boolean variable x_i or its negation $\bar{x}_i$. A *product term* (or cube) is the logical AND of a set of literals. *Exclusive-or Sum-Of-Products* (ESOPs) are a two-level boolean function representation defined as the logical XOR of a set of cubes. For example, $x_1 \oplus x_2 \oplus x_3$ is an ESOP for the three-variable parity function.

ESOPs have been extensively studied and have been shown to be more powerful than the related Sum-Of-Products (SOPs) representations [15], [16], [17]. For example, the size of the minimum ESOP of a random four-variable function is on average 10% smaller than the corresponding minimum SOP [18], a significant savings in space in the context of synthesis.

Moreover, the worst-case minimum n-variable ESOP comprises at most $29 \cdot 2^{n-7}$ cubes ($n > 6$), compared with $64 \cdot 2^{n-7}$ for the case of SOPs [19]. More importantly, for practical boolean functions, ESOPs are usually much more compact than SOPs as can be seen on Table I.

II. TCAM EXTENSION

While a traditional TCAM behaves as a natural SOP expression evaluator, it cannot immediately compute ESOPs. Fortunately, the modification required is quite simple and consists of replacing the OR gate that computes the TCAM hit signal with an XOR gate.

It is straightforward to see that this extension indeed provides the same functionality as with the unmodified TCAM case. A non-covered cube placed on the TCAM's input register, corresponding to a packet for which routing information does not exist in the routing table, will either match an even number of TCAM entries (in which case the hit signal will evaluate to 0) or none at all, as in the case of

978-1-4244-2541-9/08 $25.00 © 2008 IEEE

TABLE I

COMPARISON OF SIZES OF SOPS AND ESOPS FOR GENERAL CLASSES
AND SPECIFIC IMPORTANT BOOLEAN FUNCTIONS.

f	SOP	ESOP
arbitrary	$64 \cdot 2^{n-7}, n > 6$	$29 \cdot 2^{n-7}, n > 6$
symmetric	$64 \cdot 2^{n-7}, n > 6$	$24 \cdot \sqrt{3}^{n-7}, n > 6$
parity	2^{n-1}	n
n-bit adder	$6 \cdot 2^n - 4 \cdot n - 5$	$2 \cdot 2^n - 1$
$\bigvee_{i=1}^{n} x_i y_i$	n	$2^n - 1$
ADR4	75	31
LOG8	128	99
MLP4	126	63
SQR8	180	112
WGT8	255	59

evaluating SOPs. A covered cube will be matched with an odd number of TCAM entries, thereby forcing the hit signal to be set to one.

When then hit signal is 1, the priority encoder will return a valid matching TCAM address. While it is possible to observe a valid address at the encoder output for a non-covered cube, this event is masked out by a 0 hit signal.

A. Area Overhead

We will estimate the area overhead for a synthesized TCAM with the proposed extension. The number of internal nodes of a full binary tree having n leaves is exactly $n-1$. Even for the case where the hit signal is mapped to a parity tree of 2-input XOR gates, the additional cost can be estimated to a single XOR gate per TCAM row. Since the width of a TCAM for the specific application is in excess of 100 bits and the size of a TCAM cell is significantly larger than the size of a single XOR gate, the area overhead is negligible when compared with the unmodified TCAM.

B. Delay Overhead

The classic approach to synthesizing OR trees is by interleaving layers of NAND and NOR gates as we show on Fig. 4. Therefore, we can estimate the speed overhead by comparing the maximum of the rise and fall times of the XOR gate with the average of the maximum of the rise and fall times of NAND and NOR gates for a specific target library.

In Table II we see the rise and fall times for 2-input OR, XOR, NAND and NOR gates extracted from a high-speed 90nm industrial library. We note that an OR tree will be approximately 22% faster than a parity tree. For example, for a 32K-entry TCAM, the additional slack required will be 150ps. While the actual impact is application-dependent, it does not seem likely that such an overhead would be consequential.

C. Discussion

In a very pragmatic sense, both the OR and the XOR hit signals can coexist, providing the capability of switching between an SOP and an ESOP evaluator at run time. Moreover, since the two signals will never be required at the same time, their corresponding circuits can be folded into one if

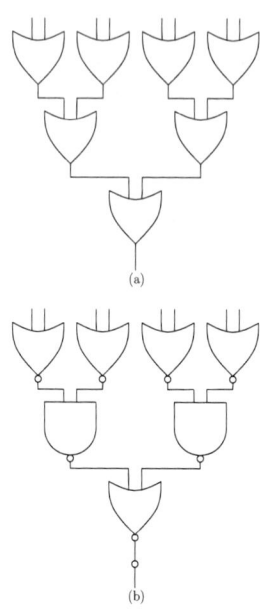

Fig. 4. An optimum way to synthesize an OR tree. (a) straightforward implementation with 2-input OR gates, (b) implementation with interleaved layers of faster 2-input NAND / NOR gates

TABLE II

SPEEDS OF 2-INPUT OR, XOR, NAND AND NOR GATES OF
COMPARABLE SIZE FROM A HIGH-SPEED 90NM INDUSTRIAL LIBRARY

gate	rise	fall
OR2	35ps	54.5ps
XOR2	46ps	40ps
NAND2	26ps	35ps
NOR2	37ps	16ps

this optimization is beneficial in terms of power consumption or required layout. We will not investigate this optimization further.

Note also that the cost of a parity tree and an OR tree in terms of both area and delay will most likely be exactly the same for the case where the circuit is mapped onto *Field Programmable Gate Array* (FPGA.)

III. ROUTING TABLE COMPACTION

A. Longest prefix Matching Algorithm

A key observation is that when multiple matches are possible on the routing table, the match with the longest prefix is returned. This routing algorithm is called *longest prefix matching*.

For example, let us examine the routing table on Table III, where the first column maintains the (IP, mask length)-tuples and the second column denotes the forwarding port. The *mask* field denotes the number of the most significant bits of the IP address that are relevant to routing.

Address 171.67.22.34 will be matched by the first and third table entry. However, the first one with the longest preamble (24) will be returned.

978-1-4244-2541-9/08 $25.00 © 2008 IEEE

TABLE III
SUBSET OF A ROUTING TABLE.

IPv4/mask	Port
171.67.22.0/24	5
169.229.0.0/16	2
171.67.0.0/16	2
128.112.0.0/16	2

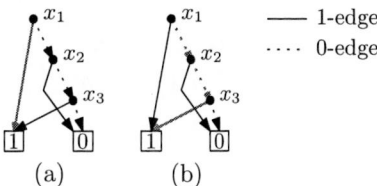

Fig. 6. Depth-first traversal of the BDD on Fig. 3 (b). (a) first path, generating cube x_1, (b) second path, generating cube $\bar{x}_1\bar{x}_2x_3$

B. Boolean Algebra Formulation

We will first cast the problem into an equivalent boolean algebra one, such that the original semantics are preserved. The routing table RT is partitioned into tables TR_j comprising entries of the same mask j. Each table TR_j is further partitioned into tables $TR_{j,k}$ that contain entries of the same output port k.

For each entry in TR_j or $TR_{j,k}$ we maintain the j most significant bits of the binary representation of its input address as a cube. For example, entry

$$169.229.0.0/16$$

on Table III corresponds to cube

$$x_{31}\bar{x}_{30}x_{29}\bar{x}_{28}x_{27}\bar{x}_{26}\bar{x}_{25}x_{24}x_{23}x_{22}x_{21}\bar{x}_{20}\bar{x}_{19}x_{18}\bar{x}_{17}x_{16},$$

which is constructed from $1010\,1001\,1110\,0101_2$.

Function g_j is defined as the logical OR of all cubes obtained from table TR_j. Similarly, function $g_{j,k}$ is defined as the logical OR of all cubes on table $TR_{j,k}$. Function $g'_{j,k}$ is then defined as

$$g'_{j,k} = g_{j,k} \bigwedge_{j'>j} \bar{g}_{j'}. \qquad (1)$$

For each output port k we define its characteristic function $f_k(\vec{y};\vec{x})$ as follows:

$$f_k(\vec{y};\vec{x}) = \bigvee_j (\vec{y}\odot[j])\cdot g'_{j,k}, \qquad (2)$$

where $[j]$ denotes the binary representation of j and $\odot$ the equivalence operation.

For example, functions $f_2(\vec{y};\vec{x})$ and $f_5(\vec{y};\vec{x})$ from Table III are constructed as shown on Fig. 5.

Each function f_k is subsequently independently minimized. For this, we have implemented two algorithms. The first algorithm is based on BDDs while the second utilizes the full power of ESOP expressions.

C. BDD Minimization Algorithm

For each f_k we construct the corresponding ROBDD. Following, we minimize the number of ROBDD nodes by applying variable reordering. For this, we use CUDD [20], an open-source, widely used BDD library developed at Colorado university.

After a minimal BDD is obtained, we extract from it a list of cubes as follows. We traverse the BDD in a depth-first manner starting from the root. Every time we reach terminal node 1, a new cube is extracted which is the product of the variables found on the particular path from the root, with polarities coinciding with the edges that have been followed.

For example, let us observe how the algorithm works on the simple BDD on Fig. 3 (b). The two paths leading to terminal node 1 are presented on Fig. 6. The generated cubes are x_1 and $\bar{x}_1\bar{x}_2x_3$ respectively.

By construction of the BDD, it holds that f_k is equal to the logical OR of the generated cubes. Note, the following property of orthogonality can be seen from the properties of ROBDDs.

The product of any two cubes e_{i_1}, e_{i_2}, corresponding to paths from the root node to terminal node 1, is always equal to 0. It is easy to see that if this were not the case, ROBDD properties (see Fig. 2) would be violated. Since $e_{i_1} \oplus e_{i_2} = e_{i_1} \vee e_{i_2}$ when $e_{i_1} \wedge e_{i_2} = 0$, the obtained cover is also an ESOP.

D. ESOP Minimization Algorithm

For ESOP minimization, we have used the algorithm presented in [17], [15]. While a full description of the algorithm is beyond the scope of this work, we will provide an outline of its operation below.

The algorithm is applied on a single output boolean function f. An initial cover F is obtained for f. The goal is to gradually reduce the size of F by selecting a number of cubes from it and transform them appropriately.

1) Initial Cover: There are numerous approaches for generating an initial cover. The simplest is the minterm cover, which is the exclusive-or of the minterms of f. Another initial cover can be the output of the BDD minimization algorithm. For this work we have selected to adopt the pseudo-Kronecker cover.

The pseudo-Kronecker cover is generated recursively. For a given boolean function f, we generate subfunctions

$$
\begin{aligned}
f_1 &= f(1,x_2,\ldots) \\
f_0 &= f(0,x_2,\ldots) \\
f_2 &= f_1 \oplus f_0
\end{aligned}
$$

and recursively obtain a minimal cover for each of them.

We can construct a cover for f from the covers of its subfunctions as follows

$$
\begin{aligned}
f &= x_1 f_1 \oplus \bar{x}_1 f_0 \\
&= f_0 \oplus x_1 f_2 \\
&= f_1 \oplus \bar{x}_1 f_2
\end{aligned}
$$

Out of the three possible choices, we select the one that leads to a smaller cover for f.

$$f_2 = y_4\bar{y}_3\bar{y}_2\bar{y}_1\bar{y}_0 \wedge$$
$$(x_{31}\bar{x}_{30}x_{29}\bar{x}_{28}x_{27}\bar{x}_{26}\bar{x}_{25}x_{24}x_{23}x_{22}x_{21}\bar{x}_{20}\bar{x}_{19}x_{18}x_{17}\bar{x}_{16} \vee$$
$$x_{31}\bar{x}_{30}x_{29}\bar{x}_{28}x_{27}\bar{x}_{26}x_{25}x_{24}\bar{x}_{23}x_{22}\bar{x}_{21}\bar{x}_{20}\bar{x}_{19}\bar{x}_{18}x_{17}x_{16} \vee$$
$$x_{31}\bar{x}_{30}\bar{x}_{29}\bar{x}_{28}\bar{x}_{27}\bar{x}_{26}\bar{x}_{25}\bar{x}_{24}\bar{x}_{23}\bar{x}_{22}\bar{x}_{21}\bar{x}_{20}x_{19}x_{18}\bar{x}_{17}\bar{x}_{16}) \wedge$$
$$\overline{x_{31}\bar{x}_{30}x_{29}\bar{x}_{28}x_{27}\bar{x}_{26}x_{25}x_{24}x_{23}x_{22}\bar{x}_{21}\bar{x}_{20}\bar{x}_{19}x_{18}x_{17}x_{16}\bar{x}_{15}\bar{x}_{14}\bar{x}_{13}x_{12}\bar{x}_{11}x_{10}x_9\bar{x}_8}$$

$$f_5 = y_4y_3\bar{y}_2\bar{y}_1\bar{y}_0 \wedge x_{31}\bar{x}_{30}x_{29}\bar{x}_{28}x_{27}\bar{x}_{26}x_{25}x_{24}\bar{x}_{23}x_{22}\bar{x}_{21}\bar{x}_{20}\bar{x}_{19}\bar{x}_{18}x_{17}x_{16}\bar{x}_{15}\bar{x}_{14}\bar{x}_{13}x_{12}\bar{x}_{11}x_{10}x_9\bar{x}_8$$

Fig. 5. Construction of functions f_2, f_5 from the entries on Table III.

2) Algorithm Outline: The algorithm iteratively performs the following steps: It scans F for pairs of Hamming distance-0 or distance-1 cubes, which are cubes that are either equal, or differ in exactly one literal, and removes them or merges them into a single cube respectively.

Subsequently, it selects two to five cubes and generates all possible ESOP subexpressions of the function they represent. If any of these new subexpressions is smaller, or generates more opportunities for distance-0 or distance-1 minimizations, it is selected and placed back into F.

The algorithm terminates after a predetermined number of iterations during which the size of the cover has not been reduced. More details can be found in [17].

E. Merging Step

After we obtain minimal ESOP covers for all possible f_k, we post-process their cubes in the following manner: For each cube c in L and all possible i, we AND c with literal y_i only if $y_i c \neq 0$. Following, each cube c in the cover of f_k is translated to "$p_c : k$" where p_c is the positional notation of cube c.

For example, cube

$$\bar{y}_3 y_1 x_5 \bar{x}_2 x_0$$

in the cover of f_2 will be transformed to

$$\bar{y}_3 y_2 y_1 y_0 x_5 \bar{x}_2 x_0$$

and subsequently translated to

$$01111XX \cdots XXXX1XX0X1 : 2.$$

All translated cubes from all covers are then together sorted non-increasingly. The parts corresponding to variables $\vec{y}$ are removed after sorting and the result is placed on the modified TCAM.

F. Discussion

We can easily verify that the boolean formulation described above respects the matching behavior of the longest prefix algorithm. The key differentiation from a flow that would optimize to SOP covers is found at eq. 1, which is used in eq. 2.

Let us observe a scenario where simply using functions $g_{j,k}$ in eq. 2 will lead to incorrect results. In the example based on Table III, functions f_2 and f_5, and therefore the covers obtained from the algorithms would both evaluate to 1 on input 171.67.22.34, forcing the parity tree on the modified TCAM to set the hit signal to 0.

In the case where f_2 is computed as shown on Fig. 5 however, it will correctly evaluate to 0 when presented with input 171.67.22.34.

IV. EXPERIMENTAL RESULTS

A. Dataset Collection

We have collected the routing table available on the AT&T WorldNet Common Backbone Route Monitor [21] that includes 219,002 entries.

According to AT&T, the information available through *route-server.ip.att.net* is offered by AT&T's Internet engineering organization to the Internet community.

The particular router has the global routing table view from a number of routers, providing a glimpse to the Internet routing table from the AT&T network's perspective.

By analyzing the routing table, we established that there exist 19 distinct mask lengths, specifically lengths 8-24 and 32. Moreover, there exist 8 distinct forwarding ports.

B. Experimental Setup

We collected our experimental results on a workstation based on Intel's Core2Duo 2.4GHz processor with 4GB of memory running SuSE Linux 10.1.

Three algorithms were benchmarked. Tool 1 is based on the BDD minimization algorithm. Tool 2 is ESPRESSO, a widely used SOP minimization tool developed at Berkeley [22]. Tool 3 is based on the ESOP minimization algorithm. The emphasis was on reducing the list of TCAM entries.

C. Results

The experimental results can be seen on Table IV. The main message obtained from these results is that the proposed routing table compaction provides *significant area benefits*.

Impressively, the compressed TCAM entry lists are 62.6%, 50.7% and 47.5% of the original TCAM entry list for the three tools respectively. Moreover, the additional gain when adopting ESOP expressions is more than 24% when compared with Tool 1 and more than 6% when compared to Tool 2.

In terms of time required to compute the resulting covers, we observe that Tool 2 is faster than both Tools 1 and 3. We note however that tool 3 reached a cover size equal to that of Tool 2 within just 70 seconds.

978-1-4244-2541-9/08 $25.00 © 2008 IEEE

TABLE IV

EXPERIMENTAL RESULTS. TOOL 1 IS BASED ON BDD MINIMIZATION.
TOOL 2 IS ESPRESSO. TOOL 3 IS BASED ON ESOP MINIMIZATION.

Tool	cover size	time (secs)
initial cover	219k	-
Tool 1 (BDD)	137k	783
Tool 2 (ESPRESSO)	111k	**114**
Tool 3 (ESOP)	**104k**	526

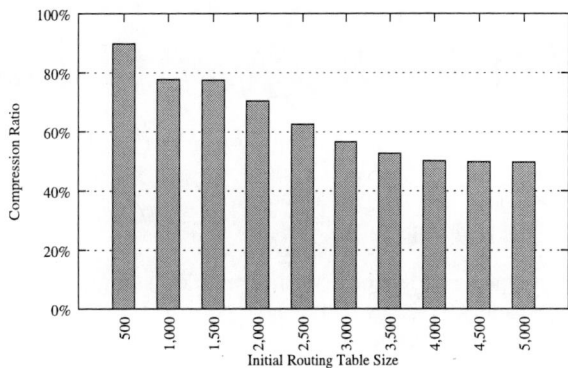

Fig. 7. Compaction Obtained from Tool 3 on Smaller Routing Tables, as a Percentage of the Initial Table Size

While the obtained execution times appear significant, we note that they are only necessary when the covers are initially generated. After that point, incremental updates will be significantly faster, especially for the ESOP minimization tool (and possibly for Tool 2,) since it will start every update from a near-optimal cover.

Such an argument is harder to establish for Tool 1, since it is not clear how the depth-first traversal part of the algorithm can be minimized for nearly isomorphic BDDs.

One may assume that significant compression ratios are only possible on very large routing tables. In order to investigate this, we minimized gradually larger TCAM tables of sizes in the range of 500 to 5,000.

On Fig. 7 we show that, in fact, the compression is more than 2X after only 5,000 initial TCAM entries, which is a small percentage of the full routing table. We note however that on small routing tables of 500 entries or less, the compression obtained is not enough to justify the overhead of the proposed methodology.

V. CONCLUSION

In this work, we investigated the applicability of *Exclusive-Or Sum-Of-Products expressions* in the context of routing table compaction for large-scale backbone internet routers.

We proposed an extension of *Content-Addressable Memories* so that they can be used as ESOP evaluators. We implemented two routing table compaction algorithms based on BDDs and two-level ESOP minimization respectively.

We evaluated our algorithms on real-life data obtained from AT&T's WorldNet Common Backbone Route Monitor and observed more than 2X compression , thus concluding that ESOPs provide significant CAM utilization savings.

We identify two possible directions for future research with this work. From a hardware viewpoint, it will be interesting to examine the trade-offs in space, power and time when more sophisticated boolean formulas, such as three-layer AND-OR-XOR expressions, are used to model TCAM based routing tables.

As a software optimization, it may be valuable to construct an application-specific ESOP minimization algorithm that takes into consideration the relevant properties of TCAMs.

REFERENCES

[1] S. Nilsson and G. Karlsson, "IP-Address lookup using LC-tries," *IEEE J. Selected Areas Comm.*, vol. 17, no. 6, pp. 1083–1092, June 1999.

[2] M. Waldvogel, G. Varghese, J. Turner, and B. Plattner, "Scalable high speed IP routing lookups," *SIGCOMM Comput. Commun. Rev.*, vol. 27, no. 4, pp. 25–36, 1997.

[3] M. Degermark, A. Brodnik, S. Carlsson, and S. Pink, "Small forwarding tables for fast routing lookups," *SIGCOMM Comput. Commun. Rev.*, vol. 27, no. 4, pp. 3–14, 1997.

[4] P. Gupta, S. Lin, and N. McKeown, "Routing lookups in hardware at memory access speeds," in *IEEE INFOCOM*, 1998, pp. 1240–1247.

[5] M. Kobayashi, T. Murase, and A. Kuriyama, "A longest prefix match search engine for multi-gigabit IP processing," *IEEE ICC*, 2000.

[6] H. Liu, "Reducing routing table size using ternary-CAM," *Hot Interconnects*, pp. 69–73, 2001.

[7] F. Zane, G. Narlikar, and A. Basu, "CoolCAM: Power-efficient tcams for forwarding engines," *INFOCOM*, April 2003.

[8] T.-B. Pei and C. Zukowski, "VLSI implementation of routing tables: tries and CAMs," *IEEE INFOCOM*, pp. 515–524, 1991.

[9] A. McAuley and P. Francis, "Fast routing table lookup using CAMs," *IEEE INFOCOM*, pp. 1382–1391, 1993.

[10] K. Pagiamtzis and A. Sheikholeslami, "Content-addressable memory (cam) circuits and architectures: a tutorial and survey," *Solid-State Circuits, IEEE Journal of*, vol. 41, no. 3, pp. 712–727, March 2006.

[11] H. Liu, "Routing table compaction in ternary CAM," *Micro, IEEE*, vol. 22, no. 1, pp. 58–64, Jan/Feb 2002.

[12] Understanding ACL on catalyst 6500 series switches. [Online]. Available: http://www.cisco.com/

[13] RFC 4271. a border gateway protocol 4 (bgp-4). [Online]. Available: http://tools.ietf.org/html/rfc4271

[14] R. Bryant, "Graph-based algorithms for boolean function manipulation," *Computers, IEEE Trans. on*, vol. C-35, no. 8, Aug 1986.

[15] S. Stergiou, K. Daskalakis, and G. Papakonstantinou, "A fast and efficient heuristic ESOP minimization algorithm," in *Great Lakes Symposium on VLSI*, April 2004, pp. 78 – 81.

[16] A. Mishchenko and M. Perkowski, "Fast heuristic minimization of exclusive-sums-of-products," in *5th International Workshop on Applications of the Reed Muller Expansion in Circuit Design*, August 2001.

[17] S. Stergiou, D. Voudouris, and G. Papakonstantinou, "Multiple-value exclusive-or sum-of-products minimization algorithms," *IEICE Transactions on Fundamentals*, vol. E.87-A, no. 5, May 2004.

[18] T. Sasao, "Exmin2: A simplification algorithm for exclusive-OR sum-of-products expressions for multiple-valued input two-valued output functions," *IEEE Trans. on CAD*, vol. 12, no. 5, May 1993.

[19] A. Gaidukov, "Algorithm to derive minimum ESOP for 6–variable function," in *5th International Workshop on Boolean Problems*, September 2002.

[20] CUDD: CU Decision Diagram Package Release 2.4.1. [Online]. Available: http://vlsi.colorado.edu/ fabio/CUDD/

[21] AT&T WorldNet Common Backbone Route Monitor. [Online]. Available: telnet://route-server.ip.att.net

[22] R. K. Brayton, A. L. Sangiovanni-Vincentelli, C. T. McMullen, and G. D. Hachtel, *Logic Minimization Algorithms for VLSI Synthesis*. Kluwer Academic Publishers, 1984.

[23] T. Hayashi and T. Miyazaki, "High-speed table lookup engine for IPv6 longest prefix match," *GLOBECOM*, pp. 1576–1581, 1999.

[24] C. Labovitz, G. Malan, and F. Jahanian, "Internet routing instability," *Networking, IEEE/ACM Trans. on*, vol. 6, no. 5, Oct 1998.

[25] V. Srinivasan and G. Varghese, "Fast address lookups using controlled prefix expansion," *ACM Trans. Comput. Syst.*, vol. 17, no. 1, pp. 1–40, 1999.

978-1-4244-2541-9/08 $25.00 © 2008 IEEE

Balancing Wrapper Chains of SoC Core Based on Best Interchange Decreasing

Maoxiang Yi
Department of Electronic Science and Technology
Hefei University of Technology, HFUT
Hefei, P.R.China
mxyi126@126.com

Huaguo Liang, Zhengfeng Huang
School of Computer and Information
Hefei University of Technology, HFUT
Hefei, P.R.China
huagulg@hfut.edu.cn, hanson_hfut@sina.com

Abstract—**An improved scheme for balancing wrapper chains partition of SoC core is proposed. Starting with the primary configuration created by LPT algorithm, we optimizes the current partition through the best interchange decreasing and iterative operation, in each step of which a pair of wrapper chains with maximum length difference is selected and the optimal two cells in the two wrapper chains are interchanged. Experiments are executed for the typical cores of the ITC'02 benchmarks. The results show that compared to the previous techniques, our scheme can create more balanced wrapper chains, decreasing the maximum scan shift length, hence the test application time of core.**

I. INTRODUCTION

It proves to be a very efficient technique to use the core-based design for shortening development cycle of the system on chip (SoC). However, ever-increasing design density and complexity of SoC core result in some key challenges in SoC testing [1,2], where increase in test data volume leads to a significant increase in investment of testing SoC. At present, the related research work focus on design for testability (DFT) [3,5,8], test data compression [4] and test scheduling [10], which aim at minimizing the test application time of SoC with some test constraints satisfied.

For SoC testing, DFT is one of the most important techniques to attack inaccessibility of internal cores and optimize test resources. DFT of SoC involves designing core wrapper and test access mechanism (TAM). The wrapper supports test isolation and mode switch for core in SoC, and implements width matching between core test ports and TAM. IEEE standard 1500 for embedded core test (SECT) [11] lefts great flexibility is for the logic implementation of wrappers, so as to meet special needs. Test wrapper/TAM co-optimization for SoC can effectively decrease the number of don't care bits in the data stored in Automatic Test Equipment (ATE) and reduce the redundant time created by transforming extra don't care bits when cores are tested, cutting the cost of SoC testing.

The cores used for designing SoC, including soft cores, firm cores and hard cores, are pre-designed and pre-verified

This work is supported in part by the Natural Science Fund of China (NSFC) under Grant 60633060.

design modules. For the last two kinds of cores, the length and number of internal scan chains depend on core providers. The balanced wrapper chains partition for the firm cores and hard cores is a NP-Hard problem [5]. Some related algorithms were presented to solve it [3,5,8,10]. In this paper, we present an algorithm called best interchange decreasing (BID) to balance the wrapper chains partition by combining with an iterative operation.

II. BASIC DEFINITIONS AND PRIOR WORK

Cores with the scan-test architecture generally involve a number of functional I/O ports and internal scan chains consisting of scan flip-flops. For a practical core, due to the limit on the number of SoC package pins, the number of its functional I/O ports and internal scan chains is much larger than the width of ATM assigned to it. Wrapper is a thin shell of DFT logic circuit around a core. Designing a core wrapper will convert each of functional I/O ports into one I/O scan cell and assign all the scan cells (including I/O scan cells and internal scan chain cells) over TAM wires to construct a set of wrapper chains. For the sake of convenience to describe, some definitions are given here.

A. Basic Definitions

Definition 1: wrapper chain (wc). Using a triple to denote a core C with scan test architecture, or $C= (FI, SC, FO)$, where FI stands for the set of functional input cells, FO is the set of functional output cells, SC stands for the set of the internal scan chain cells. Assume $WE = FI \cup SC \cup FO$ as a set consisting of all the scan cells, we define a wrapper chain wc as a scan cell sequence and the following conditions are met:

- $wc \in WE+$, where "+" means wc consists of some scan cells in WE

- $wc_i = wc_j$, if and only if $i = j$, for $i,j=1,2,\ldots,|wc|$, in other words, one scan cell can be in only one wrapper chain

- $wc = fi \mid sc \mid fo$, where "|" stands for concatenation between sub-chains, $fi \subset FI$, $sc \subset SC$, $fo \subset FO$

The length of a wrapper chain wc is

$$l(wc) = \sum_{i=1}^{|wc|} l(we_i)$$

The scan-in length of wc is $si_{wc}=l(wc)-|fo|$ and the scan-out length of wc is $soi_{wc}=l(wc)-|fi|$. When shifting in a test stimulus into a wc, we need to shift in only si_{wc} bits. Shifting out a test response is done in exactly so_{wc} clock cycles.

Definition 2: configuration K_C. K_C for C testing is defined as a set of wrapper chains, to only one of which each of the scan cells in WE belongs, expressed as

- $\forall we \in WE, \exists wc \in Kc : we \in wc$

- $we \in wc'(\in Kc) => \forall wc \in Kc \setminus \{wc'\} : we \notin wc$

$|K_C|$ presents the test width of C under K_C, which is equal to the number of TAM input wires assigned to C. The scan-in length and the scan-out length of K are, respectively:

$$si_K = \max_{wc \in K}(si_{wc}) \text{ and } so_K = \max_{wc \in K}(so_{wc})$$

Definition 3: test application time (T_K). T_K is defined as the number clock cycles needed to finish testing C in scan-test way, calculated as follows [5]:

$$T_K = \{1+\max(si_K, so_K)\} p + \min(si_K, so_K) \qquad (1)$$

where p stands for the number of test patterns for core C and is known for a given core.

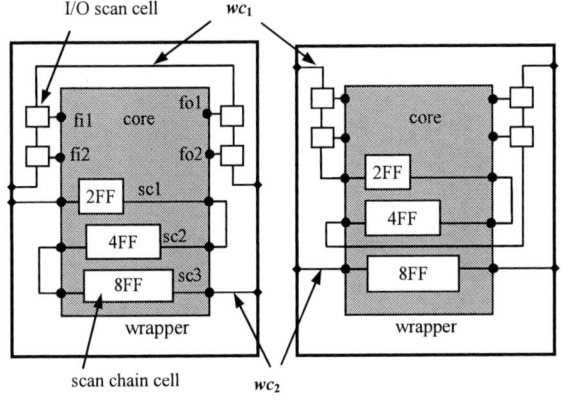

(a)   (b)
Figure 1. Wrapper chains partitions for core
（a）Configuration K_1 （b）Configuration K_2

Fig.1 shows two different configurations K_1 and K_2 for the core $C = (\{fi1,fi2\}, \{sc1,sc2,sc3\}, \{fo1,fo2\})$. Suppose that the width of TAM input wires $k = 2$ and $p =100$, for K_1, we have $si_{K1} = so_{k1}=14$ and $T_{K1} = 1514$, and for K_2, we have $si_{K2} = so_{K2} = 8$ and $T_{K2} = 908$. It is obvious that test application time is dependent on wrapper chain configuration.

Definition 4: Balanced Wrapper Chains Partition (BWCP). For given core C and the width k of TAM input wires, $|FI|=$n, $|FO|=$m, and $|SC|=$s, solving the K for C so that the test application time T_K is minimized, in other words, assigning all the $n+s+m$ scan cells over k of TAM input wires as balanced as possible.

B. Prior Work

The key to BWCP is to partition all the different length of internal scan-chain cells in SC into k wrapper chains. Literature [5] showed that BWCP is equivalent to solving multi-processor scheduling (MPS). BWCP is a NP-Hard problem because MPS is a NP-Hard [6]. Some BWCP techniques were presented [5,8,10], which are classified into heuristic and iteration-based algorithms.

Literature [5] solves BWCP by using the Largest Processor Time (LPT) heuristic algorithm that was presented for MPS. The algorithm time complexity is $O(s log s +s log k)$, but it's sub-optimum performance is low. The Best Fit Decreasing (BFD) algorithm [6] is exploited to design wrapper chains in wrapper/TAM co-optimization for SoC [8], in which, all the scan chain cells are ordered in length decreasing order, then each of them is assigned to one of the wrapper chains so that the length of the selected $wc \leq$ the length of the currently longest wc. If none of such a wc is found, the current scan chain cell is assigned to the shortest wc. In other words, the best fitting is reached when assigning a scan chain. The time complexity is $O(s log s +sk)$.

Another approach to MPS problem is MultiFit [9], which is based on iterative operation and bin-packaging algorithm. It updates the high bound and the lower bound of processor (bin) volume by bisection searching technique, and assigns each task to appropriate processor by the First Fit Decreasing (FFD) algorithm. Compared to LPT, it can reach a better result. The time complexity of this algorithm is $O(s log s +ds log k)$, where d is the iterative number. Marinissen [5] also proposed a technique called COMBINE to BWCP by combining LPT and FFD algorithms with linear searching technique, improving partition at the cost of computing time. Pouget [10] presented a partition technique, in which, an initial configuration is firstly created by treating each scan cell as one separate wc, then the two shortest wrapper chains are chained into one, forming a new configuration. The operation is done repeatedly until the width of the new configuration is equal to the number of TAM input wires.

III. OUR SCHEME AND IMPLEMENTATION

We propose a scheme to optimize BWCP based on the best interchange decreasing (BID) algorithm that is described as follows: Assume that there are two wrapper chains wc_x and wc_y and $val0 = l(wc_x) - l(wc_y) \geq 2$, let $val = int (val0/2)$, find a pair of scan cells $ex_i \in wc_x$ and $ey_j \in wc_y$ with length difference $diff = l(ex_i)-l(ey_j)$, so that $diff >0$ and $|diff-val|$ is minimum, interchange them between wc_x and wc_y.

Fig. 2 presents the code for BID algorithm, Alg_BID, which check the length difference of each pair of cells and find such a pair of cells that meets the given interchange conditions (line 3-10). If succeed, the two cells are exchanged between wc_x and wc_y whose lengths are then updated (line 11-12), returns 1, otherwise, return 0.

Fig.3 shows the code for implementing the proposed BWCP optimization procedure, called $WCPartition$, in which BID procedure in Fig.2 is used in iterative way. LPT

978-1-4244-2541-9/08 $25.00 © 2008 IEEE

Procedure Alg_BID(K, val0)

1   val = val0/2;// *val is the best exchange reference*
2   valok=-1; vdiff=val0; ixch=-1; kxch=-1; jend=0;
3   for (int i=0; i<|wc_x|; i++){ kx = -1; ndiff = -1;
4   for(int j= |wc_x|-1; j$\geqslant$jend; j--){ diff = wc_x(i)-wc_y(j);
5   if(diff $\leqslant$ 0) {jend = j-1; break;}
6   else if(diff $\geqslant$ val0) continue;
7   else{ ndiff = diff; kx=j; break;} }
8   if(kx != -1){ diffabs = |ndiff-val|;
9   if(diffabs < vdiff){ vdiff = diffabs; valok= ndiff;
10   ixch = i; kxch = kx; if(valok == val) break;} }
11   if(valok != -1){ Interchange wc_x(ixch) , wc_y(kxch)};
12   update $l(wc_x)$ and $l(wc_y)$; return 1;}
13   return 0;

Figure 2. Procedure of BID algorithm

Procedure *WCPartition (k)*

1   LPT(k);//create a primary configuration
2   sort wrapper chains for K in decreasing order
3   x = |K|; y = 0;
4   while(x < y -1){
5   while(y > x +1) { val0 = $l(wc_x)$ - $l(wc_y)$;
6   if(val0 $\geqslant$ 2 & *Alg_BID*(K, val0)) {
7   sort wrapper chains for K in decreasing order;
8   x = |K|; y = 0; }
9   else y--; }   x++; y = |K|; }

Figure 3. Procedure of optimization based on BID

algorithm is firstly invoked to cerate a primary configuration K (line 1), and |K| = k. In a double-layer repetition mechanism (lines 4-5), the longest wrapper chain wc_x and the shortest one wc_y are selected and *Alg_BID* is called if val0 = $l(wc_x)$ -$l(wc_y) \geqslant 2$ (line 7). *Alg_BID* returns 1 means that the optimal interchange occurs in it, the counters are reset and the internal repetition continues (lines 7-8). If *Alg_BID* returns 0, the next wrapper chain, which is located after and only after wc_y in increasing length, is selected to replace wc_y for the next repetition. If no interchange occurs in an internal repetition, the wc_x is removed temporarily from the current configuration and then the external repetition continues. The time complexity of the proposed approach to BWCP is: $O(s log s + s log k + d (log k+ (s / k)^2))$, where d is the iterative number.

Figure 4 demonstrates the stepwise optimization of wrapper chains partition for core 5 of ITC'02 SoC p22810, where k=4 and no I/O cell are taken into account. Based on the primary partition 4(a) created by LPT, selects the longest wc_1 and the shortest wc_3, val0=$l(wc_1)$- $l(wc_3)$ = 573-556 = 17 > 2, and uses val = int(val0 / 2) = 8 as the optimal reference. As a result, the ex_2 of length 100 in wc_1 and the ey_4 of length 92 in wc_3 are found with diff = $l(ex_2)$-$l(ey_4)$ = 8, which is

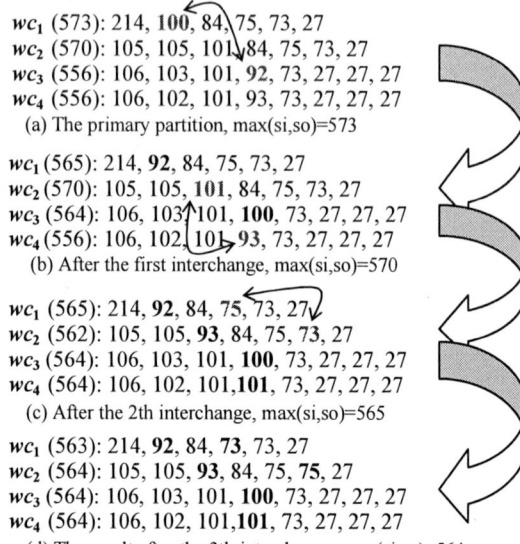

wc_1 (573): 214, **100**, 84, 75, 73, 27
wc_2 (570): 105, 105, 101, 84, 75, 73, 27
wc_3 (556): 106, 103, 101, **92**, 73, 27, 27, 27
wc_4 (556): 106, 102, 101, 93, 73, 27, 27, 27
    (a) The primary partition, max(si,so)=573

wc_1 (565): 214, **92**, 84, 75, 73, 27
wc_2 (570): 105, 105, **101**, 84, 75, 73, 27
wc_3 (564): 106, 103, 101, **100**, 73, 27, 27, 27
wc_4 (556): 106, 102, **101**, 93, 73, 27, 27, 27
    (b) After the first interchange, max(si,so)=570

wc_1 (565): 214, **92**, 84, 75, **73**, 27
wc_2 (562): 105, 105, **93**, 84, 75, 73, 27
wc_3 (564): 106, 103, 101, **100**, 73, 27, 27, 27
wc_4 (564): 106, 102, 101,**101**, 73, 27, 27, 27
    (c) After the 2th interchange, max(si,so)=565

wc_1 (563): 214, **92**, 84, **73**, 73, 27
wc_2 (564): 105, 105, **93**, 84, 75, **75**, 27
wc_3 (564): 106, 103, 101, **100**, 73, 27, 27, 27
wc_4 (564): 106, 102, 101,**101**, 73, 27, 27, 27
    (d) The result after the 3th interchange, max(si,so)=564

Figure 4. The stepwise optimization for wrapper chains partition of SoC p22810 core 5 (k=4)

most close to the optimal interchange condition and are interchanged between wc_1 and wc_3 and repositioned in length decreasing order. The new partition after the first interchange is showed in Fig.4 (b), which is further optimized in the same way until the result showed in (d) is reached after three interchanges. Through the optimization, the max(si,so) is reduced as 573 → 570 → 565 → 564.

IV. EXPERIMENTAL RESULTS

The proposed scheme is applied to all the cores of the ITC'02 SoC circuits [12-13] for BWCP. The experimental results are compared with that obtained by using the previous techniques, such as merge-based one presented in [10], BFD-based technique in [8] and COMBINE in [5]. The results show the effectiveness of our scheme.

It is the most important to assign all the scan chain cells in *SC* over TAM input wires to form as balanced wrapper chains as possible. On the other hand, no assignment of I/O cells is concerned in the known schemes [10] and [5]. Our experiments first deals with only the scan chain cells in *SC*. Considering the width k of TAM input wires varies from 3 to 8 due to that too many TAM wires for core results in no further improvement independently of used technique, we present the partial experimental results for typical cores of ITC'02 SoC p93791 and p22810, showed in Table I and II. From the 3th column, the tables give the experimental results of the maximum shift length max(si,so) by using the different schemes under various input TAM constraints.

Table III presents the results with all I/O cells considered and comparison to BFD-based technique in [8]. In our scheme, each of I/O cells is regarded as a scan chain cell with length 1 and is assigned to one of the wrapper chains created through BID algorithm, by applying LPT algorithm. A scan input cell is assigned to ensure that si_k is minimized,

978-1-4244-2541-9/08 $25.00 © 2008 IEEE

and a scan output cell is assigned to minimize so_k. The experimental results with different input TAM constraint are presented by two items, the upper one is the $\max(si,so)$, and the lower one is the test application time T.

TABLE I. COMPARING MAX(SI,SO) OF SoC P93791 CORE WITHOUT I/O CELLS CONSIDERED

Core	Alg.	k					
		3	4	5	6	7	8
C1	[10]	2522	2336	1943	1304	1218	1168
	[8]	**2267**	1747	1440	1163	1017	890
	[5]	2316	1805	1362	1204	1052	903
	Ours	**2267**	**1732**	**1361**	**1162**	**1016**	**876**
C13	[10]	3499	3408	2620	1752	1747	1744
	[8]	3281	2463	2028	1645	1428	1242
	[5]	3238	2396	1925	1688	1448	1263
	Ours	**3176**	**2382**	**1907**	**1619**	**1401**	**1227**
C20	[10]	2880	2724	1846	1440	1440	1380
	[8]	2535	1874	1526	1322	1143	998
	[5]	2501	1932	1560	1290	1122	1000
	Ours	**2486**	**1863**	**1502**	**1248**	**1070**	**972**

TABLE II. COMPARING MAX(SI,SO) OF SoC P22810 CORES WITHOUT I/O CELLS CONSIDERED

Core	Alg.	k					
		3	4	5	6	7	8
C5	[10]	852	793	610	430	422	407
	[8]	763	572	461	389	342	295
	[5]	763	577	457	**380**	333	**286**
	Ours	**753**	**564**	**452**	387	**329**	295
C9	[10]	805	776	653	405	400	395
	[8]	781	586	489	391	389	**295**
	[5]	**760**	581	477	395	**380**	**295**
	Ours	**760**	**576**	**476**	**390**	**380**	**295**
C26	[10]	5096	3197	3192	2996	2100	1600
	[8]	3939	2943	2409	2063	1809	**1530**
	[5]	3869	2991	**2328**	1931	1754	**1530**
	Ours	**3858**	**2925**	**2328**	**1930**	**1728**	**1530**

In Table I, II and III, the items in **bold** are the best archived results. The experimental results show that compared to the previous known techniques, the proposed scheme based on BID algorithm for BWCP can efficiently decrease the maximum scan shift length of testing core, hence reducing the test application time.

V. CONCLUSION

For BWCP, the existing heuristic algorithms have lower time complexity, butt he partition results are rough. Though the techniques based on iterative operation combined with heuristic algorithms improve the partition results, time complexity increases and the convergence depends on the initial conditions. The proposed scheme starts from the primary configuration created by LPT and uses BID

TABLE III. COMPARING MAX(SI,SO) AND T OF SoC CORE WITH I/O CELLS CONSIDERED

Core	Alg.	k					
		3	4	5	6	7	8
p93791 C13	[8]	3281	2463	2028	1645	1428	1242
		639989	480479	395654	320969	278654	242384
	Ours	**3237**	**2428**	**1942**	**1619**	**1401**	**1227**
		631382	**473634**	**378868**	**315899**	**273389**	**239459**
p22810 C26	[8]	3939	2945	2409	2063	1809	**1530**
		717079	536169	438619	375647	329419	278641
	Ours	**3883**	**2925**	**2330**	**1942**	**1728**	**1530**
		706876	**532531**	**424239**	**353619**	**314677**	**278641**

algorithm to optimize the current configuration in iterative operation, balancing the wrapper chains and reducing the test application time of cores. The convergence of our iterative operation is good and the time complexity is moderate. Together with the techniques related to TAM design and test scheduling, the proposed scheme can be used for actual SoC test integration.

REFERENCES

[1] Y. Zorian, S. Dey, and M. Rodgers, "Test of future system-on-chips," In Proceedings International Conference on Computer-Aided Design (ICCAD), 2000, pp. 392-398.

[2] E. J. Marinissen, Y. Zorian, R. Kapur, T. Taylor, and Lee Whetsel, "Towards a Standard for Embedded Core Test: An Example," in Proceedings IEEE International Test Conference, 1999, pp. 616-627.

[3] S. Koranne, "Design of Reconfigurable Access Wrappers for Embedded Core Based SoC Test," IEEE Transactions on Very Large Scale Integration (VLSI) System, vol. 11, no. 5, 2003, pp. 955-960.

[4] M Tehranipoor, M Nourani and K Chakrabarty. Nine-Coded Compression Technique for Testing Embedded Cores in SoCs, IEEE Trans. On VLSI Systems, vol. 13, no. 6, 2005, pp.719-731.

[5] E. J. Marinissen, S. K. Goel, and M. Lousberg, "Wrapper Design for Embedded Core Test," In Proceedings IEEE International Test Conference, Atlantic City, NJ, October 2000, pp. 911–920.

[6] M. R. Garey and D. S. Johnson, Computers and Intractability: A Guide to the Theory of NP Completeness, San Francisco, CA: W.H. Freeman and Co., 1979.

[7] R. L. Graham, "Bounds on Multiprocessing Anomalies," SIAM Journul of Applied Mathematics, 17, 1969: 416-429.

[8] V. Iyengar, and K. Chakrabarty, "Test Wrapper and Test Access Mechanism Co-Optimization for System-on-Chip," Journal of Electronic Testing: Theory and Applications, vol. 18, no. 2, 2002, pp. 213–230.

[9] D. K. Friesen, "Tighter Bounds for the Multifit Processor Scheduling Algorithm," SIAM Journul of Computing, vol. 13, no. 1, 1984 pp. 170-181.

[10] J. Pouget, E. Larsson and Z. Peng, "Multiple-Constraint Driven System-on-Chip Test Time Optimization," Journal of Electronic Testing: Theory and Applications, vol. 21, 2005, pp. 599–611.

[11] http://grouper.ieee.org/groups/1500, IEEE Standard 1500 Web Site.

[12] http://www.hitech-projects.com/itc02socbenchm/, ITC' 02 SoC Test Benchmarks Web Site.

[13] E. J. Marinissen, V. Iyengar, K. Chakrabarty, "A set of Benchmarks for Modular Testing of SoCs," In Proceedings IEEE International Test Conference, 2002, pp. 519-528.

Area-Efficient Low-Cost Low-Dropout Regulators Using MOS Capacitors

Hamed Aminzadeh, Reza Lotfi, and Khalil Mafinezhad

Integrated Systems Lab., EE Dept., Ferdowsi Univ. of Mashhad, Mashhad, I.R.Iran

E-mails: *haminzadeh@ieee.org , rlotfi@ieee.org, kh_mafi@yahoo.com*

Abstract—**Traditional design of low-dropout regulators offer the use of metal-insulator-metal (MIM) compensation capacitors to prevent instability in the absence of load capacitor with equivalent series resistance (ESR). In addition to area efficiency achieved by replacing these capacitors with MOS transistors, the location of implanted transfer function poles and zeros are adaptively changed according to the value of load current. The idea has been applied to stabilize a 1.2V, 100mA low-dropout regulator in a 0.18µm CMOS n-well process. Using the proposed technique, the regulator meets stability with a small 100pF MOS output capacitor and no ESR.**

I. INTRODUCTION

The increasing demand for portable battery-operated products presents power management designers with new challenges. To increase battery-life and to achieve better power efficiency, low-dropout regulators (LDOs) are essential. Unfortunately, the tradeoff between stability and dropout voltage of linear regulators makes uncompensated LDOs potentially unstable. Conventional LDOs employ the ESR of an off-chip capacitor to stabilize the closed-loop circuit. This resistor introduces a left-half-plane (LHP) zero to the loop-gain transfer function which counteracts the additional negative phase shift introduced by one of the two dominant poles. ESR-based compensation, however, has serious drawbacks. The large capacitor at the output can not be integrated. Its ESR strongly depends on temperature and capacitor type. There are additional transient voltage ripples originated from ESR, bond wires and so forth. Concerning these issues, compensation solutions relying on integrated components are becoming popular [1, 2].

It is very desirable to integrate the analog portion of a large mixed-signal system in standard digital CMOS technologies with no analog features. Digital circuits, as the major part of a state-of-the-art signal processing system, require only a single poly layer for the gates. However, in order to implement constant capacitors for analog applications, a second poly or extra metal layers are introduced into the process, resulting in significant increase in fabrication cost. Furthermore, although available metal layers in mixed-signal technologies can be utilized for MOM (metal-oxide-metal) capacitors, due to the relatively lower scaling rate of the oxide between these layers, the occupied physical area is noticeable [3]. Besides, a parasitic capacitance will exist between the bottom plate of an MOM capacitor and substrate.

To avoid these constraints, one of the possible solutions is to employ MOS gate junctions as capacitor. The gate oxide is thin and compared to MIMs, CMOS capacitors called MOSCAPs have larger capacitance per unit area. The main problem in largely exploiting MOSCAPs in analog applications is due to linearity issues. This is because of different regions a MOSFET experiences when its gate-bulk voltage varies. The regions, shown in Fig. 1, are accumulation, depletion and inversion [4]. For small bias voltages, the transistor is working in depletion thereby leaving the capacitor a function of the gate-bulk voltage. This degrades overall performance and mostly adds complexity to the design of analog circuits [4, 5]. Interestingly, the type of nonlinearity a MOSCAP has can be even helpful in improving the linearity of an LDO. Traditionally-designed LDOs suffer from load-dependent stability due to the variation of load current. Hence the frequency response is only optimized for a particular load current. Frequency compensation based on load-dependent zeros has therefore been proposed to fairly mitigate this issue [6]. The location of each pole or zero in loop transfer function is inversely proportional to the product of a resistance and capacitance.

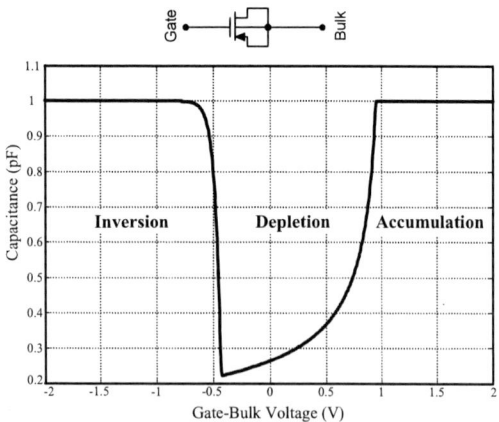

Figure 1. *C-V* charastristic of a p-Channel MOSFET

978-1-4244-2541-9/08 $25.00 © 2008 IEEE

Hence load-dependent zeros are produced using variable resistors realized by MOSFETs operating in triode [6]. However, another way to define the location of zeros (even poles) can be based on the variable capacitors. In this paper, we are going to demonstrate the capability of MOSCAPs in fully integrating LDOs in conventional digital technologies.

II. LDO DESIGN USING MOS CAPACITORS

Practically speaking, depletion-mode MOSCAPs can be widely utilized in deep sub-micron technologies [4]. Whenever required, non-linearity compensation techniques can be utilized. Fig. 2 shows a serial-compensated depletion-mode MOSCAP (SCDM) using three transistors. The gates of the two input transistors are connected to ground via a large resistance. Fig. 3 shows the C-V diagram of the result [4]. Compared to the uncompensated MOSCAP illustrated in Fig. 1, linearity is meaningfully improved but at the expense of more silicon area.

System-level architecture of an LDO consists of an error amplifier, a pass device, feedback network and a bandgap reference. Fig. 4 shows a possible circuit-level implementation of an LDO along with frequency compensation. Two compensation capacitors, namely C_{C1} and C_{C2}, are intended to properly shape the frequency response of the circuit. Fig. 5 shows the corresponding small-signal equivalent circuit. Assuming R_{C1} to be large, the effect of C_{C1} on the first stage output via i_C is negligible because of the relatively low induced ac current. The transfer function is thus obtained as follows:

$$\frac{V_{fb}}{V_{in}} \approx \frac{LG_0(1+s/z_1).(1+s/z_2)}{(1+s/p_1).(1+s/p_2).(1+s/p_3).(1+s/p_4)}, \quad (1)$$

where:

$$LG_0 = g_{mi}g_{mS}g_{mL}R_iR_S(R_L\|(R_{F1}+R_{F2}))\frac{R_{F2}}{R_{F1}+R_{F2}}, (2)$$

$$z_1 = 1/R_{C1}C_{C1}, \quad (3)$$

$$z_2 = 1/(R_{C2}-1/g_{mP})C_{C2}, \quad (4)$$

$$p_1 = 1/R_S(C_S+C_{C2}(1+g_{mP}(R_L\|(R_{F1}+R_{F2})))), \quad (5)$$

$$p_2 = 1/((R_L\|(R_{F1}+R_{F2})\|R_S)(C_S+C_L)), \quad (6)$$

$$p_3 = 1/R_iC_i, \quad (7)$$

$$p_4 = 1/(C_S+C_{C2})(R_{C1}\|R_S). \quad (8)$$

C_{C2} dominates the pole at the input of pass device (i.e. p_1) and pushes output pole (p_2) to relatively higher frequencies (well-known pole-splitting action in Miller compensation [3]). R_{C2} in series with C_{C2} creates a LHP zero (z_2) which cancels out the undesirable effect of high frequency poles (p_3, p_4). As C_{C2} is SCDM and R_{C2} is constant, the magnitude of z_2 is not affected

by the ripples of output due to load current changes. This is not however true for C_{C1} which is an uncompensated MOSCAP.

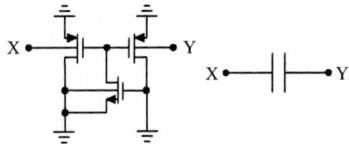

Figure 2. A serial-compensated depletion-mode (SCDM) MOS capacitor

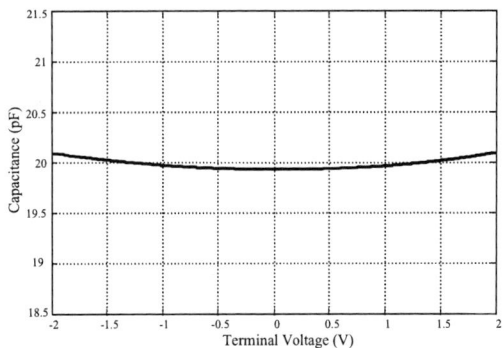

Figure 3. C-V charastristic of Fig. 2 ($W/L = 1180\mu/20\mu$)

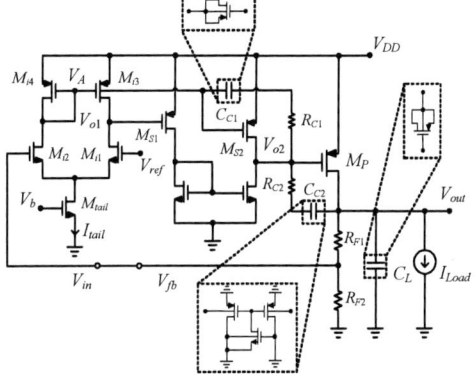

Figure 4. Detailed schematic of MOSCAP-compensated LDO

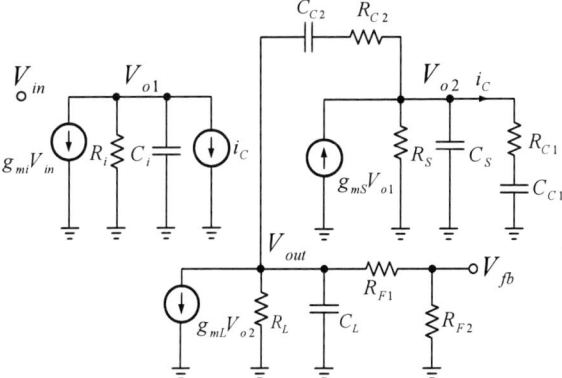

Figure 5. The equivalent small-signal circuit of the loop

978-1-4244-2541-9/08 $25.00 © 2008 IEEE

As a result, z_1 which is produced by this capacitor becomes a function of the load current. The LHP pole located at the output of LDO increases linearly with the load current according to the following well-known expression [6]:

$$p_2 \approx 1/r_{DSp}C_L = \lambda_p I_{Load}/C_L, \qquad (9)$$

where r_{DSp} and λ_p are the output resistance and channel length modulation of pass device, respectively. The zero introduced by C_{C1} is intended to counteract the phase lag introduced by this pole. Based on the well-known I-V relation of a MOSFET, the DC component of V_{o2} and V_A are respectively given by:

$$V_{o2} = V_{DD} - |V_{Tp}| - \sqrt{\frac{2I_{Load}}{\mu_p C_{ox}(W/L)_p}}, \qquad (10)$$

$$V_A = V_{DD} - V_{SG,i3} = V_{DD} - |V_{Tp}| - \sqrt{\frac{I_{tail}}{\mu_p C_{ox}(W/L)_{i4}}}. \qquad (11)$$

μ_p, C_{ox}, $(W/L)_p$ and $(W/L)_{i4}$ are the hole mobility, gate capacitance per unit area and aspect ratio of pass device and M_{i4} respectively. The gate-bulk voltage of C_{C1} (V_{gb}) is therefore expressed as:

$$V_{gb} = V_{o2} - V_A = \sqrt{\frac{I_{tail}}{\mu_p C_{ox}(W/L)_{i4}}} - \sqrt{\frac{2I_{Load}}{\mu_p C_{ox}(W/L)_p}}, \qquad (12)$$

This equation shows that V_{gb} is proportional to the square root of I_{Load}. To force C_{C1} working in depletion for the entire range of the load current, V_A can be properly set. If this is done, the capacitor value can be approximated as $a(V_{gb} + b)^2 + c$ where a, b and c are constant (see Fig. 1). Hence:

$$C_{C1} = a\left(\sqrt{\frac{I_{tail}}{\mu_p C_{ox}(W/L)_{i4}}} - \sqrt{\frac{2I_{Load}}{\mu_p C_{ox}(W/L)_p}} + b\right)^2 + c, \qquad (13)$$

When I_{Load} increases, to account for the variations of p_2, (13) shows that C_{C1} decreases to push z_1 into higher frequencies. The decrease in C_{C1} however is limited by the difference between minimum and maximum values of C_{C1} in depletion.

It is also important to investigate the effect of power supply on the location of poles and zeros because stability must be independent of V_{DD}. For a particular load current, the source-gate voltage of pass device is constant. Hence V_{o2} follows the variations of V_{DD}. C_{C2}, as an SCDM, is not indeed affected by this phenomenon because its absolute value is almost independent of the operating point (Fig. 3). This is the reason why an SCDM with minor variations is employed for realizing this capacitor. C_{C1}, on the other hand, is dependent on its operating point. Nevertheless, the terminal voltage of this capacitor is as well independent of V_{DD} because the input stage of error amplifier is biased with constant current, I_{tail}. Hence nodes V_A and V_{o2} are both V_{SG} lower than the V_{DD} in which V_{SG} is independent of power supply.

Almost all state-of-the-art LDOs require an on-chip capacitor at the output (C_L in Fig. 4) for enhancing ac and transient responses. MIMs are conventionally employed to implement this capacitor. As an alternative approach, C_L can be an uncompensated MOSCAP with higher density. Fig. 6 shows the C-V diagram of the 100pF integrated output capacitor used in the proposed LDO. Output voltage is large enough to maintain the operating point in accumulation or perhaps inversion. Furthermore, the output is always under regulation to have minor variations in magnitude. This guarantees the fact that C_L is mostly remained in voltage-independent regions under different transient conditions. Employing such a capacitor at the output is very important to significantly reduce silicon area and overall cost. No change in circuit performance of the circuit is observed when 100pF MIM capacitor of initial design is replaced with its equivalent uncompensated MOSCAP. However, the area efficiency is considerable (100000 μm^2 vs. 18000 μm^2 in our technology).

III. DESIGN EXAMPLE

The LDO shown in Fig. 4 has been simulated in a 0.18μm CMOS digital process. Table I summarizes the performance of the circuit for C_L=100pF.

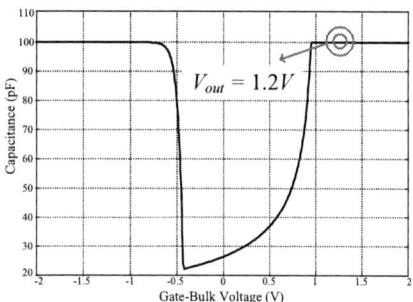

Figure 6. C-V diagram of the load capacitance (W/L = 590μ /20μ)

TABLE I.　　PERFORMANCE SUMMARY WITH C_L = 100pF

Technology	0.18μm Standard Digital Technology	
Output Voltage	1.2V	
Dropout Voltage	0.2V	
Maximum Load Current	100 mA	
DC Load Regulation (V_{DD} = 1.4V, I_{Load} = 0.01-100mA)	16 μV/mA	
DC Line Regulation (V_{DD} = 1.4-3.4V)	I_{Load} = 100μA	0.8mV/V
	I_{Load} = 100mA	1mV/V
Quiescent Current	I_{Load} = 100μA	54 μA
	I_{Load} = 100mA	83 μA
Transient Settling Time (at 0.1% error)	Load (0.01mA –100mA V_{DD} = 1.4V)	835ns +
		1350ns -
	Line (V_{DD} = 1.4-3.4V) I_{Load} = 100μA	280 ns
	Start-up (V_{DD} = 1.4V I_{Load} = 100mA)	295 ns

It requires a 0.9V bandgap reference for regulation. Fig. 7 shows the load-transient response of the circuit for a 100mA current step. The 0.1% settling error is 835ns and 1350ns for positive and negative edges. The line-transient response when V_{DD} changes from 1.4V to 2V is illustrated in Fig. 8. Fig. 9 depicts the start-up transient response for the case of dropout voltage and 100mA load current. During the start-up, the MOSCAP load capacitance changes from 26pF to 100pF when the output settles down to its final value (Fig. 6). Hence, the start-up transient response is non-linear. Nevertheless, simulations show that this variable C_L has even a positive

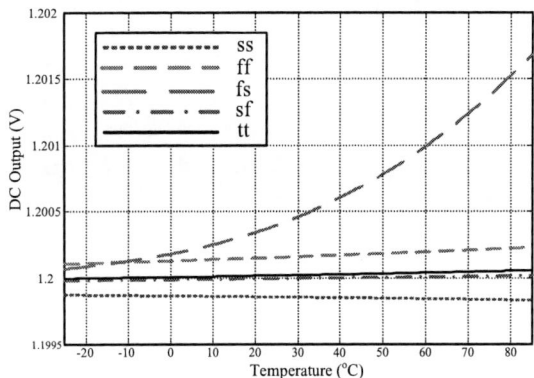

Figure 10. DC output as a function of process and temperature corners [-25°C, 85°C], where ss: slow *n*MOS slow *p*MOS; ff: fast *n*MOS fast *p*MOS fs: fast *n*MOS slow *p*MOS sf: slow *n*MOS fast *p*MOS; tt: typical condition

effect on start-up settling time because a smaller load is driven by the circuit when beginning operation. At last, the integrity of the circuit is confirmed in process and temperature corners. Fig. 10 shows the DC output in different conditions.

IV. CONCLUSION

Integrating all building blocks and reducing the physical area, while achieving the same or better performance, is an important task towards cheaper System-on-Chip (SoC) systems. An efficient way to stabilize the transient response of low-dropout regulators is proposed in this paper. It is based on MOS capacitors which have compact size and are available in all standard digital technologies. Detailed explanations to properly employing the capacitors are given. Realizing the output capacitor, by an uncompensated area-efficient MOSFET is a general viable idea proposed here. It can be applied to all integrated low-cost low-dropout regulators.

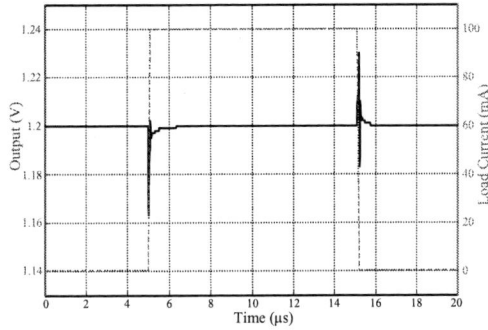

Figure 7. Load-transient response for 100 mA load current change

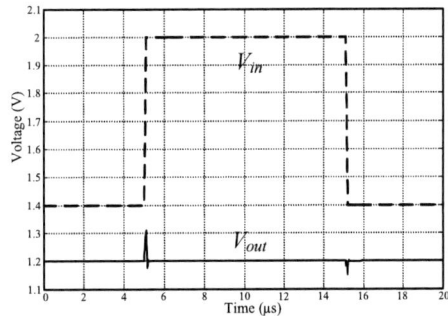

Figure 8. Line-transient response for 0.6V V_{DD} pulse (C_L = 100pF)

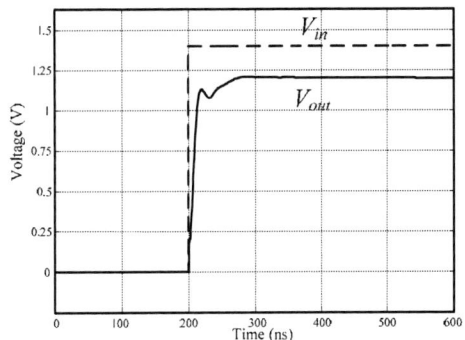

Figure 9. Start-up response for $V_{Dropout}$ = 0.2V (C_L = 26~100pF)

REFERENCES

[1] K. N. Leung and P. K. T. Mok, "A capacitor-free CMOS low-dropout regulator with damping-factor-control frequency compensation," *IEEE J. Solid-State Circuits*, vol. 38, no. 10, pp. 1691–1702, Oct. 2003.

[2] R. J. Milliken, J. S. Martinez, and E Sanchez-Sinencio, "Full on-chip low-dropout voltage regulator," *IEEE Trans. Circuits Systems*, vol. 54, no. 9, pp. 1879-1890, Sep. 2007.

[3] G. A. Rincon-Mora, "Active capacitor multiplier in Miller compensated circuits," *IEEE J. of Solid-state Circuits*, vol. 35, no. 1, pp. 26-32, Jan 2000.

[4] T. Tille, *et al.* "Design of low-voltage MOSFET-only sigma-delta modulators in standard digital CMOS technology," *IEEE Trans. Circuits Systems*, vol. 51, no. 1, pp. 96-109, Jan. 2004.

[5] H. Aminzadeh, M. Danaie, and R. Lotfi, "Design of high-resolution MOSFET-only pipelined ADCs with digital calibration," *IET Design, Automation and Test in Europe (DATE '07)*, pp. 427-432, Apr. 2007.

[6] K. C. Kwok and P. K. T. Mok, "Pole-zero tracking frequency compensation for low dropout regulator," *IEEE International Symposium on Circuits and Systems*, vol. 4, pp. 735-738, May 2002.

A 65nm CMOS Down-Sampling Micromixer with Enhanced DC Current Capability

Kurt Schweiger and Horst Zimmermann

Institute of Electrical Measurement and Circuit Design, Vienna University of Technology

Gusshausstrasse 25/35, 1040 Vienna

Austria

kurt.schweiger@tuwien.ac.at, horst.zimmermann@ieee.org

Abstract—This paper presents a low power CMOS down sampling micromixer in 65nm with an improved input stage. It is able to handle a superimposed DC current on the input port without degrading circuit performance. The mixer was fabricated in a triple-well process. A conversion gain of 17dB is achieved while only consuming $780\mu W$ from a 1.2V voltage supply up to a 3dB clock frequency of 600MHz. The gain decreases for 3dB at a superimposed DC current of $210\mu A$. The 1dB compression point and IIP3 are measured to be -22.7dBm and -16dBm, respectively.

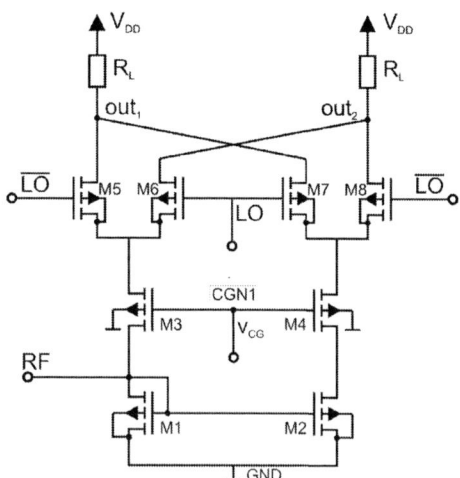

Fig. 1. Schematic of a single ended receiver chain

I. INTRODUCTION

Wireless communication is getting more important in every-day life. More and more devices are equipped with additional wireless interfaces for data exchange, logging or configuration purposes. Due to the additional integrated services in devices a faster data processing is needed as well. The increased use of radio transceiver systems leads to the demand for cheaper production of integrated circuits. The only way of achieving the digital performance and low-cost production for this type of broadened application is for the whole transceiver system to be integrated into one chip. That means that even the analog front-end circuits such as low-noise amplifier (LNA), mixer and filter need to be integrated into digital CMOS technology. CMOS transistors optimized for digital circuits have a high threshold voltage due to the need of low-power applications to present the possibility to switch off idle circuit blocks. That high threshold voltage increases the challenge for the analog design, since with a low-voltage supply the possibility of complex circuitry is limited by the supply voltage. The low supply voltage is necessary because of the very short transistor length and thin gate oxide. Higher supply voltages would cause breakthrough effects caused by too high electrical field strength. Furthermore the analog transistor properties in new CMOS technologies are very poor. For example Early voltages are in the range of only several volts.

Key system blocks for receiver systems are down-sampling mixers. These transform the high frequency modulated data signal to the baseband for further processing in the digital domain. They are placed directly after the LNA which is directly positioned after the antenna to minimize the analog circuitry preceding the digital base band processing. Fig. 1 shows the schematic of a modern receiver system with a minimum necessary analog front end. In this paper a mixer

with single ended input was designed and will be presented in detail in the following sections. Special attention will be given to the advanced input stage which can process several hundred μA DC-current superimposed on the input signal without degrading the mixer operation.

II. CIRCUIT DESIGN

In many wireless applications down-sampling mixers with single-ended input are necessary either because a single-ended antenna or a single-ended LNA is used. In general, single-ended topologies have the drawback of decreased linearity and 2nd-order distortions. Those distortions can be suppressed when a fully differential structure is chosen.

Fig. 2. Classical single ended micromixer [1]

In Fig. 2 the classical single ended design is shown, which is called micromixer in the literature. It is based on the fully differential standard Gilbert-type mixer [2]. Its RF-stage is adapted from a simple differential transconductance stage to a common-gate amplifier and a current mirror [3]. The drawback of the limited input-signal power of the differential transconductance stage can be overcome by that technique. The common-gate stage (transistor M3) can handle a virtually "unlimited" negative input signal while the current mirror (M1 and M2) can handle the "unlimited" positive signal. The combination of both shows the best requirements for a large input signal capability. Transistor M4 is added for symmetry in the two signal paths (large positive and negative signals).

In this paper a down-sampling mixer based on the micromixer is presented. The main goal of this work was to improve the circuit to be able to handle a superimposed DC current on the input node. This is useful in applications where the DC operating point is directly set from the preceding amplifier. The output DC voltage of a circuit can only be regulated to a certain accuracy with little complexity. If the following circuit block is tolerant to variations in the DC operating point, it is possible to reduce complexity which means that some chip area might be saved. As a consequence it is then possible to reduce the overall power consumption which is the main goal for mobile devices.

First of all the resistive load was replaced by an active load to increase the gain of the mixer. A common-mode feedback controller was added to regulate both outputs to approximately $V_{DD}/2$. This is necessary to provide a proper DC-voltage level for the mixer core transistors to get a well-defined operating point and to provide a fixed DC-voltage level for following circuit blocks. The proposed mixer is shown in Fig. 3. In the standard micromixer the gate voltage for the common-gate amplifier is kept constant (node CGN1). To improve the mixer to be able to process a DC current on the input a feedback circuit is added. The node voltage CGN1 is adapted for different input currents. It is regulated to be by a constant voltage higher than the input node itself. This guaranties a constant gate-source voltage for the common-gate amplifier for any input voltage level and therefore a virtually unchanged behavior. A limitation to that technique is the supply voltage itself. The gate node can only be shifted up to nearly V_{DD}, at which point the compensation will fail. Other limitation to the input current range is given by the ability of the common-mode controller to adjust the output voltage accordingly to get a valid operating point for the mixer core.

Simulations show excellent performance with currents up to several hundred μA which was also confirmed through measurements. The measurement setup and the results are shown in the next section.

III. MEASUREMENT RESULTS

The designed down-sampling mixer with single-ended input was fabricated in a 65nm low-power digital CMOS process without any analog extensions. The used low-power process has threshold voltages for NMOS and PMOS transistors of

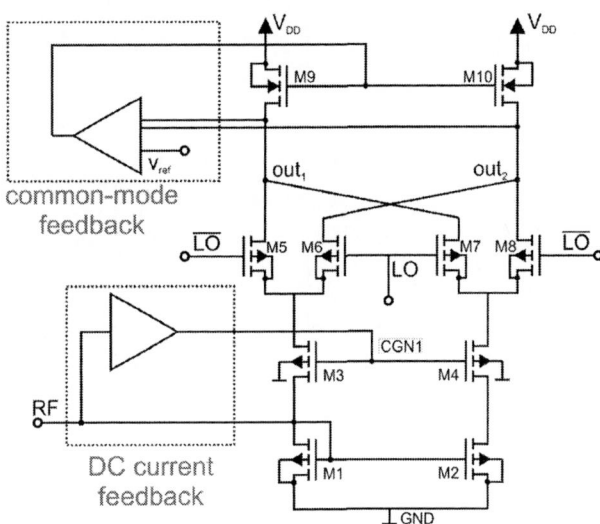

Fig. 3. Proposed single-ended mixer with input DC current compensation

about 500mV and 400mV, respectively. For measurement purpose a gold plated printed circuit board (PCB) was designed and fabricated where the chip was directly bonded. Direct bonding reduces the parasitic capacitances and inductances from any kind of package. Special care was taken to place the differential clock lines symmetrically onto the board. Furthermore the transmission lines have been designed to have 50Ω for impedance matching to the external measurement equipment. For the clock signal (LO) an external balun was used to transform the single-ended signal from signal sources to a fully differential signal for the device under test (DUT).

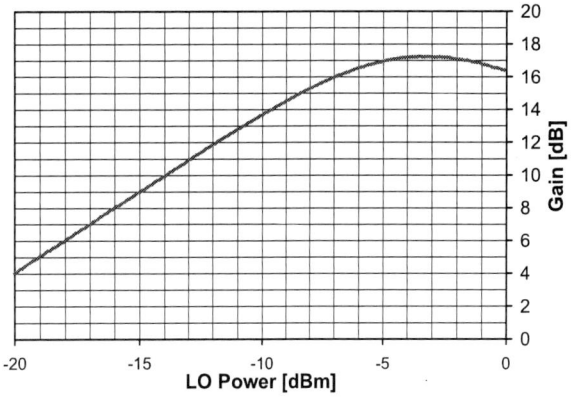

Fig. 4. Input power sweep on LO port to get the optimal LO-power value

First the optimal local oscillator power level had to be found. The gain compression on the LO-input port was measured with a network analyzer in frequency offset mode while applying a constant RF-input signal from a sinusoidal signal source. The RF-input power was chosen to be -35dBm to lie well below any compression limits. The measurement was performed for

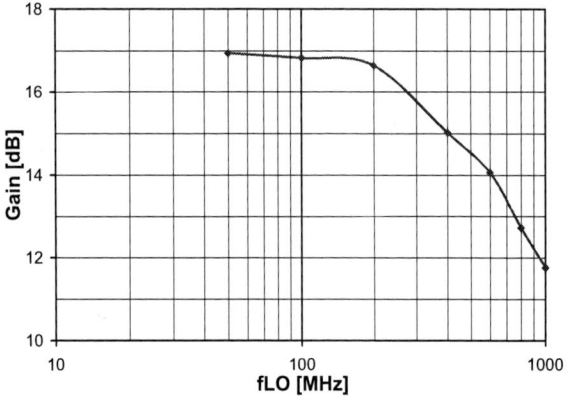

Fig. 5.   Mixer gain over LO frequency

50MHz to 1GHz and shows the same characteristic behavior for all frequencies. The results of the measurements are shown in Fig. 4. It can be seen that the optimal gain is achieved when a clock signal of -3.3dBm is applied to the clock input. All subsequent measurements were performed with the optimal LO-input signal. The conversion gain of the mixer was measured for different clock frequencies from 50MHz to 1GHz with RF signals such that the down-sampled output signal was 1MHz. The gain was measured to be +17dB up to a 3dB-corner frequency of 600MHz as shown in Fig. 5.

Measuring the feedback described above, which compensates for a superimposed DC current on the input, the conversion gain was measured for different input currents. Fig. 6 shows that a gain compression of 3dB is reached for a DC input current of $210\mu A$.

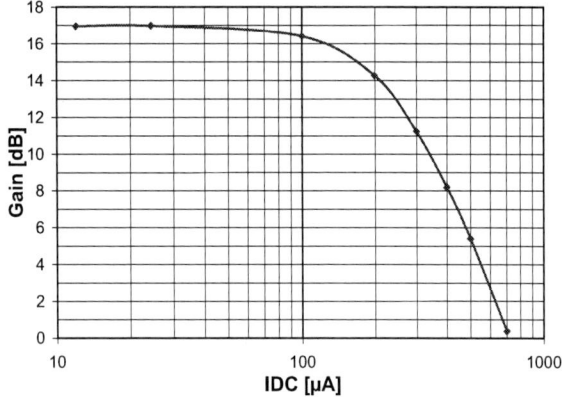

Fig. 6.   Gain dependency on superimposed DC current on the input port

The compression is caused by limitations of the control range from the common-gate stage and the limited ability to handle the DC current of the input transistor. It should be noted here, that the idle current through the input stage is in

the dimension of several ten μA to reduce the static power consumption.

Linearity measurements have been performed as well. Gain compression was measured to take all distortions into account. The input referred 1dB compression point shows the input power at which the gain is reduced by 1dB. It accounts for all distortion effects of the circuit. The measurement shows a 1dB compression point of -22.7dBm at a clock frequency of 50MHz. The 1dB compression point is rising with increased frequency and reaches -19.7dBm (3dB increase) at 550MHz. The compression was measured up to 1GHz and results to -18.9dBm. It should be noted here, that the gain compression shows the inverse behavior of the gain itself as mentioned earlier. Additionally the gain compressions caused by all distortions was measured at several input currents as well. The measurement shows, that the linearity and therefore the distortions are not getting worse. They even tend to be slightly better at higher input currents. The cause of that phenomenon needs a more detailed analysis which should be done in a redesign of the circuit.

Beside the overall distortion with single-tone measurement, also a two-tone measurement was performed. Two signals at 201MHz and 201.1MHz where applied to the input via a 6dB power combiner. A differential clock signal of 200MHz was applied which leads to a down sampled signal of 1MHz and 1.1MHz. To obtain the input referred intercept point of 3rd-order (IIP3) the amplitude of the harmonic distortions have been measured along with the fundamental signal.

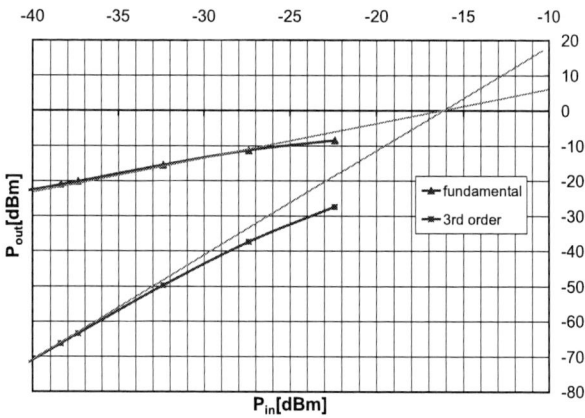

Fig. 7.   Fundamental and 3rd-order harmonic resulting to an IIP3 of -16dBm

Fig. 7 shows the resulting IIP3 of -16dBm. The measurements have been repeated for several frequencies between 50MHz and 400MHz. It can be seen that the distortions are constant at -16dBm for lower frequencies and rise to -15dBm at 400MHz. The same behavior is expected as for the 1dB compression measurement.

Fig. 8 shows a chip photograph of the bonded mixer to the gold plated PCB. The whole chip occupies an area of $450\mu m \times 340\mu m$ including bonding pads. The active area of the mixer itself only occupies an area of $52\mu m \times 52\mu m$.

978-1-4244-2541-9/08 $25.00 © 2008 IEEE   34

TABLE I
COMPARISON WITH OTHER STRUCTURES IN THE LITERATURE

reference	supply voltage	power consumption	gain	1dB Compressions	IIP3	technology
[4]	1.8V	10.26mW	1.5dB	-1.5dBm	4dBm	$0.18\mu m$
[5]	1V	1.72mW	8.28dB	-5.63dBm	4.21dBm	$0.18\mu m$
[6]	5V	60mW	15dB	-17dBm	-6.5dBm	$0.35\mu m$ SiGe-BiCMOS
[7]	1.5V	6mW	20.55dB	-9.2dBm	-0.54dBm	$0.18\mu m$
[8]	1.5V	8.25mW	13.2dB	-	4.5dBm	130nm
[9]	1V	2.8mW	9dB	-	-1dBm	$0.18\mu m$
this work	**1.2V**	**$780\mu W$**	**17dB**	**-22.7dBm**	**-16dBm**	**65nm**

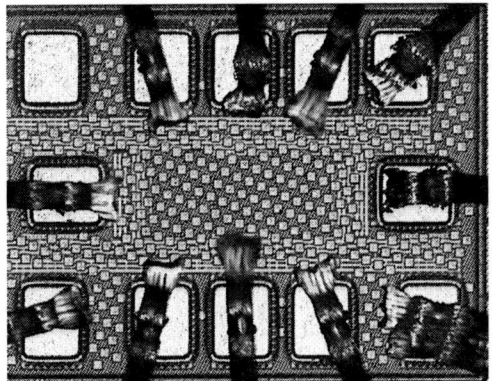

Fig. 8. Chip photograph of the bonded mixer

In Table I, a comparison to different other works in the literature is presented. It should be noted that [4]–[6] are also single-ended mixers in micromixer topology. Reference [7] shows a double balanced mixer with an enhanced transconductance stage to provide a more linear behavior. In [8] a passive mixer with transconductance stage and buffer amplifier is shown while in [9] a double balanced folded mixer is presented. Due to the enhanced input DC-current capability of the mixer it is very difficult to compare with other works since at the time of design no special care was taken for improved linearity. Yet it can be seen that only the mixer in [7] offers a higher gain but has nearly eight times the power consumption of the proposed mixer.

IV. CONCLUSION

A CMOS micromixer with a simple enhancement circuitry at the input was presented to provide the possibility to process a superimposed DC current on the input. The achieved gain of 17dB shows a 3dB compression at 600MHz clock frequency and for an input DC current of $210\mu A$. It consumes $780\mu W$ from a 1.2V voltage supply without a superimposed DC input current. Since no special care was taken to improve linearity, the IIP3 and 1dB compression point were measured to be -22.7dBm and -16dBm, respectively.

ACKNOWLEDGMENT

The authors would like to thank A. Bertl and N. DaDalt for initializing this work and for chip processing. They also thank Infineon Technologies Austria AG and the Austrian BMVIT in the FIT-IT SoC initiative for funding the project.

REFERENCES

[1] J.-Y. Choi1 and S.-G. Lee, "A Low-Noise, High-Gain Single-Ended Input Double-Balanced Mixer," *Analog Integrated Circuits and Signal Processing*, vol. 36, no. 3, pp. 263–266, 2003.

[2] W.-C. Cheng, C.-F. Chan, C.-S. Choy, and K.-P. Pun, "A 1.2 V 900 MHz CMOS mixer," *IEEE International Symposium on Circuits and Systems*, vol. 5, pp. 365–368, 2002.

[3] B. Gilbert, "The MICROMIXER: a highly linear variant of the Gilbert mixer using a bisymmetric Class-AB input stage ," *IEEE Journal of Solid-State Circuits*, vol. 32, pp. 1412–1423, 1997.

[4] B. Chi, B. Shi, and Z. Wang;, "A CMOS down-conversion micromixer for IEEE 802.11b WLAN transceivers," *IEEE International Symposium on Circuits and Systems*, pp. 3762–3765, May 2006.

[5] C.-H. Yen, W.-M. Chang, K.-H. Cheng, and C. Jou, "A New Low Voltage CMOS Micromixer for 2.45GHz Applications," *International Symposium on VLSI Design, Automation and Test*, pp. 1–3, April 2006.

[6] S.-C. Tseng, C. Meng, C.-H. Chang, C.-K. Wu, and G.-W. Hung, "Broadband Gilbert Micromixer With an LO Marchand Balun and a TIA Output Buffer," *IEEE MTT-S International Microwave Symposium Digest*, pp. 1509–1512, June 2006.

[7] S. Alam, "A 2 GHz Low Power Down-conversion Quadrature Mixer in $0.18\text{-}\mu m$ CMOS," *International Conference on VLSI Design*, pp. 146–154, 6-10 Jan. 2007.

[8] Y. Furuta, T. Heima, H. Sato, and T. Shimizu, "A Low Flicker-Noise Direct Conversion Mixer in 0.13 um CMOS with Dual-Mode DC offset Cancellation Circuits," *Topical Meeting on Silicon Monolithic Integrated Circuits in RF Systems*, pp. 265–268, 10-12 Jan. 2007.

[9] V. Vidojkovic, J. van der Tang, A. Leeuwenburgh, and A. van Roermund, "Low Voltage, Low Power Folded-Switching Mixer with Current-Reuse in $0.18\mu m$ CMOS," *International Symposium on Circuits and Systems*, vol. 1, pp. 569–572, 23-26 May 2004.

A 1V Current-Mode Filter in 65nm CMOS Using Capacitance Multiplication

Heimo Uhrmann and Horst Zimmermann
Institute of Electrical Measurements and Circuit Design
Vienna University of Technology
Gußhausstraße 25, A-1040 Vienna, Austria
Emails: heimo.uhrmann@tuwien.ac.at, horst.zimmermann@ieee.org

Abstract—**A new capacitance saving method for differential current-mode filter structures is presented. Especially filters with low cut-off frequencies need large capacitors, which comes along with large and expensive chip area. We show the opportunity to save chip-area in a 3rd-order current-mode Butterworth low-pass filter and enlarge the effective capacitance value by 30%. The proposed filter is designed to be in a transmit path in a software defined radio of a mixed-signal system on chip. It is developed and fabricated in low-power $65\,nm$ CMOS and needs an active area of $215\,\mu m \times 215\,\mu m$. The supply voltage is $1\,V$ at a current consumption of $9.6\,mA$. The filter reaches a third-order input intercept point of $1.6\,mA_p$ and a dynamic range of $73.8\,dB$ at a cut-off frequency of $1\,MHz$.**

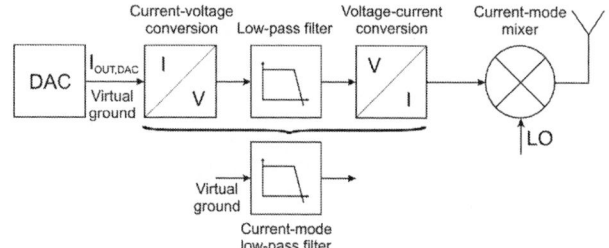

Fig. 1. Transmit path in a SDR

I. INTRODUCTION

The search for big business markets in mobile communication networks creates several new developments in mobile devices. To increase the sales figures of the mobile devices, the usability of mobile devices is raised and more and more features are implemented. The major part of those systems is the digital part and thus the CMOS technologies are optimized for digital systems. For this reason as many system operations as possible are mapped into the digital part. Nevertheless an essential part of radio communication systems are the receiver and transmitter frontends of the radio communication networks, which are mostly implemented as analog circuits. A cost-effective design forces the analog and the digital parts onto one chip, which is called mixed-signal System on Chip (SoC). The difficulties of analog circuits that come along in digital nanometer CMOS have to be solved in a cheap and robust system.

The downscaling of CMOS processes reaches gate lengths down to $32\,nm$ or even smaller. This has a direct effect on digital circuits, such as the rising integration level and the rising clock frequency. Analog circuits suffer on the characteristics of transistors in nanometer CMOS technology. Small transistor structure sizes demand a low supply voltage, to avoid the breakdown of the transistors. The low supply voltage is reflected in a small signal headroom. Other challenges are the small Early voltage in the region of a few volts that causes limited gain of the transistors, high 1/f-noise due to the small transistor dimensions and the gate leakage, which is result of the thin gate oxides. Advantageous is the fact, that small transistor structures reduce the parasitic capacitances and offer transistor transit frequencies of many tens of GHz [1], [2].

To overcome the limited voltage-mode signal headroom, current-mode circuits are used [3], [4]. In current-mode circuits, signal quantities are represented by currents, which is advantageous in some applications. Current-mode circuits provide the opportunity for a low supply voltage, have tunable input impedances and are less susceptible to power and ground disturbances. Current-mode signal processing is easy as well. Operations like additions, subtractions and multiplications by a constant can be realized easily. Adverse to voltage-mode circuits is the need for a high supply voltage and the existence of high-impedance nodes, which limit the bandwidth.

II. SYSTEM DESCRIPTION

Radio communication systems are a profitable sector in mobile communication industries. Mobile devices have many built-in possibilities for data communication. The target is to integrate all desired radio transmission methods on one chip instead of spending one separate chip for each method. Software Defined Radio (SDR) promises to comply with all desired radio transmission methods at low system costs.

The proposed filter is designed for a transmit path in a SDR. Fig. 1 depicts the scheme of a general transmit path, which is typically structured in a digital-analog converter (DAC), a filter, a mixer and an antenna. Current steering DACs are often used due to their accuracy and speed. The usage of a voltage-mode filter would force a transformation of the current-mode signal into a voltage-mode signal. The voltage-mode filter output signal is also disadvantageous, if a current-mode mixer is used. The aggregation of the two transformer blocks and the voltage-mode filter into a current-mode filter offers a power saving overall system and in addition a low voltage circuit

978-1-4244-2541-9/08 $25.00 © 2008 IEEE

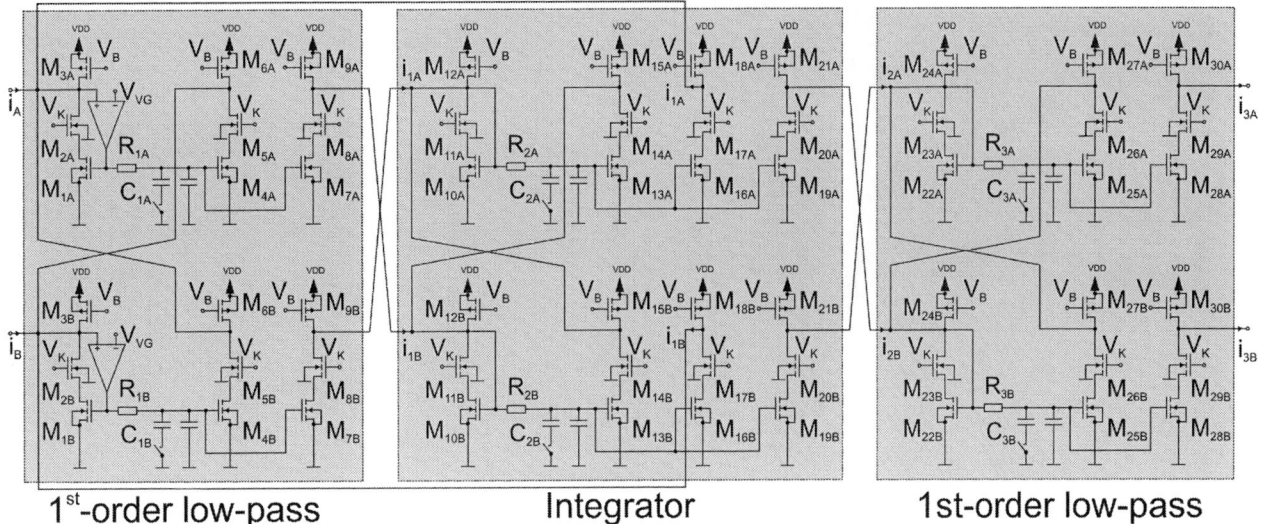

1st-order low-pass   **Integrator**   **1st-order low-pass**

Fig. 2. Overall current-mode filter structure

design is feasible. Target applications are radio systems like Bluetooth and WCDMA systems, e.g. UMTS.

III. OPERATION PRINCIPLE

The proposed filter is a fully differential current-mode third-order low-pass Butterworth filter. The circuit is shown in Fig. 2. The filter is organized in three stages. A 1st-order low-pass and an integrator, which form a 2nd-order low-pass, and a further 1st-order low-pass. The input of the filter is realized with a virtual ground regulation to regulate the input voltage to a constant potential to avoid problems due to a low Early voltage in the DAC and to lessen input impedance.

In order to fit to target applications, the cut-off frequency of the 3rd-order low-pass filter is designed to be switchable between $1\,MHz$ and $4\,MHz$. The switches shown in Fig. 2 are implemented as NMOS switches. To reach an on-resistance, which is as small as possible, the width to length ratio is very high.

One of the most important desires in chip design is, to use as little expensive chip area as possible. Low-pass filters with a low cut-off frequency, like in this case, need large capacitors. Possibilities to save capacitor area are presented in [5] and [6]. Here we introduce a new method to save on-chip capacitors for differential structures. The 1st-order low-pass in Fig. 2 is realized with a method, called capacitance multiplication. For this functionality and for symmetry reasons several assumptions have to be met: $M_{1A|B} = M_1$, $M_{3A|B} = M_3$, $M_{4A|B} = M_4$, $M_{6A|B} = M_6$, $M_{7A|B} = M_7$ and $M_{9A|B} = M_9$, where $A|B$ denotes A and B. The cascode transistors $M_{2A|B}$, $M_{5A|B}$, $M_{8A|B}$ are all equal and are used to lower the parasitic characteristics of the output transistors. The output conductance is lowered and a more ideal current-source behavior is reached. The output resistance of the loads is increased via a larger gate length. R_{1A} and R_{1B} are equal and denoted as R_1 and C_{1A} and C_{1B} are equal and denoted as C_1.

By using these values, the transfer function of the 1st-order low-pass filter calculates to

$$H_{LP}(s) = \frac{i_A - i_B}{i_{1A} - i_{1B}} = \frac{\zeta_1 \frac{g_{m7}}{g_{m1}}}{\zeta_1 C_1 (R_1 + \frac{1}{g_{m1}})s + 1}. \quad (1)$$

In this equation the factor ζ_1 is called the capacitance multiplication factor, where

$$\zeta_1 = \frac{g_{m1}}{g_{m1} - g_{m4}}. \quad (2)$$

In order to guarantee a 1st-order low-pass filter behavior the transconductances have to follow $g_{m4} < g_{m1}$. (1) shows, that a reduction of the cut-off frequency can be reached through an increment of the capacitance multiplication factor ζ_1. ζ_1 in (2) can be raised by increasing the transconductance of transistor M_4. The proposed structure uses a ζ_1 of 3. The new development results in an enlargement of the effective capacitance value of 30%. The saving of chip area is at the expense of Dynamic Range (DR), due to a smaller current in $M_{4|5|6}$ than in $M_{1|2|3}$. Nevertheless the gain in the 1st-order low-pass filter can be made larger than 1 by using g_{m7} (see (1)). A greater multiplication factor can be reached, when g_{m4} approaches g_{m1}, but the process tolerances have to be considered.

The second block in the filter design is a current-mode integrator. The transfer function is given by

$$H_I(s) = \frac{g_{m10}}{(1 + g_{m10}R_2)C_2 s} \quad (3)$$

as presented in [5] under the following assumptions: The transistors $M_{10A|B}$, $M_{13A|B}$, $M_{16A|B}$ and $M_{19A|B}$, where $A|B$ denotes A and B, are equal and named M_{10} in the formulas. The load transistors $M_{12A|B}$, $M_{15A|B}$, $M_{18A|B}$ and $M_{21A|B}$ are equal and the cascode transistors $M_{11A|B}$, $M_{14A|B}$, $M_{17A|B}$ and $M_{20A|B}$ are equal. The resistors $R_{2A|B}$

978-1-4244-2541-9/08 $25.00 © 2008 IEEE   37

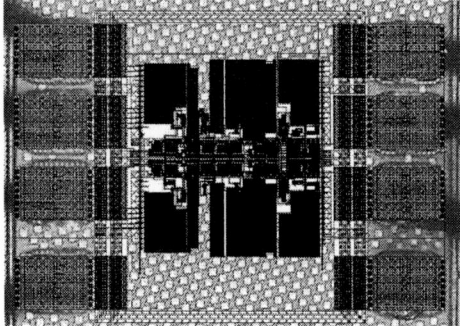

Fig. 3.   Photo of the testchip with overlaid layout plot

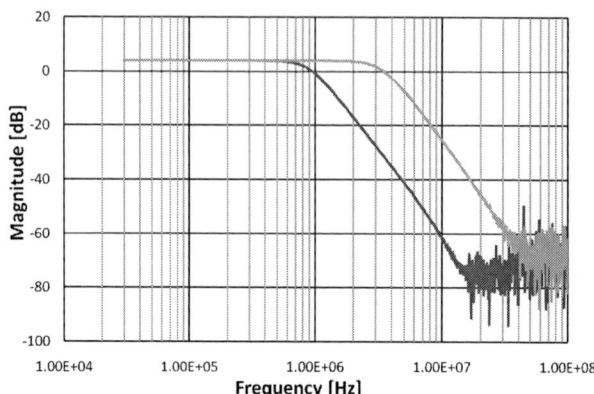

Fig. 4.   Frequency response

are equal and denoted as R_2 and the capacitors $C_{2A|B}$ are equal and are denoted as C_2.

The 2nd-order low-pass filter uses the above mentioned 1st-order low-pass filter and the integrator in feedback configuration. Its transfer function results in

$$H_{LP2} = \frac{g_{m7}g_{m10}\zeta_1}{Ds^2 + Es + F}. \qquad (4)$$

In this equation the substitutes denote as

$$
\begin{aligned}
D &= C_1 C_2 \zeta_1 (R_1 g_{m1} + 1)(1 + g_{m10}R_2), & (5) \\
E &= (1 + g_{m10}R_2)C_2 g_{m1} \text{ and} & (6) \\
F &= \zeta_1 g_{m7}g_{m10}. & (7)
\end{aligned}
$$

The entire 3rd-order current-mode low-pass filter is obtained with an additional 1st-order low-pass filter (Fig. 2). In the last low-pass filter a ζ_2 of 3 was selected as well. The transfer function of the overall 3rd-order low-pass filter using (5), (6) and (7) results in

$$H_{FILTER} = \frac{g_{m7}g_{m10}\zeta_1}{Ds^2 + Es + F} \cdot \frac{\zeta_2 \frac{g_{m28}}{g_{m22}}}{\zeta_2 C_3 (R_3 + \frac{1}{g_{m22}})s + 1}, \qquad (8)$$

where $M_{22} = M_{22A|B}$, $M_{24A} = M_{24B}$, $M_{25} = M_{25A|B}$, $M_{27A} = M_{27B}$, $M_{28} = M_{28A|B}$, $M_{30A} = M_{30B}$, the cascode transistors $M_{23A|B}$, $M_{26A|B}$, $M_{29A|B}$ are equal, $C_3 = C_{3A|B}$ and $R_3 = R_{A|B}$. ζ_2 is defined as $\frac{g_{m22}}{g_{m22}-g_{m25}}$, where $g_{m25} < g_{m22}$. g_{m7}, g_{m19} ($= g_{m10}$) and g_{m28} are chosen in such a way that the overall filter gain of $4\,dB$ results.

IV. MEASUREMENT RESULTS

The 3rd-order current-mode Butterworth filter is designed and processed in $65\,nm$ low-power CMOS technology with $V_{THn} = 500\,mV$ and $V_{THp} = 400\,mV$. The chip photo with an overlaid layout plot is depicted in Fig. 3. The chip has a dimension of $510\,\mu m \times 360\,\mu m$, by contrast the active area is $215\,\mu m \times 215\,\mu m$.

The supply voltage of the current-mode filter is only $1\,V$, where the chip consumes $9.6\,mA$. The implemented current-mode filter has a gain of about $4\,dB$. The frequency response of the filter is presented in Fig. 4. The frequency response shows a Butterworth characteristic, and two digitally switchable cut-off frequencies of $1\,MHz$ and $4\,MHz$. The frequency response was measured at different potentials of the

virtual ground regulation. At the input voltages of $400\,mV$, $500\,mV$ and $600\,mV$ the deviation of the cut-off frequency was measured. For both operation modes the deviation of the cut-off frequencies were lower than $1\,\%$.

Dual-tone measurements were performed for both cut-off frequencies. At the cut-off frequency of $1\,MHz$ two sinusoidal signals with the frequency of $500\,kHz$ and $600\,kHz$ were applied. The amplitude of the input was increased until a third-order intermodulation (IM3) of $-40\,dB$ was reached. For the virtual ground voltages of $400\,mV$, $500\,mV$ and $600\,mV$ an IM3 of $-40\,dB$ was reached, at input amplitudes of $164\,\mu A$, $168\,\mu A$ and $168\,\mu A$, which corresponds to a third-order input intercept point of approximately $1.6\,mA_p$. At the cut-off frequency of $4\,MHz$ two sinusoidal signals with the frequency of $2.4\,MHz$ and $2.5\,MHz$ were applied. Again the virtual ground voltage is adjusted to $400\,mV$, $500\,mV$ and $600\,mV$, where the IM3 of $-40\,dB$ is reached at an input amplitude of $154\,\mu A$, $159\,\mu A$ and $145\,\mu A$. This corresponds to IIP3s of approximately $1.5\,mA_p$ and $1.4\,mA_p$.

Single-tone measurements show the in-band distortions. At the cut-off frequency of $1\,MHz$ a sinusoidal signal at $100\,kHz$ was applied. At the input voltages of $400\,mV$, $500\,mV$ and $600\,mV$ total harmonic distortions of $1\,\%$ (THD$_{1\%}$) were reached at $272\,\mu A_p$, $277\,\mu A_p$ and $277\,\mu A_p$, respectively. At the cut-off frequency of $4\,MHz$ the applied sinusoidal signal had a frequency of $400\,kHz$. The THD$_{1\%}$ at the input potentials of $400\,mV$, $500\,mV$ and $600\,mV$ were attained at $272\,\mu A_p$, $277\,\mu A_p$ and $268\,\mu A_p$, respectively.

The spectral output current density is depicted in Fig. 5. The measurements are performed for both cut-off frequencies at a virtual ground voltage of $500\,mV$. The spectral output noise density in the pass-band is $40\,pA/\sqrt{Hz}$ at both cut-off frequencies which corresponds to $25\,pA/\sqrt{Hz}$ at the input.

An often used method to compare filters is the Figure of Merit (FoM) [7]. The FoM is given by

$$\text{FoM} = \frac{P}{8kT \cdot f \cdot N \cdot DR}, \qquad (9)$$

where P denotes the power consumption of the observed filter, k is Boltzmann's constant, T is the absolute temperature, f

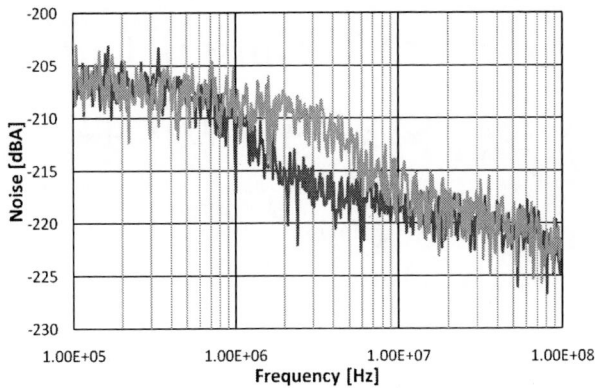

Fig. 5. Differential output current noise density

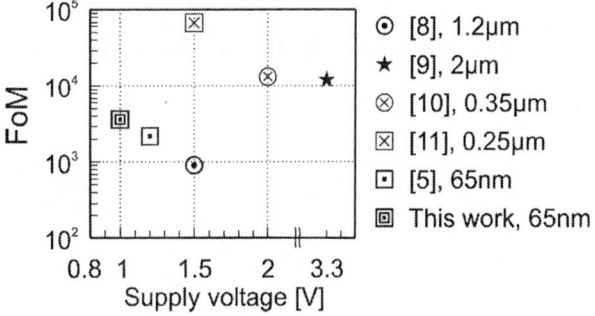

Fig. 6. FoM of current-mode filters (the lower the better)

is the cut-off frequency, N is the number of poles and DR is the Dynamic Range (DR)

$$DR = \frac{\hat{I}_{THD1\%}^2}{2 \cdot \overline{i}_{noise}^2}. \quad (10)$$

$\hat{I}_{THD1\%}$ is the amplitude of the input current, which causes a Total Harmonic Distortion (THD) of 1% and $\overline{i}_{noise}^2$ is the integrated in-band output current noise. The DR of the presented filter is calculated for an input voltage of $500\,mV$ and is $73.8\,dB$ at the cut-off frequency of $1\,MHz$ and $67.8\,dB$ at the cut-off frequency of $4\,MHz$.

Using the DR the present current-mode filter has a FoM of 3873 at a cut-off frequency of $1\,MHz$ and a FoM of 4034 at a cut-off frequency of $4\,MHz$.

The comparison of various FoMs of current-mode filters in literature is depicted in Fig. 6, where lower values are better. [8] presents a 5th-order Chebyshev low-pass filter in $1.2\,\mu m$ CMOS, where the FoM is calculated at a cut-off frequency of $525\,kHz$. A 6th-order Bessel low-pass filter in $2\,\mu m$ CMOS is published by [9]. The values for the FoM of this low-pass filter are estimated for a cut-off frequency of $10\,MHz$. [10] proposes a 3rd-order Butterworth low-pass filter in $0.35\,\mu m$ CMOS technology, while the FoM is evaluated at a cut-off frequency of $42\,MHz$. [11] presents a 2nd-order Butterworth low-pass filter. The FoM is calculated for a cut-off frequency of $22\,kHz$. In a previous work ([5]) we realized a 3rd-order Butterworth filter in $65\,nm$ CMOS without cascodes. The FoM of this filter is valid for a cut-off frequency of $1.1\,MHz$. The chart of the FoM points out, that we have the first current-mode low-pass filter at a supply voltage of $1\,V$ in the newest CMOS processes with a good FoM.

V. CONCLUSION

In the present paper we propose a 3rd-order current-mode Butterworth low-pass filter that is used in a transmit path of a software defined radio application. In this filter we apply a new method, which allows area saving for capacitors. This method is called capacitor multiplication and is advantageous for filters with small cut-off frequencies. In this way approximately $30\,\%$ capacitor area can be saved without limiting the DR

too much (DR = $73.8\,dB$). The proposed filter has two digitally switchable cut-off frequencies of $1\,MHz$ and $4\,MHz$ and is designed and produced in $65\,nm$ CMOS technology. The power consumption is $9.6\,mW$ at $1\,V$ supply voltage. Performance measurements show an IIP3 of $1.6\,mA_p$ and a DR of $73.8\,dB$ at $1\,V$ at the cut-off frequency of $1\,MHz$. A comparison of FoMs to other current-mode filters in literature shows the first current-mode low-pass at lowest supply voltage and a good filter performance.

ACKNOWLEDGMENT

The authors would like to thank A. Bertl and L. Dörrer from Infineon in Villach, Austria for the initiation of this work and for chip processing. Partial financial funding from Infineon Technologies Austria AG and from the Austrian Federal Ministry for Transport, Innovation and Technology in the FIT-IT project SOFT-RoC via FFG is gratefully acknowledged. Furthermore the authors thank R. Kolm and F. Schlögl, for support and hints that contributed to the success of this work.

REFERENCES

[1] A.-J. Annema, B. Nauta, R. van Langevelde, and H. Tuinhout, "Analog Circuits in Ultra-Deep-Submicron CMOS," *IEEE J. Solid-State Circuits*, vol. 40, no. 1, pp. 132 – 142, January 2005.

[2] M. Vertregt, "The Analog Challenge of Nanometer CMOS," *International Electron Device Meeting*, pp. 1 – 8, Dec. 2006.

[3] J. Ramirez-Angulo, M. Robinson, and E. Sanchez-Sinecio, "Current-Mode Continuous-Time Filters: Two Design Approaches," *IEEE Trans. Circuits and Systems II*, vol. 39, no. 6, pp. 337 – 341, June 1992.

[4] F. Yuan, "Low-Voltage CMOS Current-Mode Preamplifier: Analysis and Design," *IEEE Trans. Circuits and Systems I*, vol. 53, pp. 26 – 39, 2006.

[5] H. Uhrmann and H. Zimmermann, "A 3rd-Order Current-Mode Continuous-Time Filter in 65 nm CMOS," *NorChip Conf.*, Nov. 2007.

[6] J. Yan, H. Zheng, X. Zeng, and T. Tang, "Compact Current-Mode Loop Filter for PLL Applications," *Electronics Letters*, vol. 41, no. 23, pp. 1257 – 1258, Nov. 2005.

[7] U. Yodprasit and C. Enz, "A 1.5V 75dB Dynamic Range Third-Order gm-C Filter Integrated in a 0.18-μm Standard Digital CMOS Process," *IEEE J. Solid-State Circuits*, no. 7, pp. 1189–1197, 2003.

[8] R. H. Zele and D. J. Allstot, "Low-Power CMOS Continuous-Time Filters," *IEEE J. Solid-State Circuits*, vol. 31, pp. 157 – 168, Feb. 1996.

[9] S. L. Smith and E. Sánchez-Sinencio, "3V High-Frequency Current-Mode Filters," *IEEE Int. Symp. on Circuits and Systems (ISCAS)*, vol. 2, pp. 1459 – 1462, May 1993.

[10] A. Otin, S. Celma, and C. Aldea, "A Design Strategy for VHF Filters with Digital Programmability," *IEEE Symposium on Circuits and Systems*, pp. 1059–1062, 2006.

[11] N. Krishnapura and Y. Tsividis, "Micropower Low-Voltage Analog Filter in a Digital CMOS Process," *IEEE J. Solid-State Circuits*, vol. 38, no. 6, pp. 1063–1067, 2003.

978-1-4244-2541-9/08 $25.00 © 2008 IEEE

FPGA Implementation of a 2G Fibre Channel Link Encryptor with Authenticated Encryption Mode GCM

L. Henzen, F. Carbognani, N. Felber, and W. Fichtner
Integrated Systems Laboratory, ETH Zurich, Switzerland
email: henzen@iis.ee.ethz.ch

Abstract—The Galois/Counter Mode (GCM) algorithm enables fast encryption combined with per-packet message authentication. This paper presents an FPGA implementation of a complete bidirectional 2 Gbps Fibre Channel link encryptor hosting two area-optimized GCM cores for concurrent authenticated encryption and decryption. The proposed architecture fits into one Xilinx Virtex-4 device. Measurements in a working network link point out that per-packet authentication results in a speed decrease up to 20 % of the channel capacity for a reference frame length of 256 bits. Two methods of frame encryption are investigated to reduce the required GCM overhead and to exploit different network configurations.

I. INTRODUCTION

The National Institute for Standard and Technology (NIST) recommends the high-speed counter mode of operation for block ciphers to support data encryption and decryption at multi-gigabit rates [1]. However, the need of confidentiality with message authentication encouraged NIST to develop an authenticated encryption mode to protect packet flows at the rates of the fastest communication standards. A wide range of speed optimized hardware implementations of GCM [2] demonstrated that this authenticated encryption scheme can easily reach the 10 Gbps throughput [3], [4]. With the rapid growth of high-speed mass storage devices, using the 1 and 2 Gbit/s Fibre Channel (FC) standard, GCM turns out to be the only mode that is able to secure the required data transfers. The Technical Committee T11 has already adopted GCM to develop an adaption of the IPsec ESP protocol to Fibre Channel, announced as FC-Security Protocol (FC-SP [5]).

The aim of this work is to implement an FPGA hardware architecture of the GCM function for the authenticated encryption of a 2G FC network. The resulting FC link encryptor is then tested in a operating point-to-point network to demonstrate the effect of per-packet authentication on the overall throughput of the channel. The hardware constrained architecture of the GCM core leaves space for the integration of standard RSA or Diffie-Hellman protocols for key agreement in the target application. To the best of the authors' knowledge this is the first paper that presents the implementation of a bidirectional FC link encryptor with authentication and investigates the throughput decrease inherent to the message authentication process.

Outline: The reminder of this document is organized as follows. The theoretical aspects of the GCM algorithm are the topic of Sec. II. Sec. III discusses the main issues that arise in the implementation of the block cipher and the authentication core. The FPGA design of the link encryptor is explained in Sec. IV, while Sec. V illustrates the overall performances. Eventually, Sec. VI draws the conclusions.

II. GCM AUTHENTICATED ENCRYPTION

The GCM algorithm combines the Advanced Encryption Standard (AES) [6] in counter mode with data authentication. The encryption process takes as input a plaintext P, the initialization vector IV, the additional authenticated data A, and a cipher key K of appropriate length. After the computation, it generates the ciphertext C and the authentication tag T. Since AES is the underlying block cipher, P and A are divided into sequences of 128-bit strings ($P_1, P_2, \ldots, P_{n-1}, P_n^*$ and $A_1, A_2, \ldots, A_{m-1}, A_m^*$). In case a final block P_n^* or A_m^* is smaller than 128 bits, zeros are appended to make them into the right dimension (u and v denotes the dimension of this padding operation in the related equations). The authenticated encryption process is defined in the following way:

$$
\begin{aligned}
H &= \text{Enc}(K, 0^{128}) \\
Y_0 &= \begin{cases} IV\|0^{31}1 & \text{if } \text{len}(IV) = 96 \\ \text{GHASH}(H, \{\}, IV) & \text{otherwise} \end{cases} \\
Y_i &= \text{incr}(Y_{i-1}) & i = 1, 2, \ldots, n \\
C_i &= P_i \oplus (\text{Enc}(K, Y_i) & i = 1, 2, \ldots, n-1 \\
C_n^* &= P_n^* \oplus \text{MSB}_u(\text{Enc}(K, Y_n)) \\
T &= \text{MSB}_t(\text{GHASH}(H, A, C) \oplus \text{Enc}(K, Y_0))
\end{aligned}
$$
(1)

The aim of the universal hash function GHASH is the compression of the $128 \times (m + n + 1)$-bit data stream into a 128-bit hash value X_{m+n+1}. The inputs are the pre-computed variable H, resulting from the encryption of a zero block under the given secret key, the authenticated data A, and the n-block ciphertext C. The compression is performed by the multiplication over the Galois field $\text{GF}(2^{128})$ with the irreducible polynomial

$$
g(x) = x^{128} + x^7 + x^2 + x + 1.
$$
(2)

The authentication tag T is obtained from the exclusive-OR between the encryption of IV and the final hash value X_{m+n+1} of Eq. 3:

978-1-4244-2541-9/08 $25.00 © 2008 IEEE

$$X_i = \begin{cases} 0 & i = 0 \\ (X_{i-1} \oplus A_i) \cdot H & i = 1, \dots, m-1 \\ (X_{m-1} \oplus (A_m^* || 0^{128-v})) \cdot H & i = m \\ (X_{i-1} \oplus C_{i-m}) \cdot H & i = m+1, \dots, m+n-1 \\ (X_{m+n-1} \oplus (C_n^* || 0^{128-u})) \cdot H & i = m+n \\ (X_{m+n} \oplus (\text{len}(A) || \text{len}(C))) \cdot H & i = m+n+1 \end{cases}$$
$$(3)$$

$\text{len}(A)$ and $\text{len}(C)$ return the bit string length of A and C in 64-bit, respectively.

The authenticated decryption process computes the plaintext P from the ciphertext C and generates the authentication tag T'. To prove the authenticity of P, the encryption tag T and the decryption tag T' are compared. In case of mismatch, the decrypted plaintext P is discarded. Running the block cipher in the counter mode avoids the need of the inverse AES function. For decryption, P and C exchange their position in Eq. 1, except for tag generation.

III. GCM AES Hardware Architecture

The combination of GCM with a pipelined architecture of the AES cipher makes possible the achievement of the gigabit throughput. The hardware implementation hosted into the link encryptor is evaluated and designed to satisfy the maximal rates of the 2G Fibre Channel standard at minimum hardware costs.

A. AES Architecture

The demand of a block cipher for high-speed encryption leads to a fully pipelined design of the AES algorithm [7]. The architecture is composed by 14 independent instantiations of the round function block (see Fig. 1). Every round block includes the succession of the *SubBytes()*, *MixColuns()*, and *ShiftRows()* functions. Although an output block is computed every clock cycle, the fully pipelined AES requires an excessive amount of hardware resources. To reduce the overall

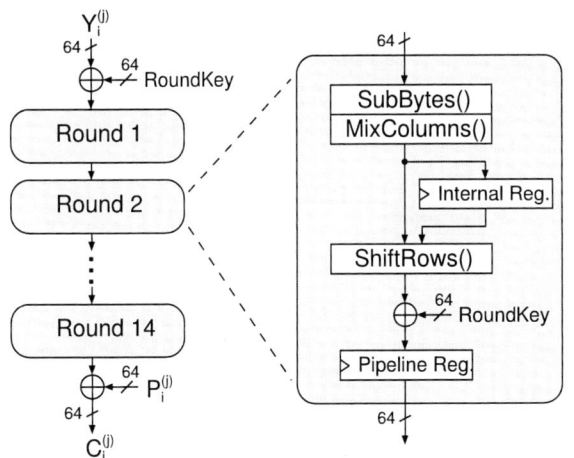

Fig. 1. AES rounds schedule for the fully pipelined implementation. Variables are $\{Y_i^{(j)}, P_i^{(j)}, C_{i-m}^{(j)}\}$ with $j = 0$ for the first 64-bit block and $j = 1$ for the last 64-bit block.

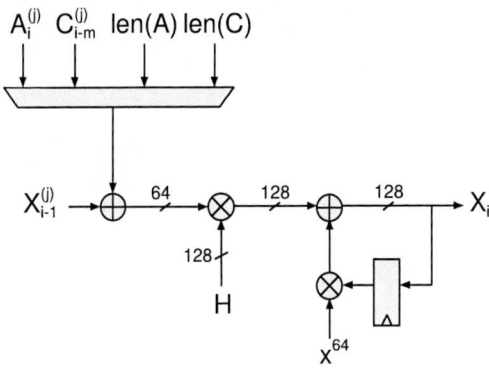

Fig. 2. BPWS multiplier of the GHASH function. Variables are $\{A_i^{(j)}, C_{i-m}^{(j)}, X_{i-1}^{(j)}\}$ with $j = 0$ for the first 64-bit block and $j = 1$ for the last 64-bit block.

dimension of the encryption core, a 64-bit data path architecture was designed instead of a conventional 128-bit data path AES. In this way, a complete round block is calculated within two clock cycles. The shortened data path halves the hardware utilization of the *SubBytes()* and the *MixColumns()* functions, while the *ShiftRows()* function is implemented as a straightforward re-routing of the signals.

In the AES the most area demanding component is the *SubBytes()* operation, which consists of a byte substitution with the multiplicative inverse in the $GF(2^8)$, the so called S-Box, followed by an affine transformation. Since in the target FPGA dual-port RAM is not a scarce resource, the S-Box is stored as lookup table in 58 blocks of this memory. The resulting *SubBytes()* function is a fast circuit using no further FPGA resources.

B. GCM Architecture

The critical computation in the authentication process of GCM is the binary multiplication over the finite field $GF(2^{128})$. All previously computed values X_{i-1} are exclusive-ORed with the ciphertext C_i and at the end multiplied by the value H. According to the 64-bit data path implementation of the AES core (see III-A), the binary multiplication in the GHASH function processes 64-bit input blocks. Thus, a basic bit-parallel structure [3] of the multiplier has been replaced by an adaptation for the $GF(2^{128})$ of the bit-parallel word-serial (BPWS) multiplier proposed in [8]. The BPWS multiplier considerably reduces the hardware complexity of the GHASH architecture, without decreasing the speed of the GCM core. The implemented multiplier, illustrated in Fig. 2, calculates the final value in 2 clock cycles.

IV. FPGA Implementation

With the aid of high-speed Multi-Gigabit Transceiver (MGT) modules [9], the bidirectional FC network interfaces are directly designed into the FPGA. The MGTs provide 8B/10B encoding scheme and translate serial data from the FC connector to 32-bit parallel data inside the FPGA. The interface region of the link encryptor (see Fig. 3) elaborates

978-1-4244-2541-9/08 $25.00 © 2008 IEEE

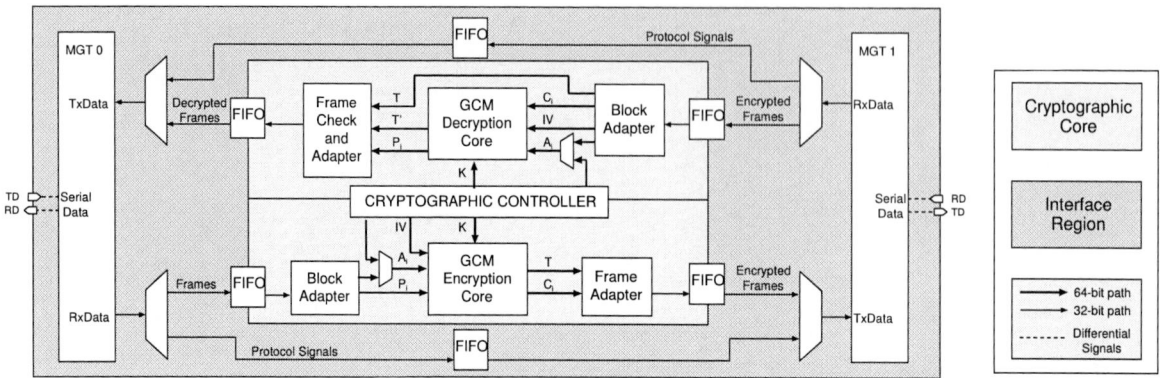

Fig. 3. Block diagram of the implemented FC link encryptor. The encryption path (bottom) detects incoming frames and sends secured frames through the FC link. The decryption path (top) decrypts the incoming secured frames and transmits the recovered frames to the destination network node.

32-bit parallelized data at 53.125 MHz to reach the specified 2G FC throughput L of 1.7 Gbps, before 8B/10B encoding. The cryptographic core hosting two GCM functions for encryption and decryption, is driven by the double frequency of 106.25 MHz in order to allow the additional processing of authentication at full rate.

In the FC standard user data are encapsulated in frames (cf. framing protocol FC-2 [10]). Special 32-bit blocks, referred to as frame delimiters, mark the beginning (SOF) and the conclusion of frames (EOF). However, to keep the nodes synchronized, a large number of primitive signals and primitive sequences is continuously transmitted through the link. The protocol signals (primitive signals and sequences) do not contain any relevant information that needs to be secured, but only information concerning the status of the nodes. They are transmitted between frames, but are not delimited with SOF and EOF. To work correctly, the nodes of a network should exchange these protocol signals without interruptions. Moreover, for primitive sequences the order of every 32-bit block should be preserved.

In the implemented link encryptor primitive signals and primitive sequences are propagated unencrypted (see Fig. 3). This way the control function of the nodes is able to guarantee the correct link synchronization.

A. Frame Encryption

Since user data are transferred in the payload field of frames, the designed link encryptor identifies SOF and EOF delimiters in the interface region and propagates the 32-bit blocks of the frame into the cryptographic core.

The first encryption method provides the entire frame as plaintext for the GCM encryption core. The initialization vector IV, the encrypted frame, i.e, the resulting ciphertext, and the tag T are encapsulated by internal SOF and EOF to compose the internal frame (see Fig. 4A). The assembled frame is then transmitted to the other link encryptor. This full encryption mode Π can be applied only in point-to-point network configurations, as the encryption of the header hides the addresses needed for a correct frame routing. Anyway, it provides a higher level of confidentiality for frames, since no

information about the content and the source/destination of the frame is visible.

A second encryption method has been therefore implemented to exploit the link encryptor in multi-point networks. The need to preserve the header field of frames unencrypted involves a different approach to secure the incoming frames. In the payload encryption mode Ψ (see Fig. 4B), only the payload field of the frame is encrypted. As the new payload of the secured frame hosts the IV, the encrypted payload, and the resulting authentication tag T, the cyclic redundancy checksum (CRC) should be re-computed over this updated field. To even improve the efficiency of the link encryptor, the GCM core uses the header field of the incoming frame as additional authenticated data A.

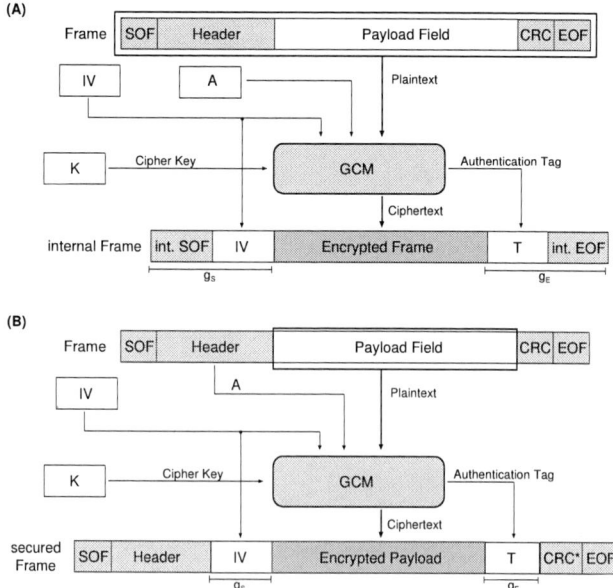

Fig. 4. Frame encryption diagram for the encryption method: full encryption $\Pi[g_S + g_E]$ (A) and payload encryption $\Psi[g_S + g_E]$ (B). g_S and g_E are the total transmission overhead caused by the per-packet authentication.

978-1-4244-2541-9/08 $25.00 © 2008 IEEE

TABLE I
RESOURCE UTILIZATION OF THE IMPLEMENTED LINK ENCRYPTOR. THE
TARGET APPLICATION IS A XILINX VIRTEX-4 FX100 FPGA.

Module	Slice	# block RAM
AES counter mode	3.1k	58
BPWS multiplier	3.8k	0
GCM core	7.1k	58
Encryption path	0.9k	9
Decryption path	1.1k	14
Link encryptor	**19.3k**	**139**

V. RESULTS AND DISCUSSION

Based on functional VHDL code, the architecture proposed in the previous section has been synthesized using Synplify Pro, placed and routed using the Xilinx ISE Design Suite, and tested in a Xilinx Virtex-4 FX100 FPGA. Tab. I resumes the hardware requirements of the principal modules inside the link encryptor.

In the performed tests, a FC traffic generator was connected in a loopback configuration with two link encryptors. Data were transmitted through the devices and then retrieved by the traffic generator. This enabled to transmit frames with variable sizes at different rates. The results depicted in Fig. 5 emphasize the effects of the per-packet frame authentication. Using larger frames, the speed of the communication can be kept at the 2G FC limit L, whereas with shorter frames, the overall throughput decreases down to 70 % of L for 128-bit frames. This effect is caused by the extension of the frames due to the appended GCM overhead (g_A and g_E, see Fig 4). In the Π configuration, the internal frame exceeds the dimension of the original frame by the internal SOF and EOF, IV, and T. For the Ψ mode the secured frame includes only the necessary IV and T of variable length t (see Eq. 1), therefore 84 % of L for 128-bit frames can be recovered. Nevertheless, the GCM overhead required for the authentication process in both encryption methods has always the same dimension for every frame size. Smaller frames proportionally carry more additional data with respect to larger frames. Thus, the throughput behavior can be predicted by the proportion between incoming frames and internal secured frames, i.e.,

$$\left(\frac{\text{User frame}}{\text{Secured frame}}\right) \text{FC limit} = \left(\frac{S}{S + (g_S + g_E)}\right) L \quad (4)$$

where S is the frame size (g_S and g_E compose the authentication overhead). Fig. 5 demonstrates the throughput decrease of the FC link for small packets, predicted by Eq. 4. A reduction of the GCM overhead leads to an increase of the performance of the link encryptor, decreasing at the same time the resulting confidentiality level. The difference between measurements and simulations for the point-to-point encryption Π are due to the padding process of frames in 128-bit blocks, described in Sec. II.

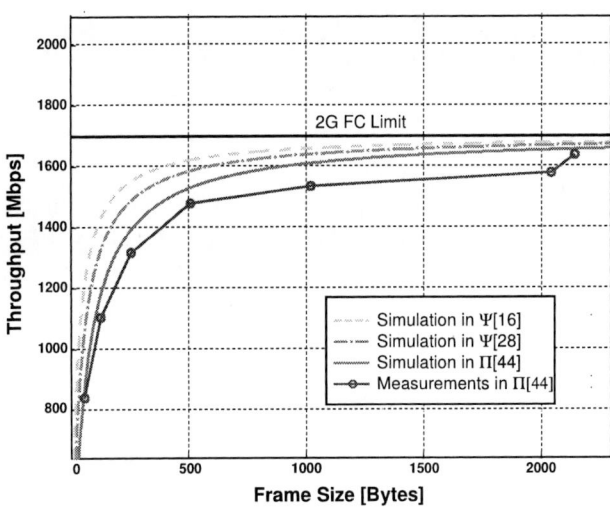

Fig. 5. Throughput/frame-length trade-off for the implemented link encryptor. Measurements in $\Pi[g_S + g_E = 44$ bytes] and simulations in $\Pi[44]$, $\Psi[28]$, and $\Psi[16]$ demonstrate how the throughput for small frames could be increased, reducing the GCM overhead (g_S and g_E) transmitted with the secured frames.

VI. CONCLUSIONS

This paper presents a compact FPGA implementation of a 2G Fibre Channel link encryptor. Authentication and en/decryption processes inside the GCM core are optimized to reach the full throughput of the 2G FC standard with the smallest possible resource utilization. Besides, an evaluation of per-packet authentication for multi-gigabit communication protocols is performed. The trade-off between security level and maximal throughput is investigated by proposing two different methods of frame encryption.

REFERENCES

[1] M. Dworkin, "Recommendation for block cipher mode of operation," Dec. 2001, NIST Special Publication 800-38A.
[2] D. McGrew and J. Viega, "The Galois/Counter Mode of operation (GCM)," May 2005, Submission to NIST Modes of Operation.
[3] G. Zhou, H. Michalik, and L. Hinsenkamp, "Efficient and high-throughput implementations of AES-GCM on FPGAs," in *Proceedings of the Int. Conf. on Field-Programmable Technology, 2007*, Kitakyushu, Dec. 2007, pp. 185–192.
[4] S. Lemsitzer, J. Wolkerstorfer, N. Felber, and M. Braendli, "Multi-gigabit GCM-AES architecture optimized for FPGAs," in *Proceedings of the Workshop on Cryptographic Hardware and Embedded Systems*, Aug. 2007, pp. 227–238.
[5] R. Snively, C. Carlson, D. Black, and C. DeSanti, "Fibre Channel Security Protocols (FC-SP)," June 2006, INCITS woking draft proposed American National Standard for Information Technology.
[6] NIST, "Advanced encryption standard (AES)," Nov. 2001, Federal Information Processing Standards (FIPS) Publication 197.
[7] A. Hodjat and I. Verbauwhede, "A 21.54 Gbits/s fully pipelined AES processor on FPGA," in *Proceedings of the Symp. on Field-Programmable Custom Computing Machines*, Apr. 2004, pp. 308–309.
[8] W. Tang, H. Wu, and M. Ahmadi, "VLSI implementation of bit-parallel word-serial multiplier in GF(2^{233})," in *Proceedings of the 3rd Int. Conf. NEWCAS*, June 2005, pp. 399–402.
[9] Xilinx, "Virtex-4 rocketio multi-gigabit transceiver," Aug. 2007, User Guide UG076 (v4.0).
[10] R. Snively, C. DeSanti, C. Carlson, W. Martin, and R. Nixon, "Fibre Channel, framing and signaling-2 (FC-FS-2)," Aug. 2006, INCITS woking draft proposed American National Standard for Information Technology.

978-1-4244-2541-9/08 $25.00 © 2008 IEEE

On the credibility of load-latency measurement of network-on-chips

Erno Salminen, Ari Kulmala, and Timo D. Hämäläinen
Tampere University of Technology, P.O. Box 553, FIN-33101 Tampere, Finland
email:erno.salminen@tut.fi

Abstract— This paper studies the impact of various simulation and network-on-chip (NoC) setups in common load-latency curves that are used for performance evaluation. The different setups yield very large variation in the observed performance yet they are too often undocumented. Vague definitions make the comparison of NoCs hard or impossible since the large uncertainties hide the actual differences between compared networks. Hence, this paper presents guidelines for performing load-latency measurements for network-on-chips to avoid these pitfalls.

Keywords: network-on-chip, benchmarking, latency, performance analysis

I. INTRODUCTION

The system-on-chip (SoC) architectures are predicted to become communication bound. Network-on-Chip (NoC) paradigm [1][2][3] brings the techniques developed for macroscale, multi-hop networks into a chip. NoC is an on-chip communication network, i.e. a (sub)system that transmits data between its terminals. However, the term is used also in wider sense as a unification of all communication-related issues on a chip. The major goal of communication-centric design methodologies and NoC is to achieve greater design productivity and performance by handling the increasing parallelism, manufacturing complexity, wiring problems, and reliability.

This paper was inspired by the observations in our earlier surveys [3][4] where we analyzed about 140 NoC publications. Networks, especially NoCs, are often compared and benchmarked via their *load vs. latency* behavior [5]. For example, this method was utilized in about 18% of the papers cited in [3][4]. Unfortunately, the measurement setup is often unclear and sometimes even erroneous which leads to unfair comparisons or prevents them. For example, the utilized transfer sizes, metrics and units are not always defined properly, the header and packet latencies may be confused, or latency values from the saturated state are sometimes given. The need for common benchmarking strategies and test cases has been motivated, for example, in [6][7]. The main concepts for benchmarking NoCs in a systematic, reproducible, and comparable way are currently addressed and standardized by an OCP-IP workgroup [8][9].

This paper presents nine different examples illustrating how the often undocumented features have a profound impact on the results. Case studies highlight that the differences between various measurement setups may appear larger than those obtained via NoC optimization. For example, omitting network interface from the measurements may account for up to $2x$ deviation in the latency results. We present our guidelines for a common measurement standard to avoid these pitfalls.

This paper is organized as follows. Section II introduces the purpose of load-latency measurement whereas Section III introduces the measurement setup. Sections IV - VII discuss the measuring units, latency breakdown, and settings of the measurement and NoC itself. Conclusions are given in Section VIII.

II. BASICS OF LOAD VS. LATENCY MEASUREMENTS

Fig. 1. shows an example of *offered traffic load vs. latency* curve for an imaginary network. It shows the transfer delay as a function of transfer rate. The X-axis shows the offered load that means here the average data rate injected to the network terminals. There is one traffic source per terminal and usually all sources are active simultaneously. The Y-axis shows the transaction latency; in this case it is the average value. As the traffic load increases beyond network-specific threshold, the injected data start stacking at the sender because network cannot accept them fast enough. Consequently, their latency increases without a bound as the situation continues, and the network is *saturated*.

The simplest traffic scenario sets equal probability for all targets (spatially uniform), injects data continuously, and keeps both message size and injection rate fixed (temporally uniform,

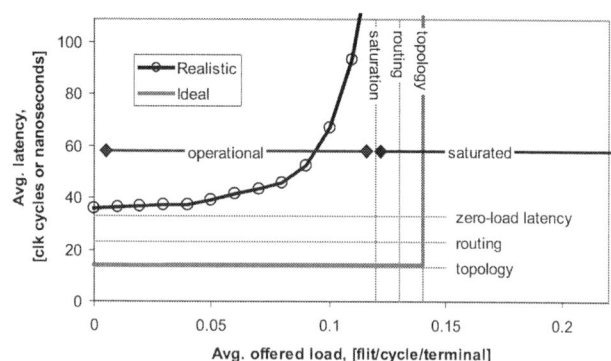

Fig. 1. Example of load vs. latency curve. X-axis shows the transfer rate and Y-axis the delay. Six analytical bounds are shown in addition to measured values.

978-1-4244-2541-9/08 $25.00 © 2008 IEEE

i.e. constant bit rate). The traffic characteristics must be documented properly to enable repeatability and comparison.

The exact curves are usually determined with simulation but certain bounds can be determined analytically [5]. The loosest bounds (bottom-right) are due to network topology. They are derived from the bisection bandwidth and the minimum number of hops (traversal of a link). Practical routing algorithms do not achieve perfect load balance and use more hops, hence defining tighter bounds. Actual load leading to *saturation* is obtained from simulation or measured from the implementation. *Zero-load latency* is measured so that only one sender is active which means that there is no contention. Due to congestion, the latency during normal operation is usually higher than zero-load latency, and increases slightly with increased load before saturation.

III. MEASUREMENT SETUP

Unfortunately, both *offered load* and *latency*, are ambiguous terms, see [3][4] for examples. This does not ruin the validity of the results *within* one paper as long as all cases use the same definition. However, comparison *between* publications becomes impossible. This Section aims to alleviate these shortcomings.

The examples are given for packet-switched $4x4$ mesh with wormhole switching [10] and utilizing spatially uniform random traffic. In general, we recommended using many types of spatial distributions but in this case the simple uniform suffices to bring out the issues we wish to address. Average latency in clock cycles is measured for transfers with 8 payload flits (flow control units). When transfer size varies, the average is still 8 payload flits. Load is given for the payload only (the headers are excluded). Resources and NoC use the same clock unless stated otherwise. Each resource initiates 500 transfers; hence there are $16 \cdot 500 = 8\,000$ transfers in total ($8\,000 \cdot 8 = 64\,000$ payload flits). The measurement settings are summarized in Table I in Section VIII with the corresponding deviations in results.

Canonical measurement setup is shown in Fig. 2(a). Traffic generator (TG) models the SoC resources, such as processors and memories, that initiate transfers. Configuration file defines the properties of the traffic:

- temporal - when and how much to send
- spatial - where to send.

Network interfaces (NI) restructure the data stream from TG to packets accepted by the NoC, and vice versa. Large (infinite, in theory) buffers must be placed between traffic source and network interfaces to avoid self-throttling during measurement [5]. Without such buffers, generator stops the data injection occasionally, and the real offered load does not match the expected. When the network saturates, these source queues start growing infinitely. Longer simulations yield larger latency for saturated traffic but the maximum observed latency cannot exceed length of the simulation. The interpretation of results must assume infinite latency during saturation, though.

There are three basic choices in load-latency measurement each with two options. That means $2^3 = 8$ possible com-

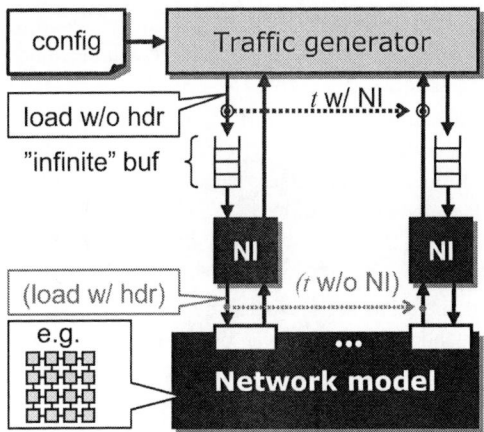

(a) Measurement setup. The non-recommended ways are in parentheses. *NI* denotes the network interface, t is the measured transfer latency, and *hdr* refers to packet header.

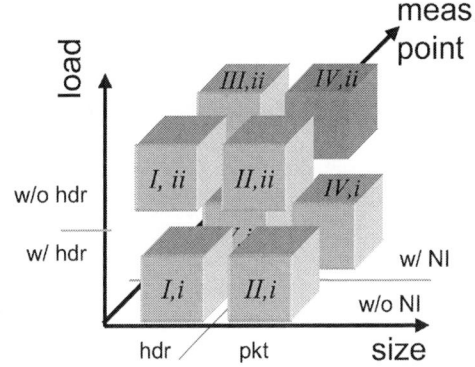

(b) 3-dimensional measurement space with 8 setups. The recommended case *IV,ii* is highlighted.

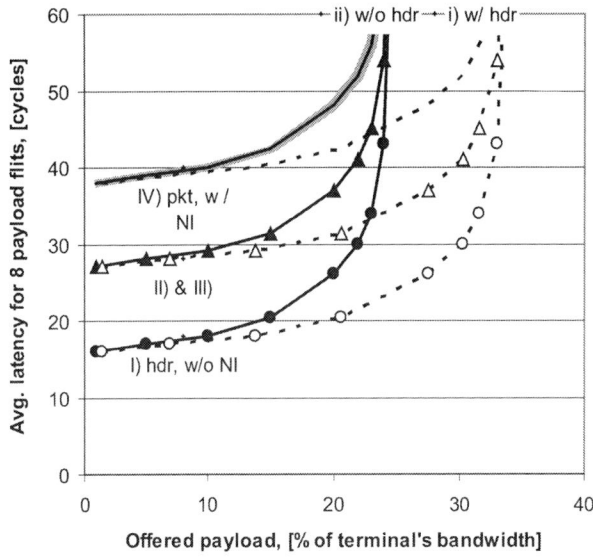

(c) Measured latencies in various setups. Note the deceitful difference between cases *IV,ii* and *I,i* although the NoC and traffic are the same.

Fig. 2. Basics of load-latency measurement: different setups and the resulting deviations in measured performance.

binations to start with. Fig. 2(b) illustrates the resulting 3-dimensional measurement space. Each choice defines one axis in the design space:

1) Size: Is latency t measured for the header or for the whole transfer?
2) Measurement point: Does latency t include network interface (NI) and waiting at the source buffer?
3) Load: Does the load r (i) include or (ii) exclude the header?

Our recommendations are presented next.

Header latency gives the lower bound for communication delay in a system but headers do not transmit any useful data in general case. Hence, latency of full packets should be measured.

The latency measurement starts when the first bit of the transfer leaves the traffic generator and ends when the last bit exits the NI at the receiver. NI must ensure that received data is in-order. Its latency and (buffer) area overheads may be substantial when reordering is required due to out-of-order delivery in the network.

The definition of load must be independent of the measured NoC. The header overhead depends on the evaluated NoC and NI, and therefore header must be excluded from the load. Otherwise, a NoC with large headers will be favored in an unfair manner, as it seems to tolerate larger load than it actually does. This applies even if source buffer and NI are implemented as part of the traffic generator.

As a summary, the latency must be measured for the whole packet including source queue, NI, and NoC latencies, and the load must include the payload only. That way the values correspond closely to the latency experienced by the actual processing elements (PEs) during real operation. The recommended setup is denoted as case IV,ii in Fig. 2(b) and 2(c). The same curve is repeated in the following graphs the sake of comparison.

Fig. 2(c) shows the impact of the various measurement options. Solid lines denote that headers are excluded from the load whereas they are included in the dashed lines. These load definitions do not affect the latency but they move the curves in X-direction. On the other hand, the measured transfer size and the impact of NI move the curves in Y-axis without affecting the saturation point. Differences are large: the latency appears to be is reduced to $0.4x$ while obtaining $1.4x$ throughput when basic choices are done "creatively" (16 vs. 38 cycles and 33 vs. 24%, respectively). Case II is here coincidentally identical to case III.

IV. UNITS OF MEASUREMENT

Another consideration regards the utilized measuring units and statistical values.

1) What values are reported? The most common and suitable choices for latency are average t_{avg} (1) and percentile $t(j)$ (2)

$$t_{avg} = \frac{1}{N} \sum_{i=0}^{N-1} t_i \qquad (1)$$

$$fraction\ j\ of\ transfers: \ t_i \leq t(j) \qquad (2)$$

$$t_i = t_{lastbit\_to\_rx\_TG} - t_{firstbit\_from\_tx\_TG} \qquad (3)$$

where t_i denotes the latency of a single transfer and N is their total number. Latency percentiles or box-plots are especially important when dealing with real-time constraints as the average performance is inadequate metric in those cases. The maximum variation or jitter $t_\Delta(j)$ is calculated as

$$t_\Delta = \frac{t_i}{min(all\ t_i)} - 1. \qquad (4)$$

Naturally, the latency equation must be reported explicitly to avoid confusion. For example percentiles $t(90\%)$, $t(99\%)$, and $t(100\%)$ are reported for latency and jitter [9]. The last one equals the maximum value of the set. Metrics t_i and $t(j)$ have the same unit (see next subsection) whereas $t_\Delta(j)$ is unitless.

2) What is the unit of latency? Latency measures how long it takes to perform certain operation; a transfer in this case. In principle, it is $time/operation$ - the inverse of data rate (throughput).

Both seconds and cycles are plausible time units. NoC frequency is implementation-dependent variable but most design-choices (routing algorithm, buffering, virtual channels etc.) can be compared with their cycle counts. Care must be taken to define "clock cycle" appropriately in a multi-clock system, e.g. using specifiers $\#cycles_{PE}$ or $\#cycles_{NoC}$.

The monitored $operation$ is case-dependent and needs rigor definition. For example, latency per packet needs information about inclusion/exclusion of header, payload length and is it varying dynamically between traffic sources.

3) What is the unit of load? We prefer measuring traffic load per terminal so that its maximum value does not depend on the network size or topology. If latency is measured at clock cycles, the recommended load unit is

$$[offered\ load] = flit/cycle/source \in [0.0, 1.0]. \qquad (5)$$

The number of traffic sources usually equals the number of terminals but not always, for example when measuring zero-load latency. In most cases, $flit/cycle/terminal$ may be opted to emphasize the fact that data is injected to all inputs. Either way, the chosen load metric has intuitive bounds that do not depend on the network size and only the payload flits are accounted. The data width and frequency of NoC terminals are implementation-dependent choices. Replacing them to (5) yields data rate in $bits/s/source$ or $Bytes/s/source$.

When the load varies between terminals or dynamically, characteristics are calculated with equations (1)-(4) by replacing latency t with data rate r. A non-constant data rate is called $bursty$ traffic. It has a profound impact on the NoC performance, as demonstrated in [11] that also presents an efficient way to generate bursty traffic loads artificially. The level of burstiness is there characterized with a unitless parameter b where $\mid b \mid \in [0, 0.5]$.

Expressing the load as a fraction (%) of terminal's bandwidth also seems to be a viable option. Unfortunately, the load is sometimes given as a fraction of network's *ideal capacity* on uniform traffic [5]. Capacity Θ means the input bandwidth that saturates the network, and the aim is to evaluate how close to ideal performance the network implementation gets. Percentage is always relative to something so it is obviously important to document clearly how it is measured, or avoid its usage. For example in Fig. 1, saturation occurs at 12% or 85% depending whether it is relative to terminal bandwidth or to network's ideal capacity. Note that it is no use to compare latency curves from different NoC topologies in a single graph if load is expressed as fraction of capacity.

V. LATENCY BREAKDOWN

Part of the latency is due to utilized network interfaces but also to inefficiencies of the traffic generator. Moreover, the evaluator must ensure the saturation is really caused by the NoC itself and not by these secondary components. Fig. 3(a) shows the results of such sanity check where the components are measured in separation. Let us define the total latency t and the saturating data rate r as

$$t = \sum t_{TG} + t_{NI} + t_{NoC} \qquad (6)$$

$$r = min(r_{TG} + r_{NI} + r_{NoC}) \qquad (7)$$

The lower bound for latency on the bottom illustrates the time needed for the mere data copying: 8 cycles in this case. Above that is the 3-cycle additional latency due to traffic generator. The generator used in this study spends 2 extra cycles between for each transfer, namely one cycle for address and one related to internal state machine. Hence, it can inject at most

$$\frac{8\,flits}{(8\,flits \cdot 1\,flit/cycle) + 2\ cycles} = 0.8\ flit/cycle.$$

Network saturation occurring beyond that would be masked away but these empty cycles have no effect in lightly loaded case. Large transfers move the NI's saturation upwards, for example 32-word transfers saturate at 94% of terminal bandwidth (not shown).

The utilized NI has latency overhead that grows with packet length. Time needed in NI for 8 payload words is 8 cycles for data, 3 for the header, and 2 empty cycles related to the current implementation. In other words, maximum load becomes $\frac{8\ words}{13\ cycles} = 0.62\ flit/cycle/terminal$ due to this particular NI. It is by chance identical to the saturation point of a single router because the control logic inside a router also wastes two cycles between the packets.

The location of the router affects its average latency but not the saturation load. For example, the corner router (0,0) has larger latency than middle router (1,1). Note also that data copying, traffic generator and NI together cause 27-cycle latency whereas the (header) latency through the NoC attributes only additional 9 cycles. Let us assume that an optimized NoC might reduce the latency by, say, one cycle. This would mean a notable 11% reduction or mere 3% depending on whether we consider the network only or the full end-to-end latency. The differences between NoC latencies diminish even further, when

the runtime overhead of SW platform running on a processor are also accounted.

The above simulation validates that the following holds

$$r = r_{NoC} < r_{bisection} < r_{router} \leq r_{NI} < r_{TG} \qquad (8)$$

In this case, the example mesh saturates at 24% load. The bisection B_{mesh} has $2k$ unidirectional links, where k denotes the number of nodes in one dimension [5]. Hence, the theoretical load bound for uniform load in 16-node mesh is $B_{mesh} \cdot \frac{2}{N} = \frac{4k}{N} = \frac{4}{\sqrt{N}} = 1.0\ flit/cycle/terminal$. This reveals room for improvement regarding routing policy, header overhead, and the control logic of the routers. Furthermore, we observe that the saturation load decreases with larger network sizes even if mesh is often dubbed "scalable".

VI. IMPACT OF THE TRANSFER COUNT AND LENGTH

The number of monitored data transfers, their length, and pseudo-random behavior affect the results. These phenomena are studied next. Fig. 3(b) shows the results when the number of transfers varies, averaged over 3 independent runs. In all cases, 8000 transfers were generated and $2000 - 8000$ were included for calculating the average. The sampled transfers were taken from beginning, middle, or from the end of the simulation. Furthermore, an example of erroneous measurement setup is shown with a case where the infinite FIFO was omitted from the source. A similar pitfall is to use a too small source FIFO.

The differences become visible only near the saturation. During saturation, the last packets will experience longer latency than the first. The knee-point of curve gets sharper with larger transfer counts and when the warmup period (i.e. the first packets) is discarded. However, all transfer counts had the same non-saturated latency and were able to recognize the saturation point. Note that the saturation point can be identified also from the load-throughput curve.

The measurements were repeated 3 times independently in order to study the impact of randomness. In this case, the generation of traffic files includes randomness whereas the simulation produces always the same results with given traffic file. The average latencies, eq. (1), of each run were compared, and the ratio between largest and smallest values is shown in Fig. 3(c). The differences are negligible at small loads and visible near saturation. Differences between runs decrease when more transfers are monitored. However, the transfer counts are here high enough that one simulation run suffices. In general, designers are encouraged to use largest transfer count that is practical regarding the simulation time. In general, the number of needed runs and transfers may be sought empirically case-by-case, for example as

$$for\ all\ loads:\ for\ all\ runs: \frac{max(t_{avg})}{min(t_{avg})} - 1 \leq \epsilon \qquad (9)$$

and let's say $\epsilon = 5\%$ for non-saturated loads.

Fig. 3(d) illustrates the variation due to different transfer lengths. Three cases are shown: transfer length being 2, 8 or 32 payload words. Similar measurements were done with single

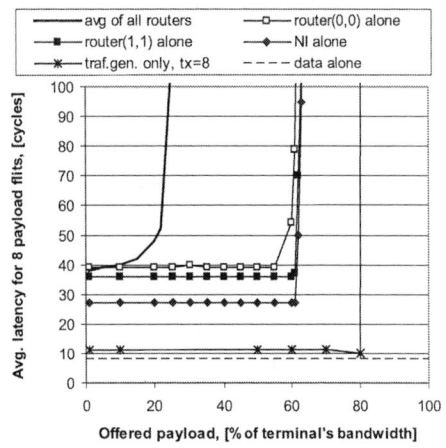

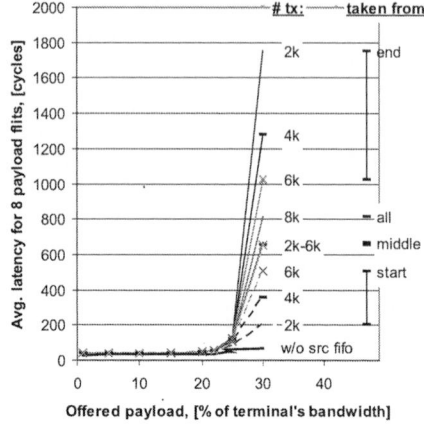

(a) Measurements done separately for each component.

(b) Varying the number of monitored transfers from 2000 to 8000. The impact of warmup period at the beginning is also visible.

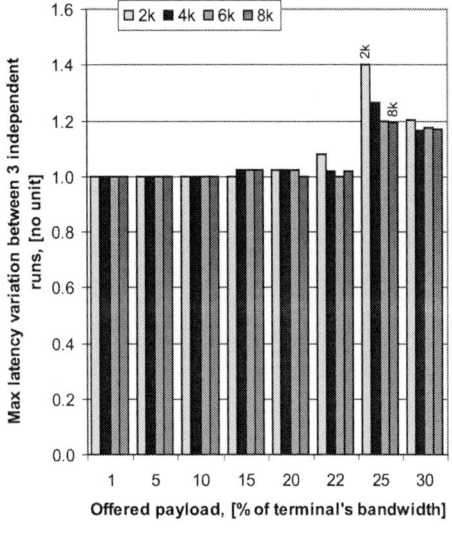

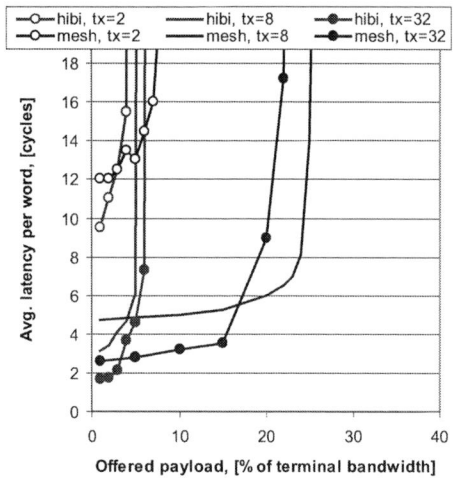

(c) Maximum variation between 3 independent runs as load and the number of monitored transfers vary.

(d) Impact of transfer length when it varies from 2 to 32 words (flits).

Fig. 3. Comparing load-latency per component and between different measurement runs.

HIBI (Heterogeneous IP Block Interconnection) segment [10]. Note that the latency is here given *per word* whereas other graphs measure the latency *per transfer*. The latency per word decreases with longer transfers because the empty cycles in traffic generator and NI occur only once per transfer. This is clearly shown with HIBI that has 1-word header (address) and varying burst length needs only single header per arbitrarily long transfer.

In addition, longer transfers move the saturation point to higher loads. Both networks have inferior results with 2-word transfers, especially the mesh that used fixed length packets of 8 payload flits and 3 header flits. Comparing results from simulations that utilize different transfer length is not meaningful. An obvious recommendation is to study multiple transfer lengths for each compared NoC and document the settings.

VII. NETWORK-SPECIFIC SETTINGS

Final remarks concern the settings related to the measured networks. The *overall performance* or *merit* is a combination of several factors, not just latency. For example, a network with the highest performance may be prohibitively costly to implement and hence out of the question. Therefore, an appropriate multiobjective trade-off is sought. Automated design space exploration tools often seek to optimize a *cost function* that considers properties such as silicon area, power consumption, runtime, and latency. Network-specific settings - such as data

978-1-4244-2541-9/08 $25.00 © 2008 IEEE

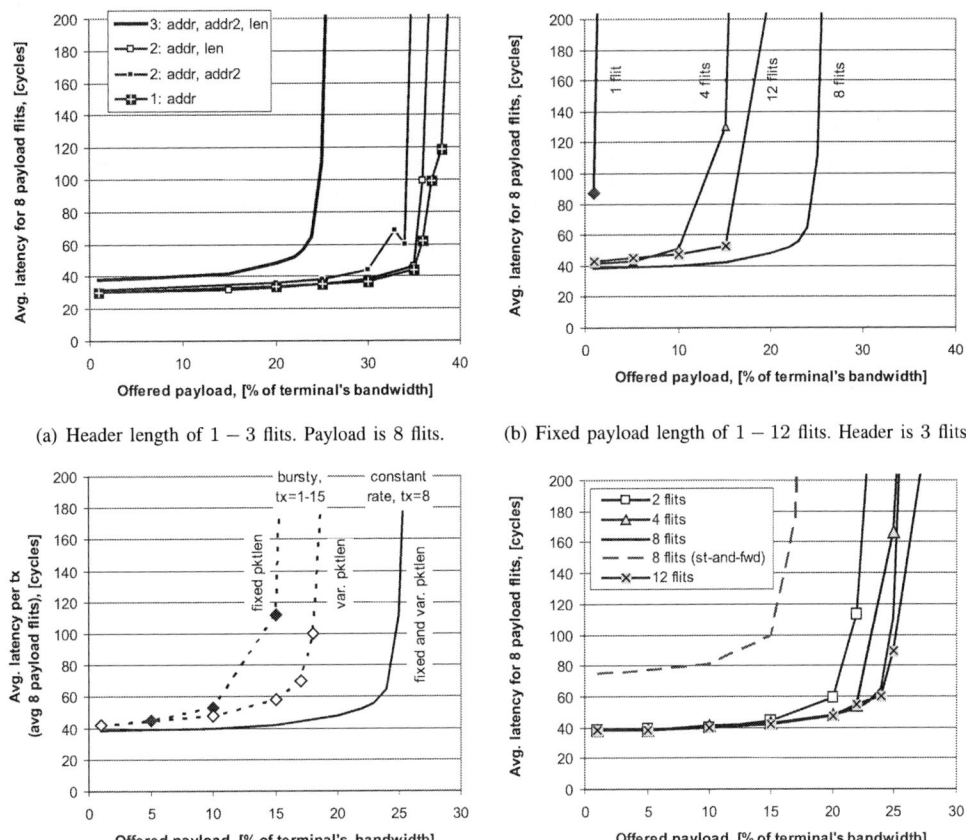

(a) Header length of $1-3$ flits. Payload is 8 flits.

(b) Fixed payload length of $1-12$ flits. Header is 3 flits.

(c) Varying versus fixed payload length. Header is 3 flits.

(d) Varying the buffer depth at the routers from 2 to 12 flits.

Fig. 4. Impact of network specific settings: lenghts of packet header and payload, and the depth of buffers at the routers.

width, frequency, buffer depth - affect all these metrics which motivates their careful selection.

Packet structure is clearly network-specific parameter whereas transfers are defined in the benchmark set. Preferably, several packet sizes must be evaluated. There are three basic choices for selecting network settings, for example packet size, for the compared NoCs:

1) Use the default packet size for each NoC, so called out-of-the-box execution. The resulting differences in performance, area, and power are accounted in cost function. A variant of this scheme uses the same packet size for each NoC.

2) Select packet size that minimizes cost function, so called full-fury optimization.

3) Select sizes that produce (nearly) equal area, and/or power, i.e. compare NoCs that conform to given constraint. The obtained performance values are now directly comparable.

Similar selection procedure can be easily generalized to other parameters, such as buffering, pipelining, frequency, and data width.

Examples are shown in Fig. 4. Utilized mesh supports up

to three header flits: mandatory NoC address, optional packet length, and optional second level address. The examples so far utilized all 3 fields. Performance with shorter headers is shown in Fig. 4(a) for four header types. The differences in latency are quite small but shorter headers provide a notable $1.45x$ increase in tolerated load. This observation clearly supports our recommendation to exclude headers from the offered load. Of course, shortest possible headers are desired and minimum length is dictated by the data width and the complexity of the network protocol.

Mesh in Fig. 4(b) uses fixed-size packets with 3-word headers and payload of $1, 4, 8$ or 12 words. The header overhead is intolerable for 1-word payload and larger payloads naturally increase performance. For example, increasing payload length from 4 flits to 8 increases the maximum data rate by a factor 1.6. However, dummy data that is used to fill packets deteriorates the performance with $12-$word payload. Dynamically varying packet overcomes this drawback.

Fig. 4(c) varies both transfer and packet size. The traffic is now *bursty* because the bit rate varies which resembles real-life applications better than constant load. The average transfer size is 8 payload words in all cases and varies uniformly within

978-1-4244-2541-9/08 $25.00 © 2008 IEEE

TABLE I

MEASUREMENT SETTINGS AND THE MAXIMUM OBSERVED DIFFERENCES ON PEFORMANCE.

	Property	Value	Unit (/note)	Max. impact in case study	Recommendation
Network	size	16 terminals		-	evaluate many
	topology	2-D mesh, bus (2 choices)		3.5x throughput	evaluate many
	header length	1 - 3 flits		1.6x throughput	shortest possible
	a) payload length	1 - 12 flits		25x throughput	evaluate many
	b) payload length	fixed or varying (2 choices)		1.2x throughput	dynamically varying
	buffer depth	2 - 12 flits		1.14 throughput	smallest adequate
	switching	wormhole, st-and-fwd (2 choices)		1.6x throughput, 2x lat.	wormhole
Traffic	spatial	uniform dst distrib. (but not to itself)		-	evaluate many
	offered load	0.0 - 1.0 flits/cycle/terminal		-	evaluate many
	a) tx length	2 - 32 payload flits		1.6x throughput	evaluate many
	b) tx length	fixed or varying (2 choices)		1.4x throughput	varying
	# tx	2000 - 8000 tx /terminal		9x latency	largest practical
Measurement	latency for	header or whole tx (2 choices)		1.6x throughput, 1.3x lat.	whole tx
	load definition	incl./excl. header - " -		1.3x latency	excl. header
	meas. point	incl./excl. NI - " -		1.3x latency	incl. NI
	infinite src queue	included/excluded - " -		exclusion is illegal	included
	latency	per transfer, per flit - " -		-	per transfer
	# src terminals	corner, middle, or all (3 choices)		3x throughput, 1.1x lat.	all
	# independent runs	3 runs		1.4x latency	largest practical
	warmup length	0 - 6000 transfers		9x latency	discard warmup

shown range. This situation is bad for fixed-size packets that need dummy data for nearly all packets. A mesh that is able vary packet size at runtime obtains categorically better results. However, the constant bit rate traffic saturates at 24% and bursty at 18%, i.e. $1.4x$ difference.

Fig. 4(d) illustrates the variation due to buffer depth in the routers as it varies from 2 to 8 flits. Furthermore, a mesh with store-and-forward switching scheme was measured and found less capable than the wormhole switched. Larger buffers clearly increase the performance but also the area cost. IN this case, increasing buffering by factor of $6x$, results in $5.5x$ area cost and $1.14x$ saturation point. Hence, performance increase is rather modest compared to cost.

VIII. CONCLUSIONS

This paper presents guidelines for performing load-latency measurements for network-on-chips. Our earlier surveys observed that the method is very common but unfortunately non-standardized and measurements are seldom adequately documented. Examples are given to 16-node 2D mesh to emphasize our claims. Very large differences are observed due to innocent-looking basic choices in measurement setup and network settings.

The guidelines are summarized as:

- An "infinite" source queue is placed between traffic generator and NI.
- The measurement points are between traffic generator and source queue.
- The latency is measured for the whole packet including source queue, NI and NoC latencies.
- Average, percentiles, and jitter are the most common values for characterizing latency. The choice must be stated explicitly to avoid confusion.
- The recommended unit for offered load is $flit/cycle/terminal$ and it includes the payload only.

- Large number of samples (monitored transfers) increases the validity.
- The min/avg/max values of packet size must be given when the size varies at runtime or across the nodes.
- Study multiple transfer lengths for each compared NoC and document the settings.
- Ensure that the bandwidth is not limited by the traffic generator or NI to effectively measure the impact of NoC parameters.
- The parameters of the studied NoC must be selected with care and considering the overall cost.

Table I summarizes the measurements, maximum impact of parameters, and recommendations.

REFERENCES

[1] W. Dally and B. Towles, Route packets, not wires: on-chip interconnection networks, in DAC, 2001, pp. 684-689.
[2] T. Bjerregaard and S. Mahadevan, A Survey of Research and Practices of Network-on-Chip, ACM Computing Surveys, Vol. 38, Iss. 1, article No. 1, 2006.
[3] E. Salminen, A. Kulmala, T.D. Hämäläinen, Survey of Network-on-chip Proposals", white paper, OCP-IP, Apr. 2008, 13 pages.
[4] E. Salminen, A. Kulmala, T.D. Hämäläinen, On Network-on-chip comparison, Euromicro conf. on Digital System Design, 2007, pp. 503-510.
[5] W. J. Dally and B. Towles, Principles and practices of interconnection networks. Morgan Kaufmann Publishers, 2004.
[6] J.D. Owens et al., Research Challenges for On-Chip Interconnection Networks, IEEE Micro, Vol. 27, Iss. 5, Sept.-Oct. 2007, pp. 96 - 108.
[7] U.Y. Ogras, Jingcao Hu, R. Marculescu, Key research problems in NoC design: a holistic perspective, CODES, 2005, pp. 69-75.
[8] C. Grecu et al., "An Initiative towards Open Network-on-Chip Benchmarks", white paper, [online]:http://www.ocpip.org/socket/whitepapers/NoC-Benchmarks-WhitePaper-15.pdf, Feb. 2007, 16 pages, OCP-IP.
[9] Zhonghai Lu, A. Jantsch, E. Salminen, C. Grecu, "Network-on-Chip Benchmarking Specification Part 2: Micro-Benchmark Specification Version 1.0", OCP-IP, May 2008, 16 pages, camera ready.
[10] E. Salminen, Benchmarking Mesh and Hierarchical Bus Networks in System-on-Chip Context", Journal of System Architectures, Aug. 2007, Vol.53, Iss. 8, pp. 477-488.
[11] R. Thid, I. Sander, and A. Jantsch, Flexible bus and NoC performance analysis with configurable synthetic workloads, in DSD, 2006, pp. 681-688.

978-1-4244-2541-9/08 $25.00 © 2008 IEEE

Realizing a flexible constraint length Viterbi decoder for software radio on a de Bruijn interconnection network

Ganesh Garga, Mythri Alle, Keshavan Varadarajan, S.K Nandy, H.S Jamadagni
Indian Institute of Science, Bangalore, India

Abstract—Building flexible constraint length Viterbi decoders requires us to be able to realize de Bruijn networks of various sizes on the physically provided interconnection network. This paper considers the case when the physical network is itself a de Bruijn network and presents a scalable technique for realizing any n-node de Bruijn network on an N-node de Bruijn network, where $n < N$. The technique ensures that the length of the longest path realized on the network is minimized and that each physical connection is utilized to send only one data item, both of which are desirable in order to reduce the hardware complexity of the network and to obtain the best possible performance.

I. INTRODUCTION

In modern wireless standards like WCDMA and cdma2000, the forward error correction specifications call for a lot of flexibility in the encoders and decoders[5]. Convolutional codes are one of the error correcting codes used in these standards. With respect to these codes, the modern standards require various constraint lengths and code rates to be supported. Convolutional codes are decoded by the well-known Viterbi algorithm[6], which is typically implemented using a multi-processor architecture[4], [1], [2] wherein each processor performs an operation called the Add-Compare-Select (ACS) operation and the interconnect pattern for the inter-processor communication is a de Bruijn graph. Our aim is to obtain a methodology for the design of Viterbi decoders capable of supporting all constraint lengths upto a certain maximum, which is also scalable in that the maximum constraint length can be set to any value. This requires us to find a scalable method to realize de Bruijn networks of different sizes(*guest* networks) on the physically provided network(*host* network). This paper presents such a method for the case when the host network is itself a large de Bruijn network, corresponding to the largest constraint length to be supported. [3] also considers implementation of flexible constraint length Viterbi decoders on a de Bruijn network, but provides a tradeoff between constraint length and throughput. Our approach is different in that we aim to obtain a constant throughput at all constraint lengths supported.

II. AVAILABLE CONNECTIONS

In an N node de Bruijn network, each source node u is connected to two destination nodes, an *even* numbered destination node given by $(2 * u) \ mod \ (N)$ and an *odd* numbered destination node given by $(2 * u + 1) \ mod \ (N)$ (see figure 1). If u is represented by an N bit binary string

Table I
THE SET OF DESTINATIONS FOR A GIVEN SOURCE AVAILABLE ON THE PHYSICAL(HOST) DE BRUIJN NETWORK AND THE CORRESPONDING BIT LEVEL TRANSFORMATION OF THE SOURCE NODE WHICH YIELDS THE DESTINATION NODE

Destination number	Transformation effected by the path	Modified node number
$u_{\log(N)-1} \ldots u_1 \mathbf{0}$	left shift and shift in a 0	$u * 2$
$u_{\log(N)-1} \ldots u_1 \mathbf{1}$	left shift and shift in a 1	$u * 2 + 1$
$\mathbf{0} u_{\log(N)} \ldots u_2$	right shift and shift in a 0	$\frac{u}{2}$
$\mathbf{1} u_{\log(N)} \ldots u_2$	right shift and shift in a 1	$\frac{u+N}{2}$

$u = u_{\log(N)} \ldots u_1$, then the even numbered destination node has the binary representation $u_{\log(N)-1} \ldots u_1 \mathbf{0}$ and the odd numbered destination node has the binary representation $u_{\log(N)-1} \ldots u_1 \mathbf{1}$.

The host de Bruijn network is assumed to have *bidirectional* links. In that case, from every source node u, there exist links to four destinations, shown in column 1 of Table I. Each link represents a specific bit level transformation on the source node numbers, as indicated in Table I.

III. REALIZATION TECHNIQUE

To realize a path of the guest network on the host network, we need to transform the set of source node numbers(SNNs) of the guest network to the set of destination node numbers(DNNs) by performing a number of left and right shifts, as seen from Table I. We categorize the paths into two groups based on the SNN from which the paths originate.

1. Paths for SNNs 0 to $(\frac{n}{2} - 1)$: These paths are always directly available in the host network. This follows from the definition of the de Bruijn graph and the fact that $n < N$.

2. Paths for SNNs $\frac{n}{2}$ to $(n-1)$: There are two choices for this group.

a) Right Shift First (RSF): In this method, we first shift out the SNN bits through right shifts and then shift in the DNN bits through left shifts. For example, consider the realization of a 4-node guest network on a 16-node host network. The paths $3 \rightarrow 2$ in the guest network will be laid out as follows: $3_d = 0011 \rightarrow 0001 \rightarrow 0000 \rightarrow 0001 \rightarrow 0010 = 2_d$. Note that there is an unnecessary link from 0001 to 0000 and back, which can be removed by shifting in the DNN bits directly after the

978-1-4244-2541-9/08 $25.00 © 2008 IEEE

penultimate left shift. In fact, for an SNN having a run of a 1s including the MSB_{guest}, the DNNs have a run of $(a-1)$ 1s including the MSB_{guest}. Hence, we need to shift out only the last $(log(n) - (a-1))$ bits of the SNN and shift in the respective bits of the DNN. Thus, in this method, we classify the SNNs into *ranges*, defined as follows:

Defn: Range m **is comprised of SNNs** $\frac{(2^m-1)n}{2^m}$ **to** $\frac{(2^{m+1}-1)n}{2^{m+1}} - 1 \ \forall \ m \in 1$ **to** $log(n)$.

All SNNs in a range m have a run of m 1s including the MSB_{guest}. For all paths in this range, we perform $(log(n) - (m-1))$ right shifts to remove the last $(log(n) - (m-1))$ bits of the SNN and then an equal number of left shifts to shift in the last $(log(n) - (m-1))$ bits of the DNN. As the same data is to be routed to both the even and odd DNNs from a given SNN, we route the path to the odd DNN along the even DNN path, except at the last step, where we shift in a 0 on the even DNN path and a 1 on the odd DNN path. This method needs a total of $(2log(n) - 2(m-1))$ shifts. Hence, the length of the longest path laid out, or the *dilation* for the RSF method is $2log(n)$, which is the length of all the paths in range 1.

b) Left Shift First (LSF): In this method, we remove the MSB_{guest} of the SNN which is unneeded in the DNN, (refer section II) by left shifting it out through the MSB_{host}, while shifting in a 0 at the first left shift on even DNN paths and a 1 on odd DNN paths. Then, we perform right shifts to get the DNN to the correct position within the node number. For example, the path $3 \rightarrow 2$ of a 4-node guest network will be realized on a 16-node host network as follows: 3_d = $0011 \rightarrow 01\mathbf{10} \rightarrow \mathbf{11}00 \rightarrow \mathbf{1}000 \rightarrow 0\mathbf{1}00 \rightarrow 00\mathbf{1}0 = 2_d$. The number of left shifts is determined by *((distance to MSB_{host} from MSB_{guest}) + 1)*, which is equal to $(log(r) + 1)$, where $r = \frac{N}{n}$ (for the above example, $r = \frac{16}{4} = 4$).

This method needs a total of $(2log(r) + 1)$ shifts, for all paths. Hence, the dilation of the LSF method is $(2log(r) + 1)$. For a given realization, we choose one of the above two method based on which method incurs the minimum possible dilation. For the RSF method, the maximum dilation is $2log(n)$ as explained earlier, while for the LSF method, it is $2log(r)+1$. Hence, we follow the rule given below:
For a given realization, If $2log(r) + 1 >= 2log(n)$, use RSF to lay out all the paths, else use LSF to lay out all the paths.

There are cases in a given realization where two paths laid out by the LSF or RSF method pass over the *same* link in the *same* direction. In other words the two paths *conflict*. The next section IV obtains the conditions for conflicts to occur and presents modifications to avoid the same, besides presenting a proof that the dilation incurred is the minimum possible.

IV. MINIMUM DILATION AND CONFLICT AVOIDANCE

Theorem 4.1: The dilation obtained by the proposed technique is the minimum possible.

Proof: The host network links allow bits to be introduced or removed only at the two extreme bit positions of the node number. Hence, the only way to change an arbitrary bit in the node number is to remove it through one of the two ends, and shift in the modified bit. As such, the dilation is proportional to the distance of the innermost bit which is different in the SNN and DNN from the two extreme bit positions of the node number.

Now, for an SNN in some range m, the innermost change that has to be effected is the removal of one of the m 1s including and immediately after the MSB_{guest}. There are only two ways to effect this removal - either by shifting one of these 1s out through the MSB_{host}, which translates to the LSF method, or by shifting one of these 1s out in through the LSB_{host}, which translates to the RSF method. As mentioned earlier, the method for a given realization is chosen based on which of the methods has the lesser dilation. ∎

Now, we begin the analysis of conflicts by considering the conflicts that arise due to the use of the LSF method.

Lemma 4.2: A conflict is possible in the LSF method iff the ratio of the DNNs of two paths is a power of 2.

Proof: During the left shifts or right shifts, we actually multiply or divide the node number by 2. Hence, a conflict will occur *iff* the numbers we start with during the left shifts, or the numbers we end with during the right shifts are such that their ratio is a power of 2. Now, considering only the left shifts' segment of any two paths, a conflict will occur if *the ratio of the SNNs is a power of 2*. However, since the SNNs for which the LSF method is applied are constrained to be in the range $\frac{n}{2}$ to $(n-1)$, the ratio of two SNNs can never be a power of 2. Hence, there cannot be a conflict in the left shifts' segment of two paths. However, conflicts are possible in the right shifts' segment of two paths, as there are paths in any particular realization for which the ratio of the DNNs is a power of 2. ∎

For any two paths in general, all potential conflicts can be avoided if the intermediate nodes reached in the right shifts' segment are such that their ratio is not a power of 2. The next lemma presents the modification required to achieve the same.

Lemma 4.3: All conflicts in the LSF method can be avoided if a 1 is shifted in at the second left shift on all paths.(refer figure 2 for an example)

Proof: Consider two SNNs i and j. We trace the path laid out from these SNNs while factoring in the above mentioned modification. At the first left shift, a 0 or 1 may be shifted in, depending on whether the DNN is even or odd. This is represented by k_1 in the expression below. $i \rightarrow (2.i + k_1) \rightarrow (4.i + 2.k_1 + 1) \rightarrow \ldots$ *(after $log(r)$ shifts)*$\ldots 2^{log(r)}.i + 2^{log(r)-1}.k_1 + 2^{log(r)-2}$. But $2^{log(r)} = \frac{N}{n}$. Hence, we get the intermediate node number after $log(r)$ left shifts as $\frac{N}{n}.i + \frac{N}{2n}.k_1 + \frac{N}{4n}$. At the next left shift, the MSB_{guest} is shifted out. The node number, after some simplification, is obtained as- $\frac{N}{n}(2.i + k_1 + \frac{1}{2} - n)$. Similarly, for the SNN j, we get the corresponding node number as-(k_2 is used instead of k_1) $\frac{N}{n}(2.j + k_2 + \frac{1}{2} - n)$. Now, consider the ratio $\frac{\frac{N}{n}(2.i+k_1+\frac{1}{2}-n)}{\frac{N}{n}(2.j+k_2+\frac{1}{2}-n)}$. This evaluates to $\frac{2(2.i+k_1-n)+1}{2(2.j+k_2-n)+1}$. This is clearly a ratio of odd numbers, which can never be equal to a power of 2. Hence, due to the modification indicated above, the numbers at the start of the right shifts segment of any two paths are always such that their ratio is *not* a power of 2. Since all the links in the right shifts' segments represent a division by 2, this ensures that the intermediate nodes reached in the right shifts's segment of any two paths will always be such that their ratio is not a power of 2. Hence, there will be no conflict in the right shift's segment of any two paths. Consequently, due to lemma 4.2, there will be no conflict between any two paths laid down by the LSF method with the above modification. ∎

Next, we consider the conflicts that arise due to the use of the RSF method.

The general approach followed in all proofs regarding the RSF method is to show that some substring within the node representation is different for the two paths at all shifts for each possible pair of ranges of shifts and/or the direction of shifting is opposite in the two paths and hence the same two node numbers cannot occur in the same order in the two paths under consideration.

We note that if the RSF method is applied without modification, two paths within any range would conflict as soon as the bit positions that are different between the two SNNs are shifted out, since 0s are shifted in from the MSB_{host} on both paths. But, analogous to the LSF method, the bits shifted in during the *right* shifts are 'dont care' bits, which we can define in such a manner as to avoid conflicts between any two paths. We can classify cases of conflicting paths into two types i.e. when the paths belong to the same range and when the paths belong to two different ranges. The next two lemmas consider these two types separately.

Lemma 4.4: All potential conflicts between any two paths laid out by the RSF method in a given range m are avoided if the complement of the even DNN (corresponding to each SNN) is shifted in, MSB first, during the right shifts, for all paths.(refer figure 3 for an example)

Proof: Consider two paths whose SNNs are such that, the pth bit is the last bit, counting from LSB_{host}, that is different. Let SNN1 have the binary representation $a_{log(n)} \ldots a_1$ and SNN2 have the binary representation $b_{log(n)} \ldots b_1$.

Let DNN1 have the binary representation $c_{log(n)} \ldots c_1$ and DNN2 have the binary representation $d_{log(n)} \ldots d_1$ (we do not specify whether the DNNs are even or odd since that does not affect our proof). The $(p+1)$th bit will be the last bit that is different (counting from LSB_{host}), between the DNNs. Now, for the first $(p-1)$ right shifts, a_p and b_p are present in the node representations on the two paths, thus making the intermediate nodes on the two paths different. Hence, the two paths cannot conflict in this part. At the next right shift, a_p and b_p are shifted out from LSB_{host}. In the worst case, when no other bits are different between SNN1 and SNN2, the same node number will be reached on both paths. However, even in this case, the same *link* will not be traversed as the bits shifted in at the next right shift, $c_{(p+1)}$ and $d_{(p+1)}$, are complements of each other. Also, these two bits remain part of the node representations on the two paths all through the remaining right shifts (at the last right shift, they are $((log(n) - (m-1)) - (p+1))$ positions from MSB_{host}), and hence the two paths will continue to be distinct throughout the remaining right shifts.

During the left shifts, for the first $((log(n) - (m-1)) - (p+1))$ left shifts, $c_{(p+1)}$ and $d_{(p+1)}$ remain part of the node representations and hence, the two paths continue to be distinct. At the next shift, these bits will be shifted out from MSB_{host}, but a_p and b_p are shifted in from LSB_{host}, which remain part of the node representation for all remaining left shifts. Hence, the node numbers will continue to be different for all the left shifts too. ∎

Lemma 4.5: There can be no conflict between two paths laid out by the RSF method which belong to two different ranges.

Proof: We can assume that $m1 < m2$, since the proof will recursively hold to cover all the ranges. Let the path in range 1 be denoted by path 1 and the path in range 2 be denoted by path 2. Now, in the right shifts' segment, the node representation on path 1 has a substring containing $log(r)$ zeroes followed by $m1$ 1s, whereas the same field in the node representation on path 2 has a substring containing $log(r)$ zeroes followed by $m2$ 1s. In all the left shifts, the node representation on path 1 has a substring containing $log(r)$ zeroes followed by $(m1 - 1)$ 1s, whereas the same field in the node representation on path 2 has a substring containing $log(r)$ zeroes followed by $(m2-1)$ 1s. As long as $m1 \neq m2$, a node number occurring in left shift (right shift) section of path1 will not occur in the left shift (right shift) section of path2. The case of the same link occurring in the left shift section of path1 and the right shift section of path2, or vice versa need not be considered, since the physical network is assumed to have bidirectional links. ∎

The following lemma formally ties up the above two lemmas.

Lemma 4.6: There can be no conflict between any two paths laid out by the modified RSF method

Proof: A potential conflict can be of two types: a conflict between two paths in the same range m or a conflict between two paths in two different ranges $m1$ and $m2$. Hence, the proof follows from lemma 4.4 and 4.5. ∎

Lemma 4.7: There can be no conflict between a path laid out by the LSF method and any of the direct paths.

Proof: One of the nodes on each link of the direct connections is always less than $\frac{n}{2}$. Hence, for a conflict to occur, an intermediate node number less than $\frac{n}{2}$ should occur in the path laid out by LSF too. In the LSF method, for all the left shifts, the intermediate node numbers are always greater than $\frac{n}{2}$ (since we start with an SNN that is greater than $\frac{n}{2}$). Hence, in this section, there can be no conflict with the direct connections. Any conflicts between the direct connections and the right shifts section of a path need not be considered, as the physical network is assumed to have bidirectional links.

∎

Lemma 4.8: There can be no conflict between a path laid out by the RSF method and any of the direct paths.

Proof: In the RSF method, for the first $(log(n) - (m-1))$ steps, we perform right shifts. Due to the assumption of bidirectional physical links, there cannot be a conflict in this section. Note that the first bit to be shifted in during the right shifts is always a 1. At the end of the right shifts, this 1 is located at the position $log(N) - (log(n) - (m-1) - 1) = (log(r) + m)$, which is always to the left of MSB_{guest}, since $(log(r) + m) \geq (log(n) + 1)$. This means the node number at the end of the right shifts will always be greater than or equal to $\frac{n}{2}$. Now, at each left shift, this node number is multiplied by 2, with or without the addition of a 1. Hence, the intermediate node numbers in the left shift section will always be greater than $\frac{n}{2}$. Hence, there can be no conflict by the same argument as that given in lemma 4.7. ∎

Theorem 4.9: There is no conflict between any of the paths laid out by the proposed realization technique.

Proof: Any potential conflict can be classified into one of the following types:

1) a conflict between two paths laid out by the LSF method
2) a conflict between two paths laid out by the RSF method
3) a conflict between a path laid out by the LSF method and a direct connection.
4) a conflict between a path laid out by the RSF method and a direct connection.

Hence, the proof follows from lemmas 4.3, 4.6, 4.7 and 4.8. ∎

V. FUTURE WORK AND CONCLUSIONS

This paper presents a scalable technique for implementing any smaller n-node guest de Bruijn network on a larger N-node host de Bruijn network, as a way to realize flexible constraint length Viterbi decoders on a physical de Bruijn network, along with relevant proofs. Due to the use of a de Bruijn network as the physical network, we are able to obtain a dilation which is lesser than the logarithm of the number of nodes(N) for all lower constraint lengths(except when $n = N/2$, when the dilation is exactly $log(N)$), while simultaneously avoiding any congestion of the physical links. This helps to improve the performance as well as to lower the energy dissipation, since the design of the interconnection network need not incorporate any kind of conflict resolution. Future work would involve comparing the results of this approach against other implementations described in the literature. Similar techniques could also be found for other physical interconnection networks.

REFERENCES

[1] M. Bóo, F. Argüello, J. D. Bruguera, R. Doallo, and E. L. Zapata. High-performance VLSI architecture for the Viterbi algorithm. *IEEE Trans. Communications*, vol. 45(2):pages 168–176, 1997.

[2] Joseph R. Cavallaro and Vaya Mani. Viturbo: A Reconfigurable Architecture for Viterbi and turbo decoding. *ICASSP*, 2003.

[3] P.H Kelly and P.M Chau. A Flexible Constraint Length, Foldable Viterbi Decoder. *GLOBECOM '93 Proceedings*, pages 631–635, 1993.

[4] C. B. Shung, H.-D. Lin, R. Cypher, Siegel P. H., and H. K. Thapar. Area-efficient architectures for the Viterbi algorithm. II. Applications. *IEEE Trans. Communications*, vol. 41(5):pages 802–807, 1993.

[5] Rudolf Tanner and Jason Woodard. WCDMA - Requirements and Practical Design. *John Wiley & Sons, Ltd*, 2004.

[6] A. Viterbi. Error bounds for convolutional codes and an asymptotically optimum decoding algorithm. *IEEE Transactions on Information Theory*, pages 260–269, April 1967.

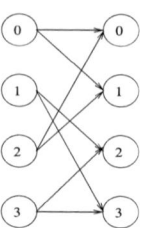

Figure 1.   4-node de Bruijn network

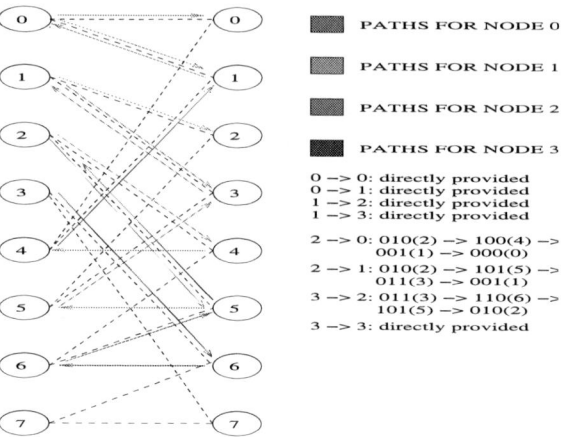

Figure 2.   Realization of a 4-node de Bruijn network on an 8-node de Bruijn network(LSF method)

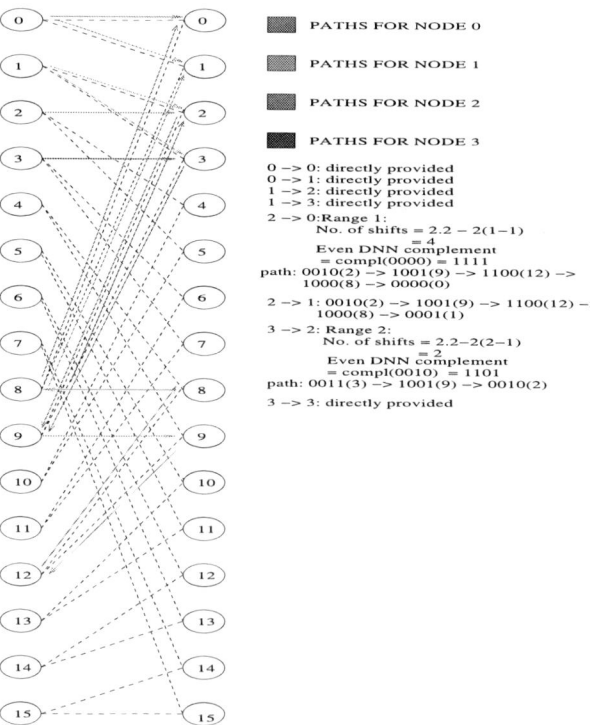

Figure 3.   Realization of a 4-node de Bruijn network on a 16-node de Bruijn network(RSF method)

TLM CO-SIMULATION FOR AN OPEN SOURCE MPSOC PLATFORM UNDER STARSOC ENVIRONMENT

Sami BOUKHECHEM and El-Bay BOURENNANE

UMR CNRS 5158 University of Burgundy
9, avenue Alain Savary B.P: 47870,
21078 Dijon CEDEX, France
Email: {sami.boukhechem, ebourenn}@u-bourgogne.fr

ABSTRACT

In the last decade, the embedded systems become more and more complex. This complexity is due to the fact that these systems contain more heterogeneous hardware and software components (CPU's, DSP, IP, etc.). Such systems called Multiprocessor-On-Chip (MPSoC) require new design approaches in order to satisfy several constraints, verification time, cost and time to market.

In this paper, we present our methodology for co-simulation at Transaction Level Modeling. This level permits modeling, verifying and validating the complete system at the beginning of the design process. Our reference platform is an open source Multiprocessor System-On-Chip labeled STARSoC (**S**ynthesis **T**ool for **A**daptive and **R**econfigurable **S**ystem-**O**n-Chip).

This work aims to propose an MPSoC platform co-simulation methodology based on an open source Instruction Set Simulator (ISS) OR1Ksim within SystemC as simulation environment. The purpose of this methodology is to allow a fast and early design space exploration.

1. INTRODUCTION

It has recently become possible to create complex embedded systems called the Multiprocessor System On Chip (MP-SoC), usually used for embedded applications. These systems contain several microprocessors, memories (shared, private), shared busses and peripherals integrated in a single die [1]. As a consequence the system designer is confronted with new challenges and difficulties related to the integration of such complex systems. So before any implementation, it is necessary to validate by simulation the system to be implemented. The chosen Transaction Level Modeling (TLM) simulation framework permits rapid exploration of several solutions containing different descriptions of the system components. Moreover, TLM co-simulation also permits the performance evaluation an validation of the whole system at the earlier stages of the design flow before build-

ing a prototype, which is faster than HDL Register Transfer Level (RTL) simulation [2, 3].

Traditionally, mixed language co-simulators are used [4] for simulation which generates a inter simulators communication overhead between different simulators, so it can cause a significant degradation in the execution time [5]. It is thus necessary to use a single language for modeling software and hardware. This motivate our choice of SystemC [6] as the modeling and simulation environment for our MPSoC platform.

In this context, the major contribution of this paper is to provide MPSoC models for an OpenRISC processor at a higher level of abstraction. This model aims to accelerate the validation, performance analysis, hardware-software partitioning evaluation and architecture exploration for our project called STARSoC.

In our case study in this paper, we have used an open source ISS of or1ksim simulation platform which is designed for mono-processor simulation. In order to use it in the case of the simulation of multiprocessor systems, we connect two ISSs with SystemC communication platform models, by using Inter-Process Communication (IPC) [7]. Thus, it is very easy to add or to remove a processor from the MPSoC design.

The interconnection models and other hardware components can be modeled in SystemC or any other Hardware Description Language. The interconnection model is based on a standard open source Wishbone bus [8][9]. Our communication mean is shared memory based. All processors are simulated at the same abstraction level, for each TLM level [10]. The rest of the paper is organized as follows: We begin in section 2 with a brief definition of STARSoC design flow and Transaction Level Modeling. In section 3, we represent the proposed simulation platform used in this work. Section 4 describe our transaction-level modeling steps of the wishbone bus. Section 5 describes the communication model adopted in our work. In Section 6 we describe the modified instruction set simulator based on Or1ksim. Section 7 presents some results of experiments we obtained by

978-1-4244-2541-9/08 $25.00 © 2008 IEEE

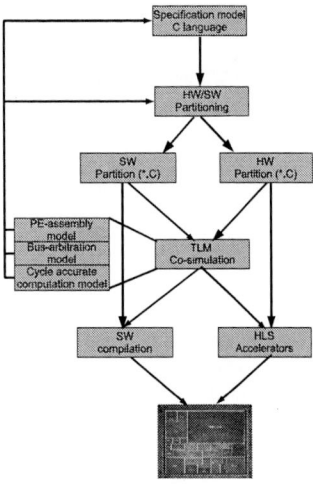

Fig. 1. STARSoC design flow.

simulation. We finish with a conclusion in section 8.

2. STARSOC DESIGN FLOW OVERVIEW

STARSoC design flow is shown in figure 1. It consists of several steps, starting by set of software and hardware processes described in C-code and ending by a synthetizable RTL architecture described in VHDL. Firstly, the application is considered as concurrency processes by using C-language [11]. Secondly, the application is manually partitioned into SW and HW Partitions, used in the simulation framework at different abstraction levels. In this work we focus on the TLM abstraction as defined in the following.

2.1. Transaction Level Modeling

Currently, transaction level modeling is widely reported in system level design literature. It permits a fast simulation and performance evaluation of a complex Systems on Chip (SoCs) earlier in the design flow, with a more modular and efficient code. This reduces the time-to-market compared with RTL and ensures practical gains for design, because TLM is less detailed than fuller realizations of the design. Timing details can be incorporated into these models to allow performance estimation and architecture exploration before the RTL HDL code is generated.

We have several definitions, concerning the exact place of TLM in the simulation level. TLM is not a single abstraction level, but involves several abstraction levels (multi-level model). We can refine the models over time to include more information. In most cases, TLM is define above the Register Transfer Level [12][13]. Cai and Gajski [10] clearly define four transaction level abstraction models, where the communication and the computation are explicitly separated.

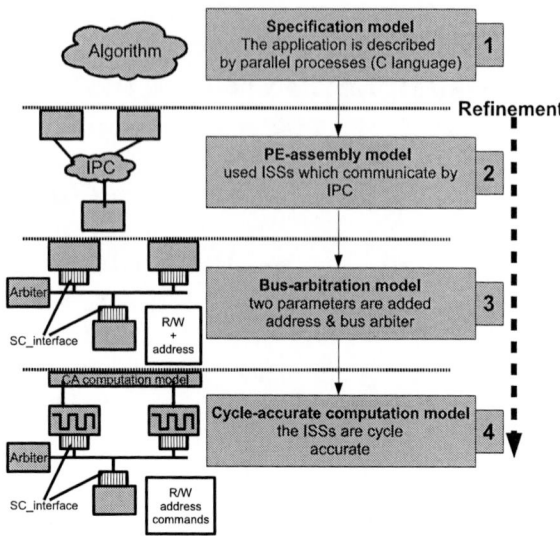

Fig. 2. STARSoC platform at different abstraction levels.

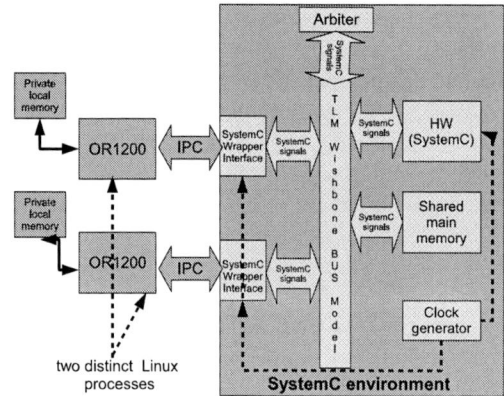

Fig. 3. STARSoC platform.

The system is represented as a set of communicating processes. These processes perform computations and communicate with other processes through an abstract channel. The different transaction levels defined by [10] are: PE-assembly model, bus-arbitration model, time-accurate communication model and cycle-accurate computation model. In our work, the use of Transaction Level Modeling in STARSoC platform (MPSoC platform) design refers to a set of abstraction levels quoted in [10].

2.2. STARSoC TLM

In figure 2 all levels belong to the TLM levels except the first level. We provide below a brief description of all these levels.

1. The first model is a *specification model* which is de-

978-1-4244-2541-9/08 $25.00 © 2008 IEEE

scribed by a parallel process (c program) without any architecture details.

2. The second model is *PE-assembly model*, implemented by using ISSs (OR1Ksim) which communicates through Inter-process Communication. We chose an implementation of IPC, called PIPE, for its capacity of data and command transfer.

3. The third model is the *Bus-arbitration model*. In this level we have added two parameters: address (for memory access) and bus arbiter (for bus access), also by using IPC also. In this level, STARSoC is a time approximate computation. In each clock cycle, the ISS performs one instruction. In our work, this level can also be called an instruction accurate execution model.

4. The fourth model is the *cycle-accurate computation model*. In this model each ISS is cycle accurate.

3. PROPOSE ARCHITECTURE

The target architecture used in this work is a Multiprocessor System On Chip (MPSoC) platform, whose communication are performed via a shared memory, as shown in figure 3.

Our platform consists of the integration of several Instruction Set Simulators wrapped under the SystemC wrapper interface (shown in figure 4), the TLM wishbone BUS model, private memories (associated with each ISSs) and a shared memory used for communication between the ISSs. The choice of shared bus as interconnection model instead Network-On-Chip (NoC) and shared memory instead Message Passing Interface (MPI). The fact we have small processors.

The ISSs used in our platform are derived from the open source simulator of the OpenRISC Processor (OR1200), written in C and called OR1Ksim [8]. Each ISS contains its own cache memory, private local memory and the minimum set of units required to provide basic functionality.

The TLM wishbone bus model is used to connect the ISSs with the rest of the system. These ISSs are executed in parallel and share the memory address space used for interprocessor communication.

The SystemC TLM Bus Model is based on the basic Wishbone communication protocol functionality, at high abstraction level. It executes the Wishbone bus transaction without timing accuracy nor pin accuracy.

The platform is entirely implemented in SystemC language, except ISSs which are wrapped under SystemC. In order to wrap ISS under SystemC, a SystemC wrapper interface is added to the ISS C model. This process involves defining a SystemC module and adding input/output ports that correspond to the ISS input/output arguments which we have implemented in the Load/Store functions from the

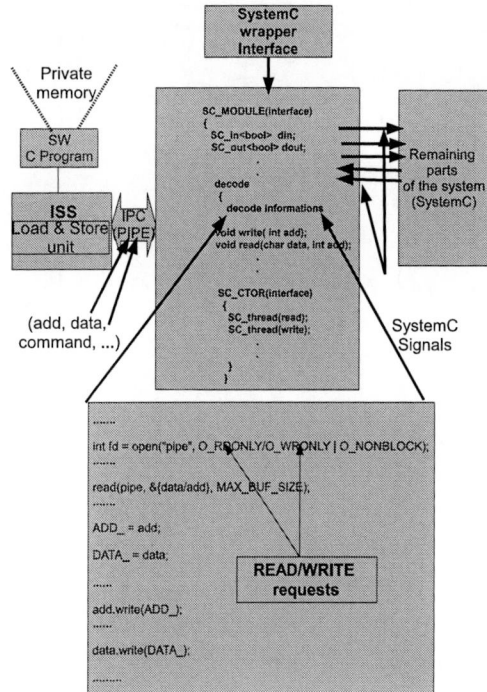

Fig. 4. SystemC ISS Wrapper Interface.

OR1k CPU directory (figure 5). This can be done by using the Inter-Process Communication as shown in figure 4. The SystemC wrapper interface is made sensitive to a positive clock edge. At every positive clock edge, the SystemC wrapper interface calls the corresponding C function inside the ISS via IPC, such as PIPE (used in our example) and Sockets [14]. SystemC wrappers and ISSs C models are both run as distinct Linux processes.

The communication between the components (Software-Software / Software-Hardware) can be effected using the shared memory, through SystemC signals.

At the beginning, we wrapped each ISS under SystemC. At every clock cycle the ISSs execute one instruction, in order to perform Instruction Accurate simulations; the pipeline effects are not considered at all. In this case we have a SystemC Instruction Accurate simulation. We then refine the simulation to a Cycle Accurate simulation: we fully simulate the processor pipeline and we take into account the number of cycles necessary for each instruction (processor pipeline stage, memory access...). We have in this case a full SystemC Cycle Accurate computation simulation.

4. TRANSACTION LEVEL MODELING OF WISHBONE BUS SYSTEM

In the case of TLM simulation, the wishbone bus protocol needs to be redefined as transaction level ports of TLM.

978-1-4244-2541-9/08 $25.00 © 2008 IEEE

```
                    or1ksim\cpu\or32\insnset.c

INSTRUCTION (l_sb)
    {
      ...
      if(add_s)
        {
          ...
          write(fd_sb_data,&PARAM1,strlen(&PARAM1));
          write(fd_sb_add,&add, strlen(&add));
          ...
        }
      ...
    }
...
INSTRUCTION (l_lbz)
    {
      ...
      if(add_l)
        {
          ...
          write(fd_lbz_add,&add, strlen(&add));
          read(fd_lbz_data,&data,strlen(&data))
          ...
        }
      ...
    }
```

Fig. 5. Example of an IPC call from the ISS.

There has been no TLM implementation for the wishbone bus to date, we have only descriptions at signal level. We implemented a transaction-level model of the wishbone bus. This model respects accurately the wishbone bus protocol. Since the bus is modeled as an abstract channel without including any specific details of the bus protocol (not pin accurate and not cycle accurate, only the computation models are cycle accurate) , it enables faster communication simulation models.

We present all the steps in our methodology to develop a wishbone bus TLM architecture, by describing our Transaction Level Modeling steps. We first used function calls (performed by calling IPC functions) to model the wishbone signals and we then used SystemC signals to implement the wishbone protocol.

In the first step, we modelize the behavior of each transaction level port. For example, in the RTL handshaking protocol, a master can immediately get an *ACK_I* (bus grant signal) from the bus, after sending an STB_O (bus request signal) if the bus is ready (free). This step is implemented as the port's transaction of a master calls *Check_bus_Grant()* and receives *true* as a return value. The arbiter selects a request from this master after applying an arbitration strategy, decodes the destination address and sends the request to the slave destination. The arbiter calls read() and write() functions implemented in the slave. The slave receives the request from the arbiter, performs any required computation, the read/write operation and optionally waits for a fixed number of cycles before sending a response back to the arbiter. The arbiter ensures eventual completion of the transaction. After that, the master (ISS) sends ADDR (address), DATA (read/write data). The transaction is a single *word/bytes* read/write transfer and receives ACK.

In the second step, we implemented wishbone signals by SystemC signals. The translation into a SystemC signal

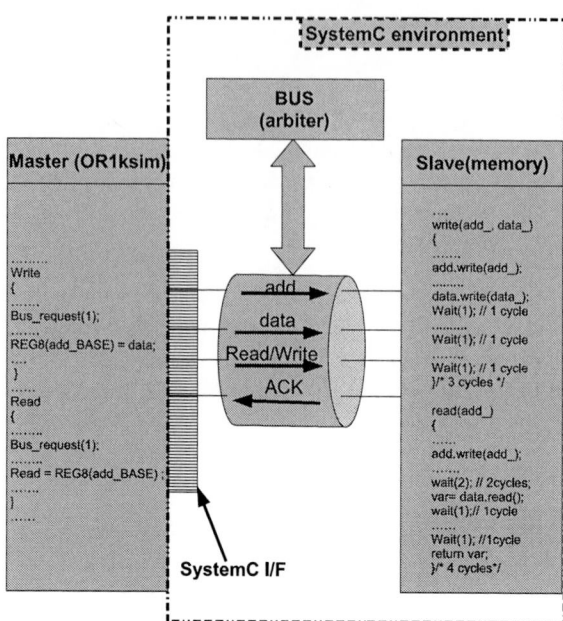

Fig. 6. TLM wishbone BUS.

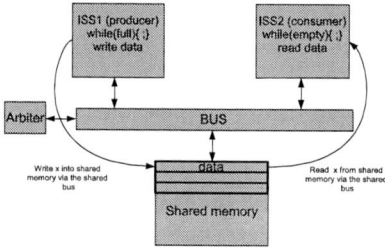

Fig. 7. Our communication model using shared memory.

is done by an *SC_Interface* module associated to each ISSs.

The realized Transaction Level Modeling, that was been conceived is shown in figure 6, where the implementation of communication is less detailed than Register Transfer Level.

The simulation bus speeds were measured at both TL Model and RTL. The TL Model is faster than the RT Level model.

5. COMMUNICATION MODEL

An important aspect in the design of multiprocessor systems is the exchange of data between SoC modules. Several communication methodologies are possible, such as shared memory and message passing. However, shared memory is the most common type of Inter Processor Communication paradigm, for multiprocessor System On-Chip (MP-SoC) platforms. where a small set of processors share a common address space.

In this section we present the model of communication

978-1-4244-2541-9/08 $25.00 © 2008 IEEE

developed and implemented in our platform architecture, which is based on the shared memory approach to perform data exchange between ISSs. We thus have two different types of memories :

A private memory space exists for each ISS which cannot be seen by any other ISS in the system except for the owner. We also have memory that is shared by all ISSs and used for communication.

An example is illustrated in figure 7 to demonstrate our communication model, established between two processors via the shared memory which is used for this purpose. The platform architecture consists of two ISSs connected by a TLM Wishbone bus model. The program running on ISS1 (producer), deposits data needed by the program running on ISS2, into shared memory and waits until this data is read by the program on ISS2 before depositing other data and vice versa. But in this case we have a well-known problem which is synchronization: it is a very critical issue in platforms based on shared memory communication. This problem arises due to the fact that the bus (in the case of a shared bus) as well as the memory are shared between different ISSs. On the one hand these ISSs exchange data through this shared memory, on the other hand these data are sent and received via the TLM shared bus. This situation generates shared resource access conflict if we have several simultaneous requests.

Thus, we need to use a simpler arbitration scheme such as round robin (RR) for the bus access and the semaphore for shared memory access, in order to ensure synchronization.

6. PROCESSING ELEMENTS

The ISS used in our framework is based on OpenRISC architecture emulator, written in C and called or1ksim, licensed under the GNU LGPL license. The goals are to emulate 32-bit OpenRISC CPUs with a high level of abstraction, capable of running real operating systems, such as NetBSD [15], RTEMS [16], eCos [17] and uClinux [18] and intended for embedded, portable and network applications.

OpenRISC 1200 is an open source core freely available from the OpenCores website [8]. This soft core is a MIPS-based 32-bit scalar RISC with Harvard microarchitecture, a 5-stage integer pipeline, virtual memory support (MMU) and basic DSP capabilities. The overview of the OR1200 architecture.

For this core, we have two descriptions at different abstraction levels. The first is a free open source description written in a synthesizable Verilog code, with a low level of abstraction (RTL abstraction) verified by several functional tests and implemented into FPGAs and ASICs.

The second description is written in C code (OR1Ksim simulator), and provides several original features [19][20] .

Abstraction levels	Number of iterations		
	10	100	1000
	Simulations times		
Instruction Accurate (IA)	0.97 (S)	1.5 (S)	8.4 (S)
Cycle Accurate (CA)	1.38 (S)	2.3 (S)	12.32 (S)
RTL	3.4 (S)	12.56 (S)	62.24 (S)

Fig. 8. Simulation time results at different abstraction levels.

7. RESULTS

In this section, we first describe the system architecture example used in our experiments, and we then present results that we carried out to validate the architeccture generation process of STARSoC and evaluate its performances . We also compare performances among the different abstraction levels.

In our experimental example, we used a shared-memory MPSoC platform generated by a STARSoC generator tool in which Multiple OpenRISC processors can be integrated and interconnected by a TLM wishbone BUS model.

We performed experiments using the SystemC design environment which we applied to a system consisting of two ISSs (wrapped under SystemC) accessing a shared memory through a TLM bus. The architecture of the system is shown in figure 7. The all system components were specified in SystemC, except for the two ISSs.

The bus arbitration mechanism is managed by the module labeled Bus arbiter in which is implemented the round robin arbitration policy.

The interface between the bus and the ISS interface consists of four signals: a read/write signal *rw*, an address *addr*, the data *data* and an acknowledge signal *ACK*. Similar signals exist between the bus and the shared memory which the access is synchronized by a semaphore.

ISS was built by the GNU cross-compiler (gcc Version 3.4) and cross-debugger (gdb Version 5.0) with the OpenRISC as a target.

The application executed by the two processors is stored in their local memory and consists of a producer-consumer application executed in parallel and synchronized by the same clock signal. The producer writes its data into the shared memory and the consumer reads these data from the same memory address. If the consumer is faster than the producer, the memory will be empty and the consumer waits until that the producer writes data. Conversely, if the producer is faster, the memory will be full, and the producer waits until that the consumer reads the data.

The same environment was employed for simulation on different abstraction levels. The simulation results were obtained by executing the platform on a Pentium IV at 3.0 GHz with a RAM memory size of 786 MBytes, based on Linux Fedora-core 3. The table in figure 8 shows three columns 10, 100, and 1000, corresponding to the fixed number of iterations (reads or writes to the memory) of the algorithm which is performed by each ISS. The simulation times shown in figure 8 correspond to the execution times of the programs at the to different simulation abstraction levels .

Three models were generated and simulated: an instruction accurate model (Bus-arbitration model)in this case the pipeline effects are not considered, a cycle-accurate system simulation model and a synthesizable RT-level model (using Verilog and simulated with Modelsim).

All of the model's descriptions were automatically generated from the STARSoC generator tool. They have also been validated by simulation using the same testbenches.

As results, we obtain in our multiprocessor system, firstly Instruction Accurate model is about 7 times faster than the RTL model and that also runs twice faster than a full cycle-accurate system simulation.

Secondly , our cycle-accurate model runs five times faster than an equivalent RTL simulation.

The simulation time analysis was adopted to compare the efficiency obtained from the TLM description in SystemC and from the underlying RTL platform.

8. CONCLUSION

The availability of a fast, high level simulation makes architecture exploration possible at different abstraction levels.

In this paper we have presented and validated our methodology for co-simulation at a high level of abstraction (TLM) within a single simulation environment based on SystemC language.

Our environment is based on the use of open source ISSs C models (OR1Ksim), wrapped under SystemC language by using UNIX Inter-Process Communication.

Comparing three different abstraction levels, namely , Instruction accurate level, Cycle accurate level and RTL level (VHDL model) we have analyzed the STARSoC generated multiprocessor SoC platform.

This model would be very useful for functional HW/SW co-simulation of large SoCs based on OpenRISC.

This motivates our choice for SystemC as a system design language, dedicated to architecture exploration in our STARSoC project which is the main contribution of this work. This gives the designer the possibility of exploring the STARSoC platform at several levels, which represents a notable advantage for STARSoC design flow.

In future work, we plan to add an embedded operating system like eCos and to integrate the heterogeneous IPs cores in our platform.

9. REFERENCES

[1] P. Paulin, C. Pilkington, M. Langevin, E. Bensoudane, D.Lyonnard, 0. Benny, B. Lavigueur, D. Lo, G. Beltrame, V.Gagne, G. Nicolescu, "Parallel Programming Models for a Multi-Processor SoC Platform Applied to Networking and Multimedia", Very Large Scale Integration (VLSI) Systems, IEEE Transactions on Volume 14, Issue 7, Page(s): 667 - 680, July 2006.

[2] A.A. Jerraya, A. Bouchhima, F. Petrot, "Programming models and HW-SW InterFaces Abstraction for Multi-Processor SoC", DAC 2006, San Francisco, California, USA, page(s): 280- 285, July 24-28, 2006.

[3] K. Hines, G. Borriello, "Dynamic Communication Models in Embedded System Co-Simulation", Anaheim, California, Page(s):395 - 400, DAC 97, 1997.

[4] A. A. Jerraya, "Systematic Design Flow For Fast Hardware/Software Prototype Generation From Bus Functional Model For MPSoC", 16th IEEE International Workshop on Rapid System Prototyping (RSP 2005), Montreal, Canada, Page(s): 218 - 224, 8-10 June 2005.

[5] J. Jung, "Performance improvement of Multi-Processor Systems Cosimulation based on SW Analysis", Munich, Germany, Page(s): 749 - 753, DATE 2001.

[6] Open SystemC Initiative, SystemC Version 2.0, Users Guide, "http://www.systemc.org", 2001.

[7] UNIX Network Programming Volume 2, Second Edition : "Inter-Process Communications", Prentice Hall, 1998.

[8] OpenCores, "http://www.opencores.org/projects/or1k".

[9] Silicore,"http://www.pldworld.com/_hdl/2/_ip/-silicore.net/wishbone.htm".

[10] L. Cai, D. Gajski, "Transaction level modeling: an overview", In Proceedings of the 1st IEEE/ACM/IFIP international conference on Hardware/Software codesign and system synthesis, page(s): 1924, ACM Press, 2003.

[11] M. Gokhal, J. M. Stone, J. Arnold and M. Kalinowski, "Streams-Oriented FPGA Computing in the Streams-C High Level Language", http://www.streams-c.lanl.gov/, Los Alamos National Laboratory and Sarnoff Corporation.

[12] Coware, "http://www.coware.com".

[13] Cadence, "http://www.cadence.com/".

[14] UNIX Network Programming Volume 1, Third Edition: "The Sockets Networking API", Addison Wesley, November 21, 2003.

[15] NetBSD, "http://www.netbsd.org".

[16] RTEMS, "http://www.rtems.com/RTEMS/rtems.html".

[17] eCos, "http://ecos.sourceware.org".

[18] uClinux, "http://www.uclinux.org".

[19] OpenCores, "http://pkgsrc.se/emulators/or1ksim".

[20] M. Bolado, H. Posadas, J. Castillo, P. Huerta, P. Sanchez, C. Sanchez, H. Fouren, F. Blasco, "Platform based on open-source cores for industrial applications", Design, Automation and Test in Europe Conference and Exhibition, Page(s): 1014 - 1019 Vol.2, 2004.

978-1-4244-2541-9/08 $25.00 © 2008 IEEE

A Flexible Modeling and Simulation Framework for Design Space Exploration

Camille Jalier and Didier Lattard
CEA-LETI, MINATEC
38054 Grenoble, FRANCE
{firstname.lastname}@cea.fr

Gilles Sassatelli
LIRMM
161 rue Ada, Montpellier, FRANCE
sassatelli@lirmm.fr

Abstract—Applications like 4G baseband modem require single-chip implementation to meet the integration and the power consumption requirements. These applications involve a high computation performance with real-time constraints, low power consumption and low cost. The concept of MPSoC is well suited to this problem. It makes it possible to adjust the architecture, by allocating the computational power where it is needed to fit the application needs. This often implies that the software has to be developed at the same time the platform is refined. Algorithm designers need accurate performance estimation to guide their decisions and system architects need to provide a design with enough calculation capacity and flexibility.

Based on the methodology used for the design of the 4G FAUST chipset, this paper presents a modeling and simulation framework for Design Space Exploration (DSE) which enables a rapid evaluation of the application-to-platform adequation. The key element of this work is a simple and flexible way of modeling application and architecture. Our SystemC-based simulation environment can support a broad range of architecture components (ASIC, DSP, NoC, bus, shared or distributed memory, ...) and application features (control flow, data exchange, interrupts, data-dependent processing, dynamic reconfiguration). Application and architecture models are separated to allow independent design space exploration. The simulation basically executes the algorithms on the architecture and monitors dynamic behavior such as communication transfers, resource conflicts, starvation, dynamic reconfiguration, etc.

I. INTRODUCTION

The increasing complexity of telecommunication algorithms in mobile terminals, requires high computational performance and low power consumption. The algorithms proposed for 4G applications are adaptive so the System On Chip (SoC) requires enough flexibility/reconfigurability to support multiple modulation schemes, Multiple Input Mutiple Output (MIMO) transmissions, etc. The implementation of such algorithms on the FAUST platform [1] has proved the efficiency of MPSoC to provide an attractive solution. This development has stressed the need for exploration tools to help the designer in the platform implementation choices:

- Processing resources: ASIC, ASIP, DSP, RISC
- Communication media: bus, point-to-point, Network-on-Chip (NoC)
- Memory topology: shared, distributed, FIFO
- Application: task partitioning, communication protocol, scheduling, mapping

The implementation of a baseband application onto a reconfigurable Multiple Processor SoC (MPSoC) adds stages in the design flow:

- Mapping and scheduling of the algorithm functionality onto each processing resources to exploit parallelism.
- Optimization of data transfer and scheduling to reduce time-overhead.
- Optimization of resource allocation to minimize the chip.
- Performance evaluation of the application on the architecture to guarantee the real-time constraints.

In order to address these issues, we propose a SystemC-based simulation framework. This framework incorporates both application and architecture aspects of the MPSoC to enable a behavioral approach. The performance is extracted through a simulation of the application behavior onto the platform. Because the application and the architecture are abstracted, it is possible to significantly speed up and ease the exploration process, to perform dynamic performance analysis and to quickly change parameters without rewriting models. In this context, this framework is useful to perform an early Design Space Exploration (DSE) and/or to provide an early execution platform for the software development. This tool might allow a faster verification of the second generation FAUST platform. The rest of this paper is organized as follows. The importance of the DSE in the design flow and the related works are discussed in section II. Section III describes the application and architecture abstraction models. Section IV describes the DSE environment.

II. SYSTEM LEVEL DESIGN

A. Framework Overview

In baseband modems, the algorithm choices widely influence the MPSoC design. In the traditional top down approach, the designer starts with an informal specification and develops a reference model of the telecom standard using a high-level language such as Matlab or C++. At this functional level of abstraction, the algorithm is developed independently from the architecture. This modeling approach is used by software designers to study different algorithms under complexity and precision aspects [2].

As shown in Fig. 1, in the next step of the design flow, this description is refined into an abstract platform. The platform

978-1-4244-2541-9/08 $25.00 © 2008 IEEE

is composed of modules, each of them executes a part of the application. They represent future hardware or software components (RISC, DSP, ASIC, etc). The transition between a functional view to a platform view is complex [3]. This implies deciding Hardware/Software partitioning, defining the architecture components and finally perform the algorithm mapping. Each architectural decision affects the global performance and the application behavior considering processing time, memory latency, data-transfer latency, possible parallel processing, resource sharing, etc. Flexible frameworks, that can model easily a broad range of platform and quickly estimate resulting performance, are required to perform an efficient DSE and converge to a satisfying solution. The exploration process at high level allows very efficient system optimizations. Our modeling and simulation framework simplifies and accelerates this process and is well suited for high complexity systems such as complex 4G baseband chipsets.

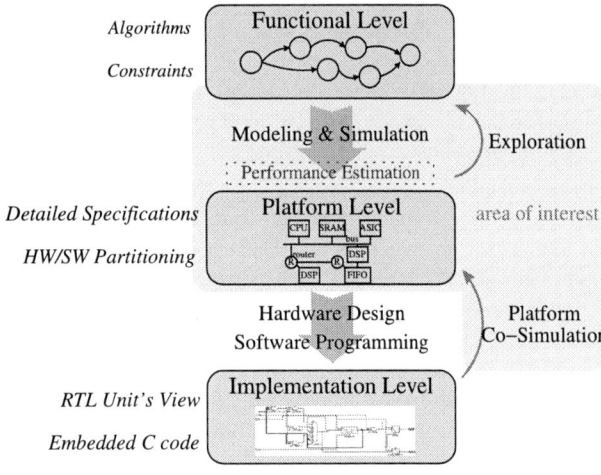

Fig. 1.   Simplified Design Flow

Once both algorithm and platform specifications are decided, every component is refined into either Register Transfer Level (RTL) model or embedded C code. The used approach is incremental for verification purposes: every low-level cycle-accurate RTL hardware component is simulated within the platform model for ensuring proper behavior.

B. Related work

Several methodologies and tools have been introduced to help the designer to model and simulate an application onto an architecture. The most relevant are mentioned in this paper.

Syndex [4] is a system level CAD software tool in which applications and architectures are modeled with independent graphs. In this approach, the application model is well separated from the architecture model to ensure an easy exploration of each aspect. However the application graph is limited to a *Data-dependence Graph*, in this case it is impossible to capture data-dependent processing or synchronization by interrupts.

ArchAn [5] is an architectural simulation environment in which applications are modeled in a task modeling language

and the abstract architecture is modeled at the cycle-accurate level. In this approach, the task modeling language includes details of the underlying architecture. Algorithms and architectural details are combined, the exploration becomes more difficult.

Artemis [6] is a modeling and simulation environment to explore the design space of heterogeneous embedded-systems architectures. This tool allows to model the architecture at multiple level of abstraction using generic building blocks provided in a library. Regarding application modeling, Artemis uses the Kahn Process Network (KPN) computational model. The KPN semantic forbids the use of interrupts, some dynamic behavior, such as run-time mapping, cannot be simply modeled.

III. Applications and Architectures Abstraction Models

A. Application abstraction models

In this work, the used approach considers the telecom algorithm (OFDM demodulation, MIMO decoding, channel decoding, etc) and also all the additional mechanisms (synchronization between resources, requests for memory space, etc) as part of the application. All these mechanisms are not defined in the telecom standard but they are mandatory to complete the platform model and they significantly influence the execution behavior, they are time-consuming. One of the motivations of this work is to keep the architecture and the application well separated so that exploration is facilitated; i.e. any functionality can be mapped to either in software (on a RISC) or in hardware (ASIC) without any expensive rewriting phase.

An application is modeled with tasks and buffers. The functionality is divided into a set of tasks. Each task may or may not contain the C++ code of the algorithm, but some specific annotations are required in order to indicate:

- communications with external tasks or buffers (data or control information)
- complexity of the data processing
- control flow (loop, conditions, ...)

These three types of annotation capture the interactions between the application and the system [7]. Data transfers between tasks are memorized in buffers. All communications can be represented as edges. The possibility for tasks to exchange some control information offers a great flexibility. The processing algorithm can be conditioned by a control event. Fig. 2 illustrates the proposed approach on a simple example that exhibits both data and control interactions among tasks. Five steps are used for describing task 3 behavior. The first one Ⓐ synchronizes the execution on a control message *START* sent by the task 5. Steps Ⓑ, Ⓓ and Ⓔ respectively model the reading, processing and writing operations of the task. Step Ⓒ is the C++ code of the interleaving algorithm used in 4G transmissions. This piece of code is not time-consuming for simulation though, it can be omitted. Finally, the complexity of the algorithm is modeled in the step Ⓓ.

978-1-4244-2541-9/08 $25.00 © 2008 IEEE

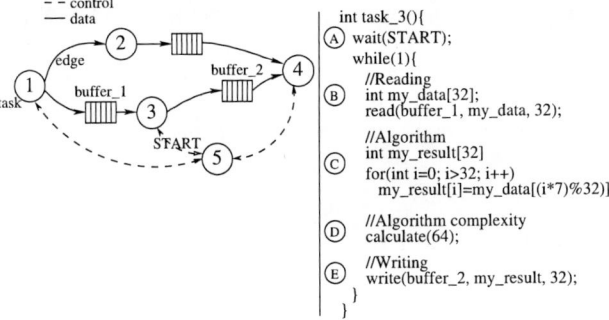

Fig. 2. Example of an application model

B. Architecture abstraction models

The platform architecture, which is the physical support for execution, is modeled with the same high-level abstraction approach as the application. To meet the well-known power, performance and cost constraints, architectures are often heterogeneous with multiple differentiated processors, complex memory hierarchy and hardware accelerators [8]. Modeling that kind of SoC involves flexible and high-level models. In the literature, two approaches are often used. Either models, like Random Access Machine [9], are employed to evaluate computing complexity but they intentionally neglect the underlying architecture details. Or low level models, like Instruction Set Simulator (ISS) or HDL descriptions, are employed but they do not allow rapid design space exploration as rewriting code is often required, they are too over-detailed.

Our approach is based on abstract models that can describe very complex and different MPSoC implementations without rewriting code. For enabling this, a set of abstract entities that can cover current platform such as FAUST and future HW/SW MPSoC are required. In the previous section III-A, the application is modeled with tasks, buffers and edges. The abstract models for the architecture are directly inspired from the application view. Three different types of abstract entities are used to describe the architecture:

- a processor model that schedules several tasks (calculate requests)
- a memory model that manages several buffers (read, write requests)
- a communication media model that manages several simultaneous communications (edges)

The abstract model for each entity is written in SystemC and is composed of a resource manager (scheduler [10]) to allow the concurrent execution of multiple application elements and a communication manager which resolves exchanges between entities. The platform architecture is specified in an XML file that describes the numbers of entities, their types and the interconnections. Each instantiated entity has its own parameters that can be tuned to predict the high level characteristics of the future physical module.

As shown in the Fig. 3, the Processor model is an abstract model used for modeling any type of processing element (DSP,

CPU, ASIC). At this stage, the partitioning between hardware and software is not defined and the architecture description is independent of the application model.

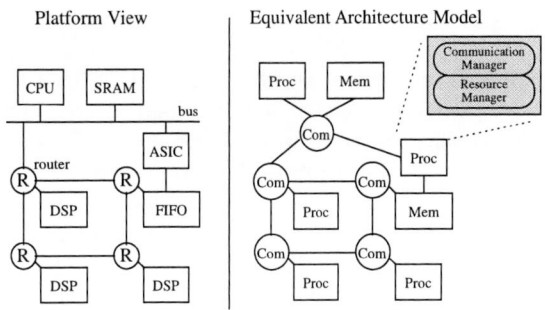

Fig. 3. A model based on Processor (Proc), Memory (Mem) and Communication (Com) entities.

C. Mapping

We have separately modeled the functionality and the architecture with independent or orthogonal models. To complete the platform description, we consider the spatial and temporal mapping. Each part of the application such as tasks (resp. buffers and edges) is placed on a Processor (resp. Memory and Communication) entity. When several tasks (resp. buffers and edges) are mapped onto a Processor (resp. Memory and Communication) module, the arbitration is done by the resource manager considering a scheduling policy. The mapping is described in an XML file to be easily changed. During this phase, the designer can efficiently distribute the application to exploit the parallelism or to share resources.

IV. DSE FRAMEWORK

Our framework follows the well-know y-chart principle [11], where a set of application models is merged with a set of architecture models in a dedicating mapping step. The platform is modeled in an orthogonal way. This mechanism is the key to enhance the flexibility during design space exploration. In our approach, we can independently explore:

- application: task partitioning, communication protocol (message passing, shared memory, etc.), synchronization model (data, interrupt, etc.), etc.
- architecture: performance for computing resources, communication topology, memory topology, number of resources, etc.
- mapping: resource allocation (to optimize performance, chip area, etc.), scheduling policy, etc.

The exploration process starts by modeling the application with tasks, buffers and edges. Then a first platform architecture is modeled with Processor models, Memory models and Communication Media models. Based on the application graph and the architecture view, the mapping consists of allocating resources for each part of the application. Using our models, the framework automatically generates a SystemC platform including the application, the architecture and the mapping.

978-1-4244-2541-9/08 $25.00 © 2008 IEEE

Based on the SystemC simulation engine (Fig. 4), the application is executed on the architecture and the execution behavior is captured. All the relevant characteristics for performance estimation can then be easily extracted: application throughput, maximum latency, utilization rate of resources, transient effects, etc.

According to the estimated performance, the designer can change the architecture and/or the algorithm architect can modify the application. This exploration process, to converge to a suitable solution, is incremental but do not need to rewrite code, only to modify XML files and/or change parameters. The designer is responsible for setting these parameters with relevant values according to his background and the technology. The predicted performance results are tightly linked to the parameters of the architectures entities.

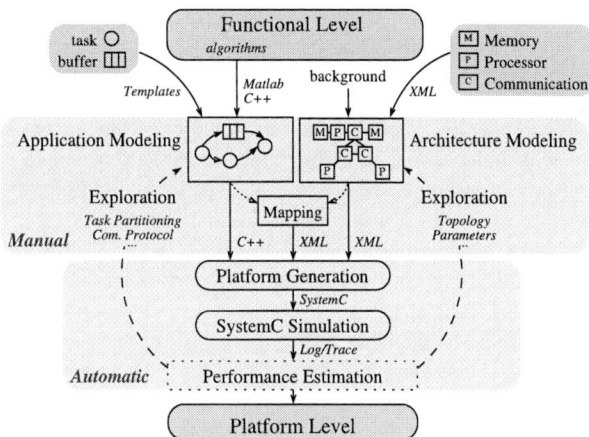

Fig. 4. A Detailed View of our Exploration Process

As all models are unified in the same framework, the algorithm architect and the designer can interact easily and work independently. In our approach, the modeling phase and the exploration is done manually by the designer relying on his skills and experience. This framework nevertheless offers an accelerated exploration process based on:

- Three parameterizable models (XML & SystemC) for architecture modeling (no code rewriting)
- XML-based description files to quickly parameterize and interconnect modules
- Automatic generation of a platform SystemC View
- Extraction of execution parameters on log files

At the end of the exploration, the designer has refined: - the application into an embedded platform software regarding communication protocols, synchronizations, etc. - the architecture together with the topology/communication architecture and the associated parameters. - the mapping.

The hardware/software partitioning is performed according to the calculation capacity and flexibility of each unit. The designer can then easily provide detailed specifications for deriving a hardware or a software implementation. The programmer can work on an embedded and optimized code

according to the target CPU and the designer can describe each hardware unit at Register Transfer Level.

As we have chosen SystemC, we can use Co-Simulation tools to plug each hardware unit, described in HDL languages, in place of our high-level model to perform a platform simulation.

V. CONCLUSION

We propose a system level simulation framework for early investigation of MPSoC platform architectures considering a specific application. The major contributions of this paper are firstly a set of flexible high level models based on XML/SystemC descriptions and secondly an orthogonal way of modeling the application, the architecture and the mapping based on a separation between the functionality and the execution support. In this context, the algorithm is modeled in an embedded manner with explicit communication protocol, processing complexity, interrupts, etc. Here the use of high-level parameterizable models enables a rapid exploration process without rewriting code. With our modeling framework, dynamic or data-dependent behavior can be easily captured.

The major advantage of our modeling approach is the combination of a quick exploration flow and an extended modeling capacity.

Based on our experience on the FAUST chip , our future work will focus on the verification of the second generation platform. This design implements more flexible components to guarantee an efficient reconfiguration. On the longer term, this framework will serve as a basis to explore homogeneous architectures in the perspective of a completely reconfigurable telecom system.

REFERENCES

[1] Y. Durand. FAUST: on-chip distributed SoC architecture for 4G baseband modem chipset. In *Proc. of the IP/SOC'05 conference*, 2005

[2] T. Kempf, M. Doerper. A modular Simulation Framework for Spatial and Temporal Task Mapping onto Multi-Processor SoC Platforms. In *Proc. of the Design, Automation and Test in Europe Conference and Exhibition (DATE'05)*, 2005

[3] P. Magarshack, P.Paulin. System-on-chip Beyond the Nanometer Wall. In *Proc. of the Design Automation Conference (DAC)*, 2003

[4] T. Grandpierre, Y. Sorel. From algorithm and architecture specification to automatic generation of distributed real-time executives: a seamless flow of graphs transformations. In *Proc. of the First ACM and IEEE International Conference on Formal Methods and Models for Codesign (MEMOCODE'03)*, June 2003

[5] A. Chatelain, Y. Mathys. Verification Strategy for Integration 3G Baseband SoC. In *Proc. of the Design Automation Conference (DAC)*, 2003

[6] A.D. Pimentel, L.O. Hertzbetger. Exploring embedded-systems architectures with Artemis. In *Computer*, vol. 34, pages 57 - 63, 2001

[7] M. Waseem, L. Apvrille. Abstract Application Modeling for System Design Space Exploration. In *Proc. of the 9th EUROMICRO Conference on Digital System Design (DSD'06)*, 2006

[8] D. Lattard, E. Beigne. A Reconfigurable Baseband Platform Based on an Asynchronous Network-on-Chip. In *Journal of Solid-State Circuits (JSSC)*, vol. 43, pages 223 - 235, Jan. 2008

[9] A. Tiskin. The bulk synchronous parallel random access machine. In Proc. of EURO-PAR'96, vol. 2, pages 327-338, August 1996

[10] J.M. Paul, A. Bobrek. Schedulers as Model-Based Design Elements in Programmable Heterogeneous Multiprocessors. In *Proc. of the Design Automation Conference (DAC)*, 2003

[11] P. Lieverse, P. van der Wolf. A Methodology for Architecture Exploration of Heterogeneous Signal Processing Systems. In *Proc. of the IEEE Int. Workshop on Signal Processing Systems (SIPS)*, 1997

978-1-4244-2541-9/08 $25.00 © 2008 IEEE

Energy Analysis of Re-Injection Based Deadlock Recovery Routing Algorithms

H. Kooti[*], M. Mirza-Aghatabar[*], S. Hessabi[+]
Computer Engineering Department
Sharif University of Technology
Tehran, Iran
[*]{kooti, aghatabar}@ce.sharif.edu, [+]hessabi@sharif.edu

A. Tavakkol
School of Computer Science
Institute for Studies in Fundamental Science (IPM)
Tehran, Iran
arasht@ipm.ir

Abstract—**There are two strategies for deadlock handling in routing algorithms in NoC: deadlock avoidance and deadlock recovery. Some deadlock recovery routing algorithms are re-injection based, such as: Compressionless (CR), Software-Based (SW_TFAR) and AFBAR. In spite of the performance comparison, none of existing researches have focused on the energy consumption of various routing algorithms. We evaluate these routing algorithms according to their energy consumption and latency. Our experimental results show the better performance and worse energy consumption of deadlock recovery routing algorithms compared to deadlock avoidance routing algorithms. In addition, the best and worst energy consumption is dedicated to AFBAR and CR, respectively.**

I. INTRODUCTION

Deadlock is the most formidable obstacle that routing algorithms in wormhole switched interconnection networks must address and overcome. Deadlock avoidance has been a traditional approach in handling deadlock problem, where routing is restricted in a way that no cyclic dependency exists between channels. As an alternative approach, deadlock detection and recovery gained attention, since it imposes no limit on routing adaptivity.

Schemes based on deadlock avoidance generally suffer from losses in adaptivity and/or increased hardware complexity which negatively impact performance. In [1,2], it was shown that deadlocks rarely occur when sufficient routing freedom is provided. Therefore, it does not make sense to limit the routing algorithm; hence, recovery schemes have gained consideration in the scientific community [3-6].

There are some deadlock recovery routing algorithms such as: CR [3], SW_TFAR [4], AFBAR [5] and Disha [6]. SW_TFAR and AFBAR are progressive while CR is regressive. In these schemes, when deadlock happens, the message will be removed from the network by ejecting it at the node containing the header flits (SW_TFAR and AFBAR) or will be killed by the source node (CR). They will re-inject the removed messages into the network at a later time.

While network performance analysis due to different deadlock recovery routing has been studied rigorously in the past [3-6], network energy analysis has not been explored. The goal of this paper is to provide a detailed evaluation of performance and energy of re-injection based deadlock recovery routing algorithms. In Section 2, we take a look at related work. Section 3, describes the architecture of our simulator and presents the evaluation metrics such as latency and energy consumption. Then, in Section 4, we will present the experimental results, and finally in Section 5, we conclude our work and give the summary.

II. RELATED WORK

One re-injection based deadlock recovery algorithm is Compressionless Routing (CR) proposed in [3]. In CR, the source node keeps track of the injected message and detects if it has reached the destination node or not. No deadlock may happen if the header flit is delivered to destination; otherwise, if the message was blocked for some time, the source breaks down the partial message path, kills the deadlocked message, and then tries sending it again later.

Another deadlock recovery routing algorithms which is re-injection based is SW_TFAR, which is introduced in [4]. It has some inevitable advantages such as: (1) requires a very small amount of hardware due to no dedicated buffer to handle deadlocks in comparison with Disha [6], (2) eliminates performance degradation at saturation point with message injection limitation, (3) uses a new deadlock detection technique which considerably reduces the probability of false deadlock detection.

AFBAR is a re-injection based algorithm too, which is an improved version of SW_TFAR. The authors of [5] improved the deadlock detection mechanism of SW_TFAR. In fact, AFBAR decreases the number of false deadlock detections; consequently, leads to higher performance.

In this paper, we will analyze the re-injection based deadlock recovery routing algorithms under different implementations and usage scenarios in terms of their

978-1-4244-2541-9/08 $25.00 © 2008 IEEE

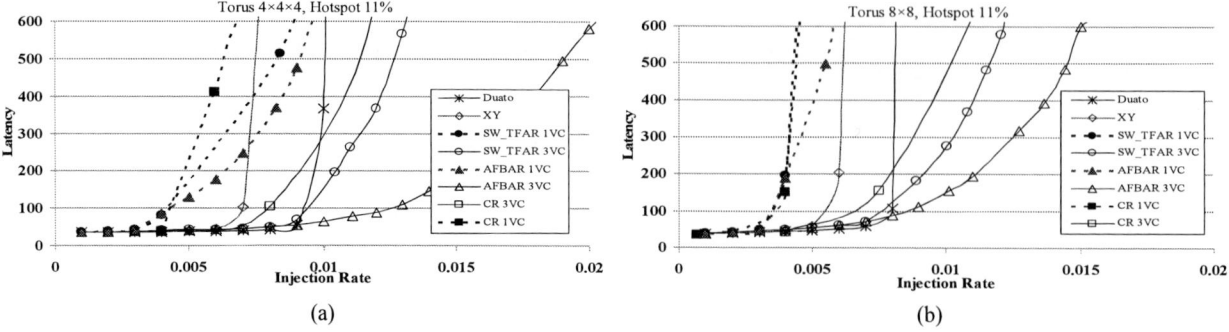

Figure 1. Latency comparison of deadlock recovery and deadlock avoidance routing algorithms

energy dissipation and performance. We will also compare our results with deadlock avoidance routing algorithms. A detailed analysis of deadlock avoidance routing algorithms is studied in [7].

III. SIMULATOR AND EVALUATION METRICS

The XMulator [8] was used to implement the proposed routing algorithm and to perform simulation experiments. XMulator is a complete, flit-level, event-based, and extensively detailed package for simulation of interconnection. Orion [9] is used in this paper to measure the energy consumption of NoC. We have performed our simulations in torus topology with two different sizes (4×4×4 and 8×8). Three conventional traffic models were selected, i.e. uniform, local and hotspot. For routing algorithms, deadlock recovery and deadlock avoidance routing algorithms were implemented. A fully adaptive deadlock avoidance routing algorithm needs at least 3 virtual channels to work properly; while, deadlock recovery routing algorithms just need 1 virtual channel [10]. Therefore, we implemented deadlock recovery algorithms with 1 and 3 virtual channels to fairly compare them with deadlock avoidance routing algorithms.

We consider the latency and energy per flit metrics to evaluate our experimental results. Latency is defined as the time that elapses between the occurrence of a message header injection into the network at the source node, including the queuing time in source, and reception of the corresponding tail flit at the destination [10]. Energy consumption in NoCs consists of two components: the energy consumed in routers, and the one associated with links. Both static and dynamic energy consumptions are considered. We used 65nm library and the clock frequency is set to 1.5GHz based on the critical path calculations.

IV. EXPERIMENTAL RESULTS

In this section, first the latency of re-injection based deadlock recovery algorithms will be presented then we will compare their latencies with deadlock avoidance routing algorithms (i.e. Duato and XY). We also analyze the energy consumption of mentioned routing algorithms under three traffic models, i.e. uniform, local 40% and hotspot 11%. In addition, we will analyze the energy consumption of these routing algorithms in the low traffic regions of the mentioned

traffic models (i.e. local and hotspot) with different percentages. Our traffic models are combined with uniform traffic model. As an exemplification, the local40% means that, 40% of the messages are distributed locally and the remaining 60% are distributed uniformly.

A. Latency Analysis

Figure 1 shows the latency comparison of deadlock recovery and avoidance algorithms under hotspot traffic model. Simulation results under other traffic models were similar with few exceptions, which we will discuss in this section. We know that a routing algorithm with more virtual channels usually gains better performance [10]. It is obvious from Figure 1 that all deadlock recovery routing algorithms with 3 virtual channels have a better performance than the ones with 1 virtual channel under all traffic models. A key point of this paper is the dependency of latency to the network's diameter. The diameter of torus 4×4×4 topology is 6 and the diameter of torus 8×8 topology is 8. Although both topologies have 64 nodes, we claim that unequal diameter lengths lead to different latency behaviors in some situations.

AFBAR and SW_TFAR have a better performance than Duato routing algorithm with 3 virtual channels in a 2 dimensional torus topology under all traffic models. But, in 4×4×4 topology this is not correct and the best latency belongs to Duato with 3 virtual channels under local and uniform traffic models. Shorter diameter in 4×4×4 topology leads to less blocking time; furthermore, each node has 6 input/output ports which enhance the adaptivity of each node to route packets. On the other hand, the killing or ejection and re-injection procedures of deadlock recovery routing algorithms lead to performance degradation. In fact, Duato routing algorithm dose not suffer from this process and also benefits from high node adaptivity in 4×4×4 topology which lead to slightly better performance. Under hotspot traffic model, due to high traffic around the hot node, the blocking time is high. Hence, Duato cannot benefit from more adaptivity of each node in 4×4×4 topology. Figure 1 (a) shows the better performance of deadlock recovery routing algorithms with 3 virtual channels in comparison with Duato. Unlike 4×4×4 topology, in 8×8 topology each node has 4 input/output ports that cause lower adaptivity for routing. Hence, the more efficient usage of virtual channels in deadlock recovery routing algorithms leads to their better

978-1-4244-2541-9/08 $25.00 © 2008 IEEE

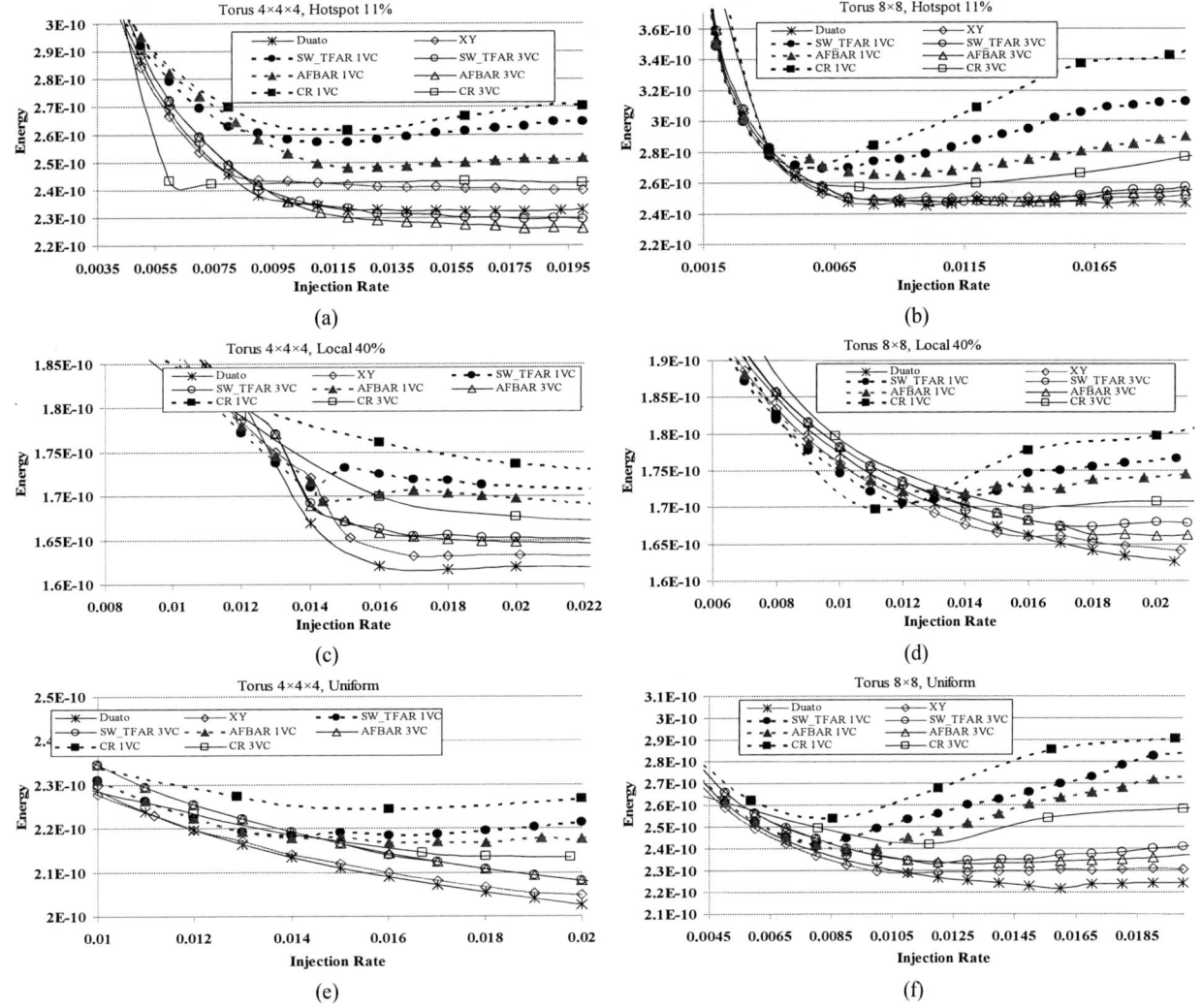

Figure 2. Energy consumption comparison of deadlock recovery and deadlock avoidance routing algorithms

performance in comparison with Duato routing algorithm.

In all conditions, the best performance among deadlock recovery routing algorithms belongs to AFBAR. This gap is considerable under hotspot traffic model, since AFBAR uses a more efficient deadlock detection mechanism [5].

Another key point is the worse performance of CR in comparison with AFBAR and SW_TFAR with 1 and 3 virtual channels in all circumstances, since deadlock recovery overhead in CR is more than that of the other ones.

B. Energy Analysis

Figure 2 shows the energy consumption of the mentioned routing algorithms under hotspot, local and uniform traffic models in torus 4×4×4 and 8×8. The important point that the authors would like to make is that the worst energy consumption belongs to CR with 1 virtual channel, since the worst delay is associated with CR (Figure 1). In addition, CR deadlock recovery procedure leads to more number of

transferred flits per cycle, which increases the power consumption as well.

Another important point is that in most cases, the best energy consumption is gained by Duato routing algorithm among all routings algorithms. Using three virtual channels by Duato, results in better performance and lower delay. Additionally, this algorithm does not kill or re-inject any packets to the network due to its deadlock avoidance nature. Hence, this algorithm has lower power consumption as it would not increase the number of transferred flits per cycle in comparison with deadlock recovery routing algorithms.

There is an exception were Duato does not have the best energy consumption. This case appears in torus 4×4×4 with hotspot traffic (Figure 2 (a)). As aforementioned, in this topology, the lower diameter reduces the number of blocking packets and so SW_TFAR and AFBAR will detect lower number of packets engaging in deadlock cycle. Therefore, the number of re-injected packets and number of transferred

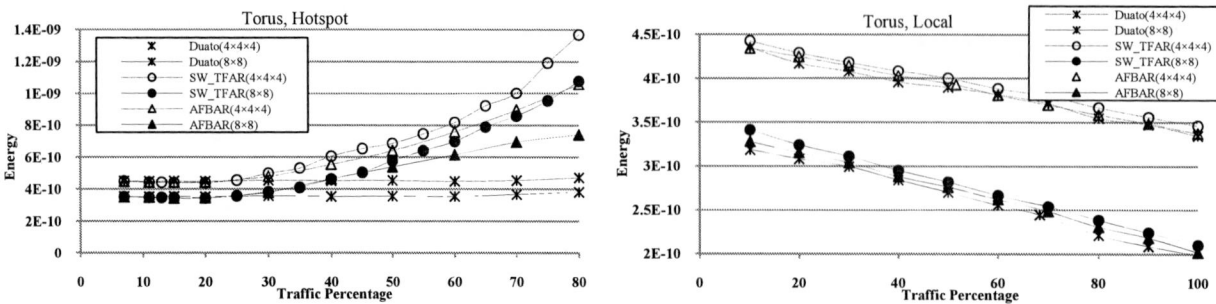

Figure 3. Energy consumption of deadlock recovery and deadlock avoidance routing algorithms at different traffic percentages

flits per cycle will reduce which leads to lower power delay product or energy consumption.

Another point is that deadlock recovery routing algorithms with 1 virtual channel consume more energy compared to deadlock recovery routing algorithms with 3 virtual channels in all cases. Although more virtual channels cause more power consumption, it does not increase energy consumption. It means more efficient usage of virtual channels by deadlock recovery routing algorithms, which leads to better performance, overcomes the more power consumption based on more virtual channels.

Our experimental results show that less energy is consumed by AFBAR than SW_TFAR. The energy consumption gap for AFBAR and SW_TFAR is more obvious with 1 virtual channel than 3 virtual channels, as the better deadlock detection mechanism of AFBAR is more efficient in lower number of virtual channels.

We consider energy consumption of these routing algorithms in low traffic under local and hotspot traffic models with different percentages. Figure 3 shows the results. The interesting result is that by increasing the percentage of local traffic, the energy consumption will decrease. This is due to the fact that locality increment reduces the average distance, and hence, leads to lower delay and lower power-delay product or energy consumption. On the other hand, by increasing the percentage of hotspot traffic, the energy consumption will increase. This is due to the fact that more blocking time leads to more channel monitoring [11] which increases the power consumption, and hence, leads to more energy consumption.

V. SUMMARY AND CONCLUSIONS

In this paper, we considered the deadlock recovery routing algorithms. We compared them with each other and with deadlock avoidance routing algorithms from latency and energy consumption points of view.

AFBAR has the best performance among all routing algorithms, while CR has the worst performance. Another important result was that increasing the number of virtual channels in deadlock recovery routing algorithms leads to lower energy consumption. Duato has the best energy consumption due to its good performance and power consumption. In addition, high percentage of hotspot traffic

model leads to more energy consumption; while, high percentage of local traffic model leads to lower energy consumption.

As a conclusion, we can say that whenever the energy consumption is a critical parameter for a designer, the Duato deadlock avoidance routing algorithm is a better selection, and whenever the performance or delay is the critical parameter, the AFBAR deadlock recovery routing algorithm is the best choice.

REFERENCES

[1] T.M. Pinkston and S. Warnakulasuriya, "On deadlocks in interconnection networks", the 24th International Symposium on Computer Architecture, June 1997

[2] S. Warnakulasuriya and T.M. Pinkston, "Characterization of deadlocks in interconnection networks," In Proc. of the 11th International Parallel Processing Symposium, April 1997

[3] J. Kim, Z. Liu and A. Chien, "Compressionless Routing: a framework for adaptive and fault-tolerant routing," In Proc. of the 21st International Symposium on Computer Architecture, pp. 289-300, April 1994

[4] J.M. Martinez, P. Lopez, J. Duato and T.M. Pinkston, "Software-Based deadlock recovery technique for true fully adaptive routing in wormhole networks," 1997 International Conference Parallel Processing, August. 1997

[5] M. Mirza-Aghatabar, A. Tavakkol, H. Sarbazi Azad, "An adaptive software-based deadlock recovery technique," IEEE International Conference on Advanced Information Networking and Applications, pp. 514-519, March 2008

[6] K. V. Anjan and T. M. Pinkston, "DISHA: a deadlock recovery scheme for fully adaptive routing," In Proc. of the 9th International Parallel Processing Symposium, pp. 537-543, April 1995

[7] M. Mirza-Aghatabar, S. Koohi, S. Hessabi, and Massoud Pedram, "An empirical investigation of mesh and torus NoC topologies under different routing algorithms and traffic models," in Proceedings of the 10th IEEE Euromicro Conference on Digital System Design, pp. 19-26, 2007

[8] A. Nayebi, S. Meraji, A. Shamaei, H. Sarbazi-Azad, "XMulator: an object oriented XML-Based Simulator," in Asia International Conference on Modeling & Simulation, pp. 128–132, 2007

[9] H. S. Wang, X. Zhu, L. S. Peh and S. Malik, "Orion: a power-performance simulator for interconnection networks," In Proceedings of MICRO 35, Istanbul, Turkey, November 2002

[10] L. Ni and C. Glass, "The Turn Model for adaptive routing," In Proc. of the 19th International Symposium on Computer Architecture, IEEE Computer Society, Vol. 20, No. 2, pp. 278-287, May, 1992

[11] S. Koohi, M. Mirza-Aghatabar, S. Hessabi, "Evaluation of traffic pattern effect on power consumption in mesh and torus-based Network-on-Chips," International Symposium on Integrated Circuits, Singapore, 2007

978-1-4244-2541-9/08 $25.00 © 2008 IEEE

UML profile for Estimating Application Worst Case Execution Time on System-On-Chip

Fateh Boutekkouk[1,2], Sébastien Bilavarn[1], Michel Auguin[1], and Mohammed Benmohammed[2]

[1] Laboratoire d'Electronique, Antennes et Télécommunications (LEAT)
Université de Nice Sophia Antipolis – CNRS, Nice, France
[2] Department of Computer Science, University of Constantine, Algeria

Abstract— **Systems-on-Chip (SOC) design is confronted with the problem of the so-called productivity gap. In order to cope with this problem, authors emphasize on using the Unified Modeling Language (UML) as a system level language, so higher level of abstraction is achieved. In this context, we present a UML profile and a methodology for estimating application Worst Case Execution Time (WCET) on SOC. The proposed profile allows the designer to express hierarchy among application tasks, and most of parallelism forms that exist in typical embedded applications such as task parallelism, pipelining, and data parallelism, while making control and communication over tasks explicit. In order to estimate application WCET, the hardware platform on which application is mapped on, should be abstracted too. Consequently, each hardware component is parameterized by a set of parameters matching the abstraction level of the application. A particularity of our flow is that it starts by establishing a sequential object model using UML sequence diagram, from which a task-level model is extracted. We think that the sequential model is strongly preferred from the system designer's perspective for two reasons. First, because it facilitates the modelling task relieving the designer of the burden of concurrency modelling. Secondly, starting from an existing sequential model (e.g. legacy C code) which is generally considered as the reference model, we can then parallelize it, and explore the design space. We show how we have used our profile for H264 decoder modeling.**

Key words —: SOC; Y-chart approach; UML; WCET

I. INTRODUCTION

Modern Embedded Systems (ESs) are becoming increasingly complex. Despite the prevalence of such devices, we can remark the scarcity of abstract and visual programming models supporting features related to these systems. In order to manage design complexity, ESs designers have resorting to software engineering and borrow from it many well practiced concepts and methodologies. On the other hand, the emergence of the Unified Modelling Language (UML) as a standard for object-oriented modelling may improve the design quality of ESs.

Beyond visual modelling and documentation capabilities, UML can also be used in performance analysis like time and power consumption. The remarkable maturity of UML-based tools for code generation will push designers to concentrate on higher level of abstraction rather than coding.

Since UML does not dictate any particular development process to be used, it is for designers to define a design flow. We think that the Y-chart approach is the most appropriate.

The Y-chart approach [7] puts strong emphasis on the Platform-Based Design (PBD) [6]. According to our knowledge, the PBD is one of the best-validated industrial approaches for achieving high reuse in SOC (System-On-a Chip) design.

SOC can be defined as a complex integrated circuit that integrates the major functional elements of a complete end-product into a single chip or chipset. In general, SOC design incorporates at least one programmable processor, On-Chip memory, accelerating functional modules implemented in hardware, and a communication infrastructure ranging from a simple shared bus to a very complex network (NOC for Network-On-Chip). It also interfaces with peripheral devices, and/or the real world.

Due to the extra high integration density, and the strict time-to market window, SOC design is confronted with the problem of the so-called productivity gap. In order to cope with this problem, authors emphasize on raising the level of abstraction of both application and hardware platform. Although, abstraction reduces the complexity, fasters the simulation speed, and allows a larger design space exploration, it comes at the price of inaccurate estimations. For this reason, a deep analysis on the pertinent factors that have great impacts on the system performance is required. In the same context, we present a UML profile for estimating application Worst Case Execution time on SOC.

The rest of the paper is organized as follows: section 2 reviews the related work. Section 3 presents our proposed approach including application, architecture, mapping modeling. Section 4 describes a set of guidelines for parallelism extraction from sequence diagram. Section 5 presents a set of mapping and optimization guidelines. The WCET estimation technique is detailed in section 6. Section 7 shows our tool and a case study before concluding.

II. RELATED WORK

Due to extension mechanisms offered by UML, UML can be tuned to a particular domain by the definition of a profile. With regard to the application of UML to the ESs and SOCs fields, the literature is very rich. However we can mention some pertinent works.

MARTE [11] is an UML2.0 profile that targets real time embedded systems. It offers a facility for modelling, and analyzing real time applications. UML-SOC profile [12] intends to describe SOC specific information using UML2.0. It integrates concepts from SOCs and allows automatic code

978-1-4244-2541-9/08 $25.00 © 2008 IEEE

generation for the hardware part. UML-SystemC profile [9] captures both the structural and the behavioral features of the SystemC language and allows high level modelling of SOC with straightforward translation to SystemC code. TUT profile [8] provides an automated path from UML design entry to FPGA prototyping including the functional verification and the automated architecture exploration focusing on automatic profiling and performance values back annotation. Gaspard2 [2] is an UML2.0 profile, targeting intensive signal processing (ISP) domain. It allows the expression of task and data parallelisms using Array-OL language. Gaspard2 defines stereotypes for application, hardware platform, and mapping.

Since, most profiles focus on the process (task) paradigm, they lack of capabilities for the higher level object oriented service-based application. Although, the process paradigm is more suitable for synthesis, hardware/software partitioning and performance analysis, it lacks reusability and abstraction. MARTE targets mainly real time embedded systems. A more profound discussion of abstraction and hierarchy of both application and hardware platform modelling would be needed [5]. UML-SOC and UML-SystemC target hardware related aspects. Consequently, they show limitations toward software part. TUT profile lacks expressiveness for tasks pipelining, and data parallelism modeling. Finally with Gaspard2, designers deal with complex index expressions for multi-dimensional data.

III. OUR APPROACH

Our flow starts at an early stage of development at which designer models his/her application using UML sequence diagram annotated by temporal constraints. In other words, our initial model is a pure sequential object paradigm. We think that such a paradigm brings many benefits. First because, an object-oriented paradigm is preferable in term of abstraction and reuse. Secondly, we think that a sequential model facilitates the modelling task relieving the designer of the burden of concurrency modeling. Thirdly, starting from an existing sequential model (e.g. legacy C/C++ code) which is generally considered as the reference model for embedded applications, we can then extract many types of parallelism that exist in typical ESs, explore, and compare between different alternatives.

The second step in our approach is the transformation of the sequence diagram to the design model comprising a set of communicating tasks with explicit task parallelism, data parallelism, pipelining, hierarchy and virtual communication channels. In order to perform high level performance estimation, the hardware architecture on which application will be mapped should also be abstracted to match the abstraction level of the application. In our case, The SOC architecture is a set of abstract components. Each architecture component is modeled as an abstraction of its fine grain model and it is generic so that the whole architecture could potentially be used for modeling all types of SOCs.

A. computation modeling

- **Leaf behavior**

A Leaf Behavior (LB) represents the elementary schedulable computation of an application. LB inputs and outputs are attached to its ports as tagged values. The size of the data to be transmitted over the channel is expressed in term of abstract tokens number. An abstract token is the elementary datum communicated or processed by LBs. The advantage of such abstraction is the large opportunity to apply static formal analysis and fast simulations can be easily performed. We define a new stereotype named *"behavior"* with the following tagged values:
- The WCET (Worst Case Execution Time) expressed in term of cycles number.
- The maximum number of iterations.
- The maximum number of read and/or write access to a shared data.

- **Behaviors sequencing**

In order to model behaviors executing in a sequence fashion, we introduce a new stereotype called *"sequence"*.

- **Behaviors pipelining**

We introduce a new stereotype called *"pipeline"*. This stereotype composes behaviors in sequence, with the output of one connected to the input of the next. The *"pipeline"* stereotype contains two tagged values that are the *"PipeDepth"*, and the maximum number of iterations. The former specifies the number of the pipe stages. Each behavior included in the "pipeline" stereotype is allocated to a different CPU. Pipeline execution implies the iterative execution of children. The pipeline stereotype also supports communication buffering modeling between the pipeline stages [4].

- **Data parallelism**

To satisfy streaming applications needs, we introduce a new stereotype called *"datapartition"*. This stereotype distributes data to a set of parallel streams, which are then joined together. Each data stream is executed by the same code and requires a memory buffer. Hence this stereotype duplicates behavior into many synchronized behaviors. The "datapartition" stereotype contains three tagged values: *PartitionsNumber* to specify the number of data partitions, *PartitioningMechanism* to specify how incoming data are scattered over behaviors, and *CollectionMechanism* to specify the mechanism of data gathering. Each data partition is stereotyped by *"partition"* with one tagged value: *PartitionSize* in term of tokens number. A behavior stereotyped by "datapartition" plays the role of a controller (master). It is responsible of creating, synchronization between the slave behaviors, data scattering and collecting.

- **Hierarchy**

We introduce a new stereotype called *"structure"*. This stereotype defines a hierarchical behavior. We use this stereotype to manage complexity and to enable hierarchic descriptions.

- **Mutually exclusive behaviors**

In some cases, it is preferable to expose behaviors which are IF-dependent (only one behavior is executed at a time). In this case, we will define a new stereotype called *"exclusive"*. A behavior stereotyped by *"exclusive"* plays the role of a controller. Depending on the condition value, it triggers the appropriate behavior. Hence, the behavior of the controller is modeled as an FSM (Finite State Machine).
Exposing these kinds of dependencies leads to a more accurate estimation and increases optimization opportunity. For instance and in order to minimize the number of hardware

978-1-4244-2541-9/08 $25.00 © 2008 IEEE

resources, we map sequential and exclusive behaviors on the same hardware components.

B. Communication modeling

Two communication models are supported, signal passing and shared memory.

- **Signal passing**

In this case, behaviors communicate via abstract channels. Each channel is connected to two ports. We define a stereotype called *"channel"*. This stereotype is applied on SysML flows [10]. It has two tagged values: the *SampleMax* specifying the maximum size of the data FIFO (expressed in term of tokens number) attached to the channel, and the communication style which is either Blocking Read-Blocking Write (BRBW), Blocking Read- Non blocking Write (BRNW), Non blocking Read- Non blocking Write (NRNW), or Non blocking Read- Blocking write (NRBW).

- **Shared memory**

In this case, behaviors communicate via shared data. Here, two styles are possible, synchronous and asynchronous. In synchronous mode, a lock is associated with each shared memory block and only one behavior can access the memory at one specific time. Meanwhile, the asynchronous mode does not have a lock associated with the memory, and therefore concurrent accesses can happen. We define a stereotype called *"shared data"* with two tagged values: *SizeData* specifying the size of shared data in term of token numbers and *Communication Mode* which is either synchronous or asynchronous.

C. Architecture Modeling

- **The CPU model**

This model concerns both General Purpose Processors (GPP) and Application Specific Instruction Processors ASIP (e.g. DSP). The CPU can execute one or many behaviors. In the latter case, it needs a scheduler. Each CPU is parameterized by five parameters that are the cost, the speed factor (*SF*), the local data memory size, the scheduling policy and the context switching overhead. We denote a cycle to be the amount of time required to execute an elementary instruction. The speed factor is a number showing the relative speed of the CPU. For a GPP, $SF = 1$. For a CPU faster than GPP (e.g. DSP), $SF < 1$.

- **The IP model**

This model concerns pre-characterized blocks. Each IP is parameterized by its cost, and *SF*.

- **The FPGA model**

The FPGA model is parameterized by its cost, the *SF*, the local memory size, and the reconfiguration time in term of cycles number.

- **The BUS model**

At this level of abstraction, we assume that bus communicate directly with other components without using interfaces. Each bus is parameterized by three parameters that are: the cost, the transfer rate *TRB* in term of number of tokens transferred per cycle, and the bus type (shared or dedicated). In the case of a shared bus, we must specify the arbitration mechanism to solve the problem of concurrent transfers. If two hardware components need a fast link between them without arbitration, the designer may configure the bus as dedicated.

- **The Memory model**

The memory model is characterized by its cost, the transfer rate *TRR* in term of number of read tokens per cycle, the transfer rate *TRW* in term of number of written tokens per cycle, the number of simultaneous reads *SRN*, and the number of simultaneous writes *SWN*.

D. Mapping modeling

Mapping consists in allocation and scheduling of application components to architecture components, so behaviors are mapped to computing resources (CPU, IP, FPGA), communication channels are mapped to buses, and data to memories. We define a new stereotype called *"AllocatedTo"*. This stereotype is applied on the UML constraint and it has two stereotypes specifying the hardware resource to which logical component is allocated, and the execution order of the behavior on the hardware resource. The latter information is specified whenever more than one behavior is mapped to the same hardware resource.

IV. GUIDELINES FOR PARALLELISM EXTRACTION FROM SEQUENCE DIAGRAM

Our proposed flow starts by establishing a pure sequential model of the application. For this purpose we use the sequence diagram (SD). The latter is a good choice to model sequential (eventually hierarchic) interactions between objects. Furthermore, it exposes control and data dependencies, loops, and conditions explicitly. We enrich the SD with temporal constraints (WCETs). Fig. 1 shows an example of a sequence diagram that supports hierarchy. There are five objects named MAIN, O1, O2, O3, and O4. These objects interact via message sending. Since we are dealing with a pure sequential application model, all objects are considered passive and all messages are supposed synchronous.

A. The first step: identification of hierarchic, exclusive, concurrent, sequenced and pipelined behaviors

End-to-end scenarios may be concurrent behaviors. An end-to-end scenario is a sequence of dependant methods triggered by a method call or external event. In the example of Fig. 1 we can identify four concurrent behaviors named B1, B2, B3,

978-1-4244-2541-9/08 $25.00 © 2008 IEEE

At this stage, we can also extract mutually exclusive behaviors. For instance B1 and B2 belong to different branches of the operator "alt", so we encompass them in one behavior stereotyped by "exclusive". Finally B4 is a hierarchic behavior. Using the sequence diagram, we can for each behavior extract, a set of tagged values. For example B1 WCET = 40 + 10 + (5 + 20) * 40 = 1050 cycles. Here loop <1,40> specifies the min and the max iterations number. B1 iterations number = 1, Max Reads access number to shared data = 1 (Get_attrib), Max writes access number to shared data = 40.

B. The second step: identification of data partition behaviors

This kind of parallelism cannot be identified from sequence diagram. It requires knowledge on method internal data structures and the fashion the method manipulates these structures. Generally, methods with a large execution time are good candidates for splitting into less intensive concurrent behaviors.

V. MAPPING AND OPTIMIZATION GUIDELINES

A. Mapping guidelines

Mapping guidelines deal with two contradictory goals: communication overhead decreasing and behaviors execution time minimization.

1. In general, computing-intensive behaviors are implemented in hardware. But, due to limited hardware resources, some resources should be shared. To overcome this problem, we map sequential and mutually exclusive behaviors on the same CPU.
2. If a behavior is control-dominated, then it will not be mapped to FPGA, rather than it is preferable to map it on a GPP (General Purpose Processor).
3. If a behavior is data-dominated with a very high computational load, then it will not be mapped to GPP, unless there is no available hardware resource.
4. If the first objective is the communication overhead decreasing then map behaviors with a high communication workload on the same CPU, even they are concurrent.
5. To minimize bus congestion, map channels with high traffic to fast links.

B. Optimization guidelines

Indeed, there are many factors that affect the overall system performance. The most important ones are the granularity of behaviors, the granularity at which data is communicated, the amount of data processed by each behavior, the data scattering and gathering mechanisms in the case of data partition, the management of the shared data, the communication scheme, and the depth of the pipeline.

These parameters have to be chosen carefully. The overhead caused by synchronization may counteract the benefits of parallelism, coarse granularity behaviors (which mean small number of behaviors) are better than fine grained behaviors in term of communication overhead. However,

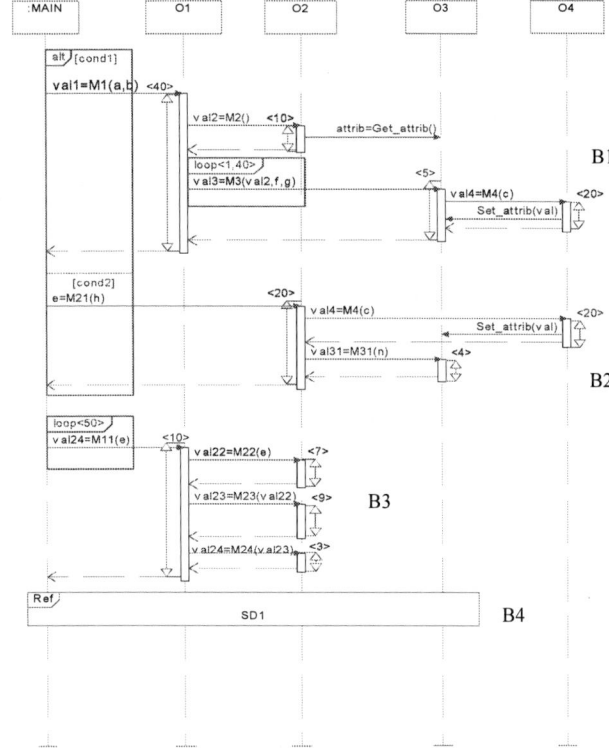

Figure 1. Hierarchic Sequence Diagram

and B4. B1=(M1,M2,Get_attrib,M3,M4, Set_attrib), B2=(M21,M4,Set_attrib,,M31), B3 = (M11, M22, M23, M24), B4=(SD1). Communication between these behaviors is achieved through shared memory. In this example, B1 and B2 access to the attribute "attrib" of object O3 via methods Get_attrib (read), and Set_attrib (write). So we create a new data object called "attrib" stereotyped by "shared data".

Of course, we can transform the shared memory based communication scheme to an equivalent signal (message) passing based communication scheme. In this case, we do not create data objects, rather than, we replace all shared data accesses by point-to-point communication.

Another form of parallelism that we can extract from SD is pipeline. In Fig. 1, methods M22, M23, and M24 of behavior B3 can be executed in a pipeline fashion. In the same example, we remark that the returned value of B2 (e) serves as input for B3. Consequently, B2 and B3 are in sequence. Since messages are synchronous, the caller must wait (blocked) for returned values. Not all methods receive or return data. In this case we will introduce two zero delay control events: the **Request** event and the **Ret** event.

behaviors may have to wait longer. Finally, depending on the application, data partitioning will give communication overhead for data dependencies between partitions. According to estimation results, the designer may:

1. Merge behaviors with high communication workload into one behavior.
2. For each computational bottleneck behavior, refine it following guidelines of section 4.
3. In order to minimize the effect of synchronization overhead due to data partitioning, it is preferable that each data partition processes larger data blocks [1].

VI. WCET ESTIMATION

Our proposed approach deals with two situations:
1. The application under construction does not exist. In this case, based on the previous experience, and similar existing designs, the designer can introduce for each object method its WCET.
2. The application model exists (i.e. legacy code in C). In this case, the WCETs of methods are obtained via profiling.

The estimation model is based on analytic formula. Since we are dealing with higher level of abstraction the analytic analysis seems more appropriate. Of course our formula is inexact, but it serves as a good first attempt to model aspects related to time at higher level of abstraction. But before presenting our estimation technique, we make three assumptions:

1. All concurrent executions, and transfers, to a shared resource (CPU, BUS) are processed following the order imposed by the designer (during the mapping).
2. The Communication via shared data is asynchronous.
3. The communication mode on channels is supposed NRNW (FIFO with infinite size).
4. Numbers of concurrent memory reads and writes never surpass SRN (SWN).

Let t be the WCET of behavior B and SF the Speed Factor of the Computing Resource CR. If B is allocated to a CPU, then the estimated time Et for B is $Et = t * SF$. If B is mapped to an FPGA, we include the overhead due to reconfiguration: $Et = Et + T_{reconfig}$. (The overhead is added one time). If there are other concurrent behaviors which are allocated to the same CPU, then $Et = Et + T_{cpu}$ where T_{cpu} is the overhead due to behaviors sequencing. If B access shared data, then $Et = Et + T_{data}$ where T_{data} is the overhead due to shared data access (in this case T_{data} is introduced by the designer, otherwise it is equal to zero according to the second assumption). $T_{data} = T_{read} + T_{write}$. If B is the only behavior that access to the shared data then $T_{read} = NB_{read} * (T_{trans} + T_{rmem})$ where T_{trans} is the time of bus transfer, T_{rmem} is the actual reading time and NB_{read} is the number of read access. (we assume that shared data access is done via a shared Bus). $T_{trans} = DataSize/TRB$, and $T_{rmem} = DataSize/TRR$. $T_{write} = NB_{write} * (T_{trans} + T_{wmem})$ where T_{wmem} is the actual writing time, and NB_{write} is the number of write access. $T_{wmem} = DataSize/TRW$. If there are many behaviors that access to the same shared data concurrently, then $T_{trans} =$ $T_{trans} + T_{bus}$ where T_{bus} is the overhead due to transfers sequencing. Furthermore if B is executed iteratively, then $TEt = Et*Iter$ where $Iter$ is the maximum number of iterations and TEt is the total execution time of B.

Let $t1$, $t2$ be the WCETs of behaviors B1, B2 respectively.

1. B1 and B2 are in sequence
If B1 and B2 are mapped to the same CPU, then $T = TEt1 + TEt2$ (we neglect the communication time between B1 and B2). If CPU is an FPGA, then $T_{reconfig}$ is added one time. IF B1 and B2 are mapped to two distinct CPUs and the two CPUs are linked by a fastlink (non shared bus), then $T = TEt1 + TEt2 + T_{com}$ with $T_{com} = Iter*DataSize/TRB$: is the communication time between B1 and B2 where $DataSize$ is the size of the transferred data between B1 and B2, and $Iter$ is the maximum iterations number of B1. If the link between CPUs is a shared bus then $T_{com} = T_{com} + T_{bus}$.

2. B1 and B2 are mutually exclusive
$T = MAX (TEt1, TEt2)$.

3. B1 and B2 are concurrent
If B1 and B2 are mapped to two distinct CPUs then $T = MAX (TEt1, TEt2)$. If B1 and B2 are mapped to the same CPU, then we compute T as B1 and B2 are in sequence.

4. B1 and B2 are pipelined
Let n be the number of iterations of the pipeline. B1 is mapped to CPU1 and B2 is mapped to CPU2. Only behavior B1 will be executed in the first iteration. In the second iteration, B1 and B2 will be executed concurrently. In the third and all following iterations, both behaviors are executed in parallel. After the n iteration, only B2 will be executed (in the $n+1$ iteration). $T = TEt1 + (n-1) * MAX (TEt1, TEt2) + TEt2$. More generally, if the depth of the pipeline is m and the iterations number is n, then we can estimate T by the formula: $T = (n-m-1)*MAX (TEt_1, ... TEt_m) + TEt_1 + MAX (TEt_1, TEt_2) +...+MAX (TEt_1, TEt_2, ... TEt_{m-1}) + TEt_m + MAX (TEt_m, TEt_{m-1}) + MAX (TEt_m, TEt_{m-1}, TEt_{m-2})+...+ MAX (TEt_m, ..., TEt_2)$ [4].

5. B is a datapartition
Let n be the number of data partitions and S1, S2... Sn their data partitions sizes respectively. A datapartition behavior B plays the role of a controller. It creates n behaviors B1, B2...Bn executing concurrently. It also splits and collects data. Assuming that the execution time is proportional to the amount of processed data. For each behavior Bi, we can estimate its WCET by the formula $ti = t*Si/S$ where t is the WCET of B, and Si is the data partition size of behavior Bi, and S is the sum of all data partitions sizes. When Bi is allocated to CPUi then $Eti = SFi*t*Si/S$. Before slaves execution starts, the master should transmit data partitions to his slaves. We can estimate transmission time by formula $T_{part} = MAX (S1/T_{trans}, S2/T_{trans}, ..., Sn/T_{trans})$ [1].

The controller itself executes concurrently with its slaves and takes time for splitting and data collection. $T = T_{part} + MAX (TEt1+T_{coll}, TEt2+T_{col2}..., TEtn+T_{coln})$ where T_{coli} is the overhead due to collection data. $T_{coli} = Iter*DataSize_i/T_{trans}$ where $Iter$ is the maximum executions number of the slave Bi, and $DataSize_i$ is the size of Bi data outputs. Supplementary overhead can be added, if designer has accurate information on data collection and functionality of the controller.

VII. CASE STYDY

We have tested our methodology on the H264 decoder for video compression. We were focused on the C code furnished by Chemnitz [13] (with a minor modification). Using the reverse engineering tool integrated in the Rhapsody [14], we recuperated the set of H264 classes. After that, we established the sequence diagram from which we extracted a stereotyped hierarchic tasks graph (Fig. 2) following the guidelines of section 4.

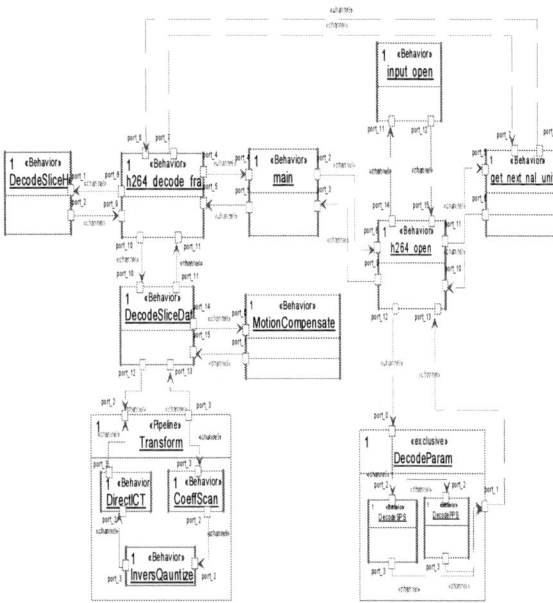

Figure 2: H264 decoder concurrent model

VIII. CONCLUSION

In this paper, we present a new UML profile for application WCET estimation on SOC. Our proposed flow starts from a sequential model from which, a hierarchical task graph exposing many forms of tasks and data parallelism is generated. In order to estimate application WCET, the architecture on which the application is executed is abstracted too. As a perspective we plan to automate the passage from the sequence diagram to the concurrent model, take into account more sophisticated communication infrastructures such as NOC, and estimate power consumption.

IX. REFERENCES

[1] I.Ahmed, Y. He, and M. L. Liou. Video compression with parallel processing. In Parallel Computing Journal 28, pp. 1039-1078, 2002.

[2] R. Ben Atitallah, P. Boulet, A. Cuccuru, J.L. Dekeyser, A. Honré, O. Labbani, S. Le Bleu, P. Marquet, E. Piel, J. Taillard, and H. Yu. INRIA. Rapport technique, Gaspard2 UML profile documentation.. N° 0342. September 2007.

[3] F.Boutekkouk, and M. Benmohammed. A Novel UML2.0-based approach for System On a Chip modeling and Co-design. In VLSI-SOC PhD Forum, Nice, France, 2006.

[4] R. Domer. System-level modeling and design with the SpecC language. Dissertation zur Erlangung des Grades eines. Doktors der Naturwissenschaften der Universitat Dortmund am Fachbereich Informatik. Dortmund, 2000.

[5] ITEA. Information Technology For European Advancement. MARTES. Model-Based Approach for Real-Time Embedded Systems development. Title: Current limitations of best practices. Deliverable ID: 1.1, Version: 1.0. Editor Kari Tiensyrja. Status: Final. Confidentiality: Public. Date: 31/03/2006.

[6] K. keutzer, S. Malik, R. Newton, j. Rabaey, and A. Sangiovanni-Vincentelli. Sytem level design: orthgonalization of concerns and Platform-Based Design. In IEEE transactions on computer-aided design of circuits and systems, Vol. 19. , No. 12 ,December 2000.

[7] B. Kienhuis, Ed F. Deprettere, P.V. Wolf, and K. Vissers. A Methodology to design Programmable Embedded Systems. The Y-chart approach. In LNCS series vol. 2268, page 18-37 by Springer Verlag © 2001.

[8] P. Kukkala, J. Riihimaki, M. Hannikainen, T.D. Hamalainen, and K.Kronlof UML2.0 Profile for Embedded System Design. In Proceedings of the Design, Automation and Test in Europe Conference end Exhibition (DATE'05).

[9] E.Riccobene, P. Scandura, A. Rosti, and S. Bocchino. A SOC Design Methodology Involving a UML2.0 Profile for SystemC. In DATE'05.

[10] OMG. Systems Modeling Language (SysML) Specification. *OMG document: ad/2006-03-08-01*, version 1. Draft, April 2006.

[11] OMG. UML Profile for MARTE, Beta 1. *OMG Adopted Specification, ptc/07-08-04*, August 2007.

[12] OMG. UML Profile for System on a Chip (SOC). *OMG Available Specification, version 1.0.1 formal /06-08-01*, August 2006.

[13] http://rtg.informatik.tu-chemnitz.de.

[14] www.ilogix.com.

Specification of GNSS Application for Multiprocessor Platform

Heikki Hurskainen, Jussi Raasakka, and Jari Nurmi

Department of Computer Systems
Tampere University of Technology
Tampere, Finland
{firstname.lastname@tut.fi}

Abstract—In this paper we present a satellite navigation application for a multiprocessor platform. We give a preliminary description to the multiprocessor platform with an introduction to the project where it will be designed and implemented to a fabric die. The platform contains a matrix of processing tiles, connected together via a Network-on-Chip. The satellite navigation application is one of the streaming applications designed to be executed on this platform. Detailed descriptions of the signal processing functions for acquiring and tracking the satellite signals for navigation are given. The estimations of their requirements for computational complexity, measured as multiply-accumulate counts, are also presented in this paper. The results indicate that one processing tile on the platform could barely perform the required signal processing functions and thus more tiles are preferred for a better user experience. This work is carried out in the EU FP7 project called CRISP (Cutting edge Reconfigurable ICs for Stream Processing).

I. INTRODUCTION

CRISP (Cutting edge Reconfigurable ICs for Stream Processing) is a project funded from European Union 7th Framework Program (EU FP7) funds. The CRISP consortium is a good mixture of industrial and academic know-how and competence. The consortium members are: Recore Systems (NL), University of Twente (NL), Atmel (DE), Thales Netherlands (NL), Tampere University of Technology (FI), and NXP (NL).

The main objective of this project is to research the optimal utilization, efficient programming and dependability of a reconfigurable multiprocessor platform for streaming applications. During the project a General Stream Processor (GSP, not to be mixed with the GPS (NAVSTAR Global Positioning System)), a system with multiple cores, will be specified, designed and eventually implemented onto a single fabric die. This multi-core system will consist of a number of Tile Processors (TP), embedded memories and other dedicated blocks, connected via a Network-on-Chip (NoC). The manufactured die will be accompanied by a General Purpose Processor (GPP) die in a System-in-Package (SiP). The processing tiles are reconfigurable.

The project has four main themes. The Streaming Applications theme is about specification, implementation and integration of the streaming applications to the GSP. The streaming applications selected for the project for proof-of-concept purposes are a beamforming application in electronically steered radar which is out of the scope of this paper, and a Global Navigation Satellite System (GNSS) application.

The streaming applications are developed in parallel with the development of the multiprocessor platform. The goals for developing the streaming applications within the project are to obtain experience in the software development for reconfigurable multi-core systems; improve the design methodology for developing complete streaming applications; obtain realistic design specifications for the GSP; and use realistic streaming applications for evaluating project results. The other themes in the CRISP project are General Stream Processor (GSP), Run-time Mapping and Dependability.

From a GNSS applications point of view the anticipated result of the whole project is to create a software navigation application running on the multi-core platform (General Stream Processor). Expected output of the application is a real-time PVT (Position, Velocity and Time) solution. The application should take full advantage of the available computational resources. To our knowledge the GNSS receiver application running on multiple cores is novel. Examples of implementations of parts of the GNSS/GPS receiver functionality to reconfigurable machines can be found in literature, e.g. in [1] and [2].

This paper is divided to the following sections. The planned CRISP platform is described in Section II, where we discuss the components of the system. Since the specification work of the multiprocessor platform has not been finalized we cannot give detailed information about the platform. In this paper our focus is in the description of the GNSS application and what it demands from the platform. The

---

This research is conducted within the FP7 Cutting edge Reconfigurable ICs for Stream Processing (CRISP) project (ICT-215881) supported by the European Commission.

978-1-4244-2541-9/08 $25.00 © 2008 IEEE

functions of the GNSS application are described in Section III and their computational complexity is estimated in Section IV. The mapping of GNSS functionality to the CRISP platform is given in Section V. After that the conclusions with the direction of our future work follows.

II. CRISP PLATFORM

In this section we describe the planned CRISP platform. The information about the platform is preliminary at the time of writing.

A. GSP + GPP

The CRISP platform will be a combination of an off-the-shelf General Purpose Processor and a General Stream Processor. The GSP is a reconfigurable matrix of a number of Montium Tile Processors, memories and dedicated I/Os. The detailed architecture nor the amount of the TPs in the platform are not finalized at the time of writing and thus no more detailed information can be given here.

The off-the-shelf GPP is provided by Atmel, one of the consortium members. The selection of the GPP has not been finalized, but it will belong to the Atmel's AVR processor product families.

B. Montium Tile Processor

The Montium Tile Processor is developed by Recore Systems (NL), one of the consortium members and the leader of the project. It is a programmable architecture that has significantly lower energy consumption than DSPs for fixed-point digital signal processing algorithms [3].

The Montium TP is illustrated in Figure 1. The architecture of the Montium TP consists of 5 processing units, accompanied with 10 parallel memories to meet the high demand of memory bandwidth of the parallel units. The datapath width of the core and memory are customizable at design-time [3], [4]. The processing units, or Arithmetic Logic Units (ALUs), can compute both signed integer and signed fixed-point values. One ALU can perform one multiplication and accumulation operation (MAC) in one clock cycle [4]. One MAC can be expressed mathematically as $a + b*c = d$, where a, b, and c are input values and d is an output value.

Since there are five parallel ALUs, the Montium can perform 5 MACs per clock cycle, and if the processor is clocked, e.g., at 200MHz the computational performance of the TP is theoretically 1GMAC/second.

C. NoC

The matrix of Montium TPs, memories and dedicated I/Os are connected together by a Network-on-Chip. The data between tiles is planned to be transmitted with packet switched approach. The specification of the NoC is not finalized at the time of the writing, but will be based on work from University of Twente (NL) and Tampere University of Technology (FI) [5], [13].

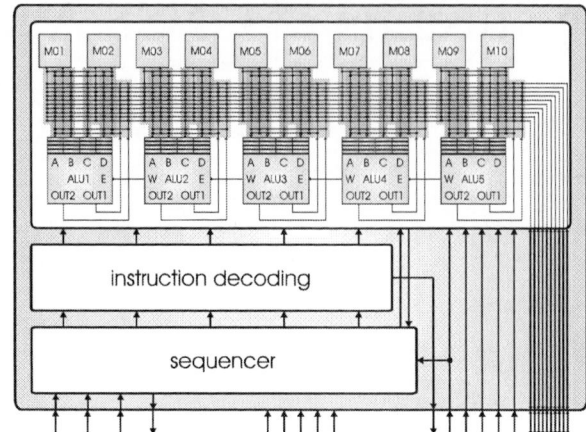

Figure 1.   Montium Tile Processor. [3]

III. GNSS APPLICATION

In this Section we describe the GNSS application. Shortly, the GNSS application receives signals from multiple satellites and uses dedicated methods for extracting data from the received signal. After that the data, combined with the timing information is used to compute the pseudoranges between the user and a number of satellites. With four or more measured pseudoranges (ρ_i), and known satellite locations (x_i, y_i, z_i), the user's position (x_u, y_u, z_u) and clock bias (b_u) between the receiver and the system can be computed by the following equation:

$$\rho_i = \sqrt{(x_i - x_u)^2 + (y_i - y_u)^2 + (z_i - z_u)^2} + b_u \qquad (1)$$

Besides position the GNSS application can also solve the user's velocity and time. Usually the "output" of the GNSS application is called as PVT solution, standing for Position, Velocity and Time.

A. Global Navigation Satellite Systems

Most of the GNSS satellites in the space belong to the American NAVSTAR Global Positioning System (GPS). For a long time GPS has been the only functional GNSS. During last few years the European Galileo system has been rising to compete with the GPS and complement the GNSS. Galileo will reach the full operational capability by 2013. There are other systems, partially existing GLONASS from Russia and the planned Chinese Compass/Beidou system, but the GPS and Galileo are the only ones with agreed interoperation and common transmission frequencies. Figure 2. shows the common frequencies between GPS and Galileo in both L5 (centered at 1176.45MHz) and L1 (1575.42MHz) frequency bands [6], [7].

In the CRISP project our focus is to specify, implement and integrate the GNSS application, supporting the GPS and Galileo L1 signals. The L1 is selected since it is the only frequency band with guaranteed existence of signals at

978-1-4244-2541-9/08 $25.00 © 2008 IEEE

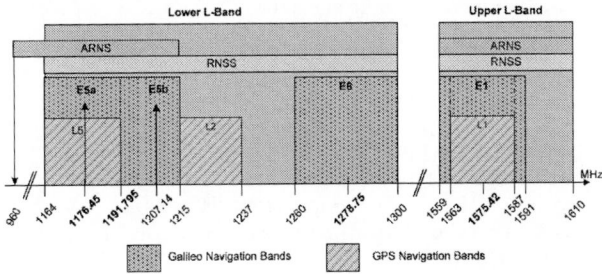

Figure 2. Galileo and GPS frequency plan. [6]

CRISP project timeframes. The interface specifications for new systems are not fixed but still changing, and thus they are one of the main drivers for programmable receiver implementations.

B. Functions of the GNSS Application

The input for the GNSS Application is the digital signal stream coming from the radio front end, where the L1 signal is amplified, filtered, down-converted to intermediate frequency (IF) and then digitalized. A Typical rate for such a digital signal stream could be 1-bit data with 16.367 MHz sampling rate and IF around 4 MHz [8].

In usual case the user has no knowledge about the incoming signal; the sources (each satellite has a unique transmission sequence), delays, and Doppler shifts are all unknown. The signal is also totally buried under noise. For these reasons specialized functions are needed to find, and keep track of the satellite signals. These receiver baseband processes are called acquisition and tracking and due to their requirements for computations they are traditionally implemented as hardware [10].

The programmable platform baseband implementation allows the receiver algorithms to be tuned to meet the surrounding signal environment better. E.g. the tracking can be optimized either for accuracy when the signal is good or for robustness when the signal is weak.

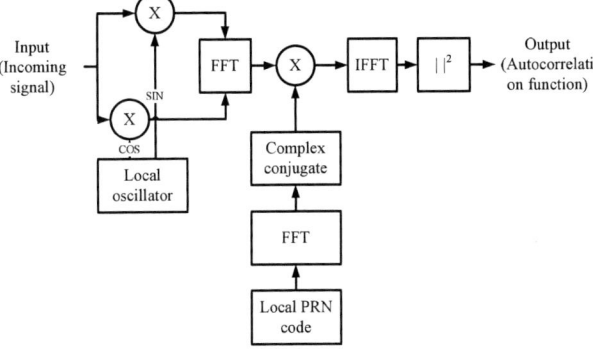

Figure 3. FFT Acquisition process.

1) Acquisition

The function of the acquisition is to find and identify the satellite signals from the received stream. The outputs of the acquisition are satellite ID, phase/delay and Doppler frequency information of the signal, which are used to start the tracking at the correct point. The acquisition is usually needed only in the initialization phase (i.e. finding satellites), but in case of high dynamics (acceleration), bad signal environment (multipath presence, urban canyons, indoors) or other disturbance (noise, interference) it can be applied to perform re-acquisition for the lost signals.

Acquisition, as well as tracking, is based on the autocorrelation function (ACF) properties of a finite length binary sequence. Each satellite has a unique, finite length pseudorandom number (PRN) code, which identifies it. The main parameters of GNSS L1 PRN codes are presented in TABLE I. When a replica PRN with the right delay and frequency is generated in the user's receiver, the ACF achieves a peak value, whereas the correlation result will be low-amplitude noise in other cases. Acquisition can be done in serial (PRN multiplied chip by chip) or parallel (whole epoch is multiplied at once) mode [9].

One way to perform the parallel acquisition is to use Fast Fourier Transform. A Block diagram of FFT-based code acquisition is presented in Figure 3. The advantage of this acquisition method is that to cover the full search space for one satellite, only 41 frequency domain searches are needed (when Doppler is ±10kHz and search bin space is 500Hz) [9].

TABLE I.   GNSS L1 PRN PARAMETERS

GNSS L1 PRN code parameters [6], [7]		
Parameter	*GPS*	*Galileo*
PRN size (chips)	1023	4092
PRN frequency (MHz)	1.023	1.023
Epoch time (ms)	1	4
Modulation	BPSK	CBOC

Unlike other methods, the FFT reveals all code delays at once in one PRN code epoch time. The incoming signal is multiplied with the locally generated IF carrier (includes Doppler). The resulting complex value is fed to FFT of which output is multiplied with the complex conjugate of the FFT of the locally generated PRN code. The multiplication result is then fed to an inverse FFT and the result of that is squared. The squaring result is the autocorrelation function of the acquisition process, having the maximum peak at the position indicating true code delay (if the Doppler and satellite ID were correct). The acquisition is compatible with both GPS and Galileo. Galileo only needs larger FFTs due to longer PRN codes and different modulation. GPS uses Binary Phase-Shift Keying (BPSK) [7], and according to the newest specification [6] Composite Binary Offset Carrier (CBOC) modulation will be used for Galileo.

978-1-4244-2541-9/08 $25.00 © 2008 IEEE

2) Tracking

The function of the tracking is to follow the received signal phase by replicating it as accurately as possible and using feedback loops to steer the replication. The typical GPS tracking process, with three correlators, is illustrated in Figure 4. First the incoming signal is multiplied with the locally generated IF carrier, which is generated by the carrier plant. The heart of the plant is a Numerically Controlled Oscillator (NCO). After this carrier wipe-off process the resulting baseband signal is correlated with several versions of the local PRN replica.

The PRN replica is also generated by a NCO and code generator/memory. The early, prompt and late versions of the correlation are usually steered so that he maximum peak of the ACF is found at the prompt phase. The correlation results are integrated and the discriminator functions are used to compute the feedback, which are filtered before steering the plants. The tracking process is similar in Galileo case, only the minimum number of code correlation fingers is five, since the ACF of CBOC modulated code has side peaks on both sides of the main ACF peak and the extra fingers can be used to remove the ambiguities [11].

When tracking is locked, the receiver can demodulate the navigation data out of the incoming stream. The navigation data is having a low rate, 50 Hz for GPS and 250 Hz for Galileo [6], [7].The sign of the integration result over a data symbol period is one navigation data bit. The outputs of tracking are navigation data and pseudoranges computed from the timing information inside the tracking channels (NCO phases). For each satellite one tracking channel is needed. Since at least four satellites are needed to solve the users position and clock bias, the number of tracking channels should be four or preferably much greater. [9], [10]

3) PVT computation

The PVT solution process collects and timestamps the navigation data which is extracted from the input data stream by the tracking channels. The data integrity is checked and ephemeris parameters collected. According to [6], four types of data must be received for positioning:

- Ephemeris which are needed to indicate the position of the satellite

- Time and clock correction parameters which are needed to compute pseudo-ranges

- Service parameters which are needed to identify the set of navigation data, satellites, and indicators of the signal health

- Almanac which are needed to indicate the position of all the satellites in the constellation with a reduced accuracy

The detailed equations how to solve the PVT solution are out of the scope of this paper and they can be found e.g. in [9] and [10]. The PVT computation is planned to be implemented on the GPP of the CRISP platform. The GPP should be powerful enough to execute the PVT computation in real-time with a reasonable update rate (e.g. 1 Hz).

IV. ESTIMATING THE APPLICATION COMPLEXITY

In this section we estimate the computational complexity of the GNSS application core functions; acquisition and tracking. We use the number of MAC operations as a baseline for our estimations.

A. Acquisition estimates

For acquisition complexity estimations we divided the process to subprocesses. For each subprocess we estimated a MAC count as a function of the FFT size (marked as N_{fft}) over one frequency search bin. E.g. for carrier wipe-off the NCO is one MAC and two MACs are required for in-phase and quadrature phase multiplications, totaling 3 MAC operations for the subprocess. The estimated MAC figures for acquisition subprocesses are given in TABLE II. The generation of local PRN, its FFT and complex conjugate can be neglected in estimations since they are generated only once in the beginning of each new satellite search.

For the size of FFT (N_{fft}) we used 16K (2^{14}) for GPS and 64K (2^{16}) for Galileo, reflecting the number of discrete samples over one PRN epoch with sampling rate of 16.367 MHz, rounded up to the closest power of two value. The sampling rate is taken from an existing front end [8] capable to process both GPS and Galileo signals. In literature the computational complexity of radix-2 FFT is estimated with $N_{fft}*log2(N_{fft})$ [12]. In Montium the FFTs are implemented as radix-2 Decimation-in-Time FFTs. There the number of complex butterfly units is $(N_{fft}/2)*log2(N_{fft})$. Per complex butterfly, 4 real multiplications are required and thus we used $2*N_{fft}*log2(N_{fft})$ to estimate the FFT complexity.

TABLE II. MAC COUNTS FOR ACQUISITION PROCESS

Estimations over one frequency search bin			
Sub process	*MACs*	*GPS*	*Galileo*
Carrier wipe-off	$3*N_{fft}$	49,152	196,608
FFT (incoming signal)	$2*N_{fft}*log2(N_{fft})$	458,752	2,097,152
Complex multiplication	$4*N_{fft}$	65,536	262,144
Inverse FFT	$2*N_{fft}*log2(N_{fft})$	458,752	2,097,152
Squaring	$2*N_{fft}$	32,768	131,072
TOTAL		1,064,960	4,784,128

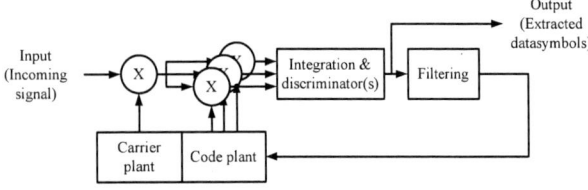

Figure 4.   Typical GPS tracking process.

The total values reported in the table estimate the MAC operations for one acquisition frequency search bin, to cover full frequency space (41 bins) in one second, 43.66 MMAC/s for GPS and 196.15 MMAC/s for Galileo are needed.

B. Tracking estimates

The tracking process was also divided to subprocesses. Unlike parallel FFT acquisition the tracking is a serial process. We estimated the MAC operation counts per sample in tracking. The estimations are given in TABLE III. The NCOs were estimated as 1 MAC/sample, and carrier wipe-off as 2 MAC/sample since the incoming signal is multiplied with in-phase and quadrature phase replica carrier resulting in I and Q branches for baseband processing. The variable $N_{correlators}$ represents the number of correlation fingers, which is in GPS case 3 and in Galileo 5.

To meet the real-time requirements of the signal tracking the TP should be able to perform computations over 16.367 Mega samples in second, this yields 163.67 MMAC/s for GPS and 229.14 MMAC/s for Galileo. These figures represent the requirements for one tracking channel. The computation of discriminators and feedback happens at rates between 1 kHz to datasymbol rates, which are 50 Hz for GPS and 250 Hz for Galileo and thus they add only a few kMAC/s to these counts.

TABLE III.   MAC COUNTS FOR TRACKING PROCESS

Estimations per sample			
Sub process	*MAC/sample*	*GPS*	*Galileo*
Carrier plant (NCO)	*1*	1	1
Carrier wipe-off	*2*	2	2
Code plant (NCO)	*1*	1	1
Correlation & Integration	$2*N_{correlators}$	6	10
TOTAL		10	14

V.   MAPPING THE APPLICATION TO THE PLATFORM

The figures and numbers given in the previous section are only estimates of computational power needed for executing the acquisition and tracking parts of our GNSS application on CRISP GSP. We sum up these estimations in TABLE IV. We assume the Montium TP running at 200 MHz, and thus having 1 GMAC/s computational performance. The estimations indicate that we could implement either 22 parallel acquisition processes for GPS, 5 acquisitions for Galileo, 6 tracking channels for GPS or 4 for Galileo in a single TP. In a real situation the resources are divided between processes, i.e. we could have four GPS acquisitions and tracking channels in one TP. The results also indicate that theoretically we could perform the baseband processes of GNSS application with only one Montium TP, but more TPs for execution of the acquisition and tracking are preferred for a better user experience. A four-channel receiver will not be sufficient for constant, reliable positioning of the user device.

The reduction of MAC/s counts is also possible by decimating the input signal, for acquisition purposes we could live with sampling rates around 4 to 8 MHz. Another motivation for signal decimation is the memory available in the Montium TP. The maximum size of the FFT is limited by the size of the memories. The earlier integrations of Montium TP are reported to have memories with a size of 1024 addresses [13]. So, if we want to implement large FFTs the memory size needs to be increased or the FFT functionality needs to be distributed over several Montium TPs.

GNSS application requirements for the NoC are quite low in the output end, if we close the feedback loops inside a single TP; we only need to forward the low rate navigation data and pseudoranges to the GPP. The more critical is the real-time requirement of the input signal. It can be alleviated by grouping several input bits into a single NoC packet. Thus, this application does not stress the network very much.

TABLE IV.   TP RESOURCE USAGE

Montium TP at 200 MHz, 1 GMAC/s			
Process	*Usage (MMAC/s)*	*Usage of TP (%)*	*Max number of processes per TP*
Acquisition (GPS)	43.66	4.4	22
Acquisition (Galileo)	196.15	19.6	5
Tracking (GPS)	163.67	16.4	6
Tracking (Galileo)	229.14	22.9	4

VI.   CONCLUSIONS

In this paper we introduced the CRISP project and consortium, presented the outline for the CRISP platform and gave detailed information about the GNSS application. We describe the processes for acquisition and tracking. The computational complexity of these GNSS application's core functions is also given. The results indicate that a single Montium TP is barely capable to execute the required acquisition and tracking processes, but having several tiles available will improve the user experience. The programmable receiver implementation allows faster adaption of new signal specifications.

VII.   FUTURE WORK

In future we will follow closely the finalization of the CRISP GSP and NoC specifications and we will start doing the implementations for the functions of GNSS application. Our intention is to report the implementation phase, followed by the application integration to the fabric die as the CRISP project proceeds.

ACKNOWLEDGMENT

The authors want to thank Gerard Rauwerda from Recore Systems (NL) for his useful comments.

REFERENCES

[1] T. Ristimäki, "Reconfigurable IP Blocks : A MIMD Approach" PhD Dissertation. Tampere University of Technology, 2005. ISBN 952-15-1491-4.

[2] F. Garzia, C. Brunelli, L. Nieminen, R. Mastria, and J. Nurmi. "Implementation of a tracking channel of a GPS receiver to a reconfigurable machine" Proceedings of the EUROCON 2007, Sep. 9-12. Warsaw, Poland.

[3] P.M. Heysters, G.K. Rauwerda, and L.T. Smit, "A Flexible, Low Power, High Performance DSP IP Core for Programmable Systems-on-Chip". Proceedings of the IP/SoC 2005, Dec 7-8. France.

[4] P.M. Heysters. "Coarse-Grained Reconfigurable Computing for Power Aware Applications" The 2006 International Conference on Engineering of Reconfigurable Systems & Algorithms (ERSA'06), June 7-9. Las Vegas, Nevada.

[5] T. Ahonen, "Designing network-based single-chip system architectures," Ph.D. dissertation, Tampere University of Technology (TUT), Department of Information Technology, Institute of Digital and Computer Systems (IDCS), Tampere, Finland, October 2006, 242 pages, TUT Publication 625, ISSN: 1459-2045, ISBN: 952-15-1666-6.

[6] "Galileo Open Service Signal in Space Interface Control Document (OS SIS ICD)", Draft 1, Feb 14, 2008.

[7] "NAVSTAR Global Positioning System Interface Specification. (IS-GPS-200)", Revision D, Dec 7, 2004.

[8] Atmel. ATR0603 GPS Front End IC. Datasheet.

[9] K. Borre, D.M. Akos, N. Bertelsen, P. Rinder, S.H. Jensen, "A Software-defined GPS and Galileo Receiver – A Single-Frequency Approach", Birkhäuser Boston, 2007. ISBN 0-8176-4390-7.

[10] E.D. Kaplan, and C.J. Hegarty (Eds), "Understanding GPS – Principles and Applications", 2nd edition, Artech House, 2006. ISBN 1-58053-894-0.

[11] P. Fine, and W. Wilson, "Tracking Algorithm for GPS Offset Carrier Signals", Proceedings of the 1999 National Technical Meeting of the Institute of Navigation, Jan 25 - 27, 1999, San Diego, CA.

[12] Smith, J.O. "Radix 2 FFT Complexity is N Log N", in Mathematics of the Discrete Fourier Transform (DFT) with Audio Applications, Second Edition, http://ccrma.stanford.edu/~jos/mdft/Radix_2_FFT_Complexity.html, 2007, online book, accessed Apr 29, 2008.

[13] Gerard J. M. Smit, André B. J. Kokkeler, Pascal T. Wolkotte, Philip K. F. Hölzenspies, Marcel D. van de Burgwal, and Paul M. Heysters, "The Chameleon Architecture for Streaming DSP Applications," EURASIP Journal on Embedded Systems, vol. 2007, Article ID 78082, 10 pages, 2007. doi:10.1155/2007/78082

Inherent Reliability Evaluation of Networks-on-Chip Based on Analytical Models

Mojtaba Valinataj, Siamak Mohammadi, Saeed Safari

Dept. of Electrical and Computer Engineering, University of Tehran
Tehran, IRAN
m.valinataj@ece.ut.ac.ir; {smohamadi,saeed}@ut.ac.ir

Abstract—**Reliability evaluation based on analytical models is a precise method for dependability analysis before and after designing the fault-tolerant systems. In this paper, we present the precise formulations for the inherent reliability of mesh-based NoCs that also depend on the employed routing algorithm and traffic model. Based on this analysis, the effects of some permanent failures in the links, switches or cores on the packet delivery of NoCs are determined. The models can be extended to evaluate the fault-tolerant methods in addition to other topologies and routing algorithms.**

I. INTRODUCTION

As CMOS technology scales down into the nano-technology domain and the complexity of evolving integrated circuits design increases, VLSI systems become more and more vulnerable to permanent faults in addition of transient faults. The Network-on-Chip design paradigm [1] has been proposed as the best scalable communication infrastructure for Multi-Processor System-on-Chips (MPSoCs) designs. Thus, it's essential to design the reliable NoCs. The reliability assessment which can be done analyti--cally or by simulation is a key method for dependability evaluation that can be used in design of reliable systems.

While performance analysis for large scale interconnects and NoCs has been reported extensively, reliability assessment for NoCs has not received much attention. In NoC domain, [2] provides a model for determining the probability that an NoC link fails due to manufacturing variation. In [3] a fault-tolerant analysis of different mesh NoC architectures including the topology, the router structure and the number of network interfaces is presented. Another work [4] proposes an analytical model to assess the reliability of a mesh NoC against transient faults effects on switches and routing algorithms. In this paper, some preliminary probabilistic and analytical models are presented to evaluate the intrinsic vulnerability of mesh-connected NoCs based on the computation of average path length for all the packets traversing the network with a specific routing algorithm and traffic model. The main investigated routing algorithms are XY and XY-YX. But, this analysis is extensible to other topologies and adaptive algorithms. The used traffic patterns are the uniform, transpose 1 and 2 and hot-spot traffics. This analysis is performed for permanent faults that lead to the permanent failures in the switches, the links and the cores of MP-SoCs.

The rest of the paper is organized as follows. In Section II, some preliminaries needed for the analyses are presented.

The analytical reliability models for mesh NoCs are described in Section III. In Section IV, some extensions to analytical models and in Section V the evaluation of the models and inherent reliability is discussed based on analytical and simulation results. Finally, some concluding remarks are drawn in Section VI.

II. BACKGROUND

A 4×4 mesh NoC and some simple paths are depicted in Fig. 1. The proposed analytical models are based on the following assumptions:
1) Each node consists of a switch or router, a core and a network interface.
2) For simplicity, the failure rates of the links are assumed the same. This applies to the failure rates of the switches and the failure rates of the cores.
3) There exist enough buffers in the input ports of the switches so no packet is dropped due to congestion.

To compare the NoCs with different sizes, routing algorith--ms and traffic models, we define the following parameters:

Path length: It's defined as the number of hop-by-hop links that exist on a path between a source and a destination node.

Path Reliability (PR): It's the reliability of a path that a packet traverses, and its failure rate is a function of the failure rates of the network components in the path.

Packet Completion Probability (PCP): It's defined as the number of received packets divided by the total number of injected packets into the network, and is the complement of *Packet Drop Probability (PDP)*.

In XY routing algorithm, there is one path between a source and a destination. A packet is first routed in the X direction and then in the Y direction to reach the destination. But, in XY-YX routing there are two paths for most of source-destination pairs. Thus, if a path fails, the packets are sent from another path, if it exists, to the destination.

III. ANALYTICAL RELIABILITY MODELS

To perform analytical reliability assessment for mesh-connected NoCs, we use the average values of parameters mentioned in Section II. Therefore, these parameters can be computed if, for a specific traffic model and network size, we compute the *Average Path Length* of all the packets:

Average Path Length (APL) is defined as the sum of all shortest paths lengths between any source and destination nodes according to the traffic model, divided by the total number of paths. It should be noted that for any source and

978-1-4244-2541-9/08 $25.00 © 2008 IEEE

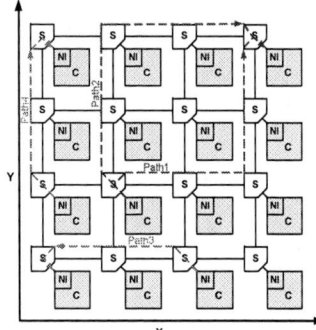

Figure 1. A 4×4 mesh NoC with some simple paths

destination pairs only one minimal path is considered and the *APL* is independent of the routing algorithms.

Assume the failure rates of the switches, links and cores equal λ_s, λ_l and λ_c, respectively, then the reliability functions of these components are stated as $R_S(t)$, $R_L(t)$ and $R_C(t)$ in which $R_i(t) = e^{-\lambda_i t}$. We can use the multiplication of $R_{PE}(t)$ and $R_{NI}(t)$ as $R_C(t)$ in which *PE* and *NI* stand for the processor element and the network interface, respectively.

Lemma 1: *The average Path Reliability for all the packets traversing the network is computed by the general equation:*

$$PR(t) = R_S^{(APL+1)}(t) \cdot R_L^{APL}(t) \cdot R_C^2(t) \qquad (1)$$

Proof: To correctly send a packet to the destination, the cores and switches in the source and destination nodes must be healthy, in addition to the middle links and switches. The average number of links and switches in a path are *APL* and *APL+1*, respectively. Thus, (1) is valid.

In (1), it's assumed that the reliability functions are constant in period of time in which a packet is transmitted from a source to a destination.

A. Uniform Traffic

In this traffic model, a core sends the packets to any other cores in the network with equal probability. The *APL* parameter for this traffic model is computed using (2):

$$APL = \sum_{i=1}^{MN(MN-1)} L(Path_i)/[MN(MN-1)] = (M+N)/3 \qquad (2)$$

where *M, N* are the number of rows and columns in the mesh and $L(Path_i)$ is the length of *i*th path. The term *MN(MN-1)* is the total number of paths in this mesh with uniform traffic model. The average path reliability for XY routing is computed by (1) using (2). In XY-YX routing, if the source and destination are in the same row or column, there is one path between them that called *1-way path* (Path3, Path4 in Fig. 1), otherwise there are two paths called *2-way path* (Path1, Path2 in Fig. 1). Thus, the average path reliability for this routing is computed by (3):

$$PR_{XY-YX} = \frac{M+N-2}{MN-1} \cdot PR_{1-way} + \frac{(M-1)(N-1)}{MN-1} \cdot PR_{2-way} \qquad (3)$$

In equation above, the factors of the first and second terms are fractions of the total paths that are 1-way and 2-way paths, respectively. Hereafter, we only consider $N \times N$

meshes to obtain simple equations. Thus, PR_{1-way} in (3) is obtained using (1) by replacing *APL* by APL_{1-way} in which APL_{1-way} is the average path length of 1-way paths in the mesh with uniform traffic and is computed by (4):

$$APL_{1-way} = \sum_{i=1}^{2N^2(N-1)} L(Path_{1-way_i})/[2N^2(N-1)] = (N+1)/3 \qquad (4)$$

PR_{2-way} is computed by (5):

$$PR_{2-way} = R_S^2 . R_C^2 \left[1 - (1 - R_S^{(APL_{2-way}-1)} . R_L^{APL_{2-way}})^2 \right] \qquad (5)$$

where APL_{2-way} is obtained using (6):

$$APL_{2-way} = \frac{N+1}{N-1} \cdot APL - \frac{2}{N-1} \cdot APL_{1-way} = \frac{2}{3}(N+1) \qquad (6)$$

Since $R_i(t)$ is not greater than one, based on (1) and (3), when the mesh size increases *PCP* is decreased.

One component failure:

We can analyze the NoC reliability from another perspective. To do so, we compute the *PDP*s when one or more permanent failures occur in the NoC architecture. For example, when a link failed in an $N \times N$ mesh, the average probability that a packet does not reach the destination with XY routing equals $PDP_{XY,1-L} = APL/[2N(N-1)]$ in which the denominator is the number of total links. Similarly, we have $PDP_{XY,1-S} = (APL+1)/N^2$ for a switch failure and $PDP_{XY,1-C} = 2/N^2$ for a core failure. For XY-YX routing the following equations are used:

$$PDP_{XY-YX,1-L} = \frac{2}{N+1} \cdot \frac{APL_{1-way}}{2N(N-1)} = \frac{1}{3N(N-1)} \qquad (7)$$

$$PDP_{XY-YX,1-S} = \frac{2}{N+1} \cdot \frac{APL_{1-way}+1}{N^2} + \frac{N-1}{N+1} \cdot \frac{2}{N^2} = \frac{2(4N+1)}{3N^2(N+1)} \qquad (8)$$

In (7) and (8) the factors *2/(N+1)* and *(N-1)/(N+1)* are fractions of the total paths that are 1-way and 2-way paths, respectively. Equation (7) is obtained through the fact that one link failure can only stop the packets on 1-way paths; but for one switch failure this can occur in the source and destination nodes of 2-way paths extra to the 1-way paths.

Two component failures:

For two link and two switch failures, using the probability computations, we estimate the *PDP*s as follows:

$$PDP_{XY,2-L} = 2PDP_{XY,1-L} - PDP_{XY,1-L}^2 \qquad (9)$$

$$PDP_{XY,2-S} = 2PDP_{XY,1-S} - PDP_{XY,1-S}^2 \qquad (10)$$

$$PDP_{XY-YX,2-L} = \frac{2}{N+1}(2q_1 - q_1^2) + \frac{N-1}{N+1} \cdot 2q_2^2 + 2 \times \frac{2}{N+1} q_1 \times \frac{N-1}{N+1} q_2 \quad (11)$$

$$PDP_{XY-YX,2-S} = \frac{2}{N+1}(2q_3 - q_3^2) + \frac{N-1}{N+1}(2q_4^2 + q_5) + 2 \times \frac{2}{N+1} q_3 \times \frac{N-1}{N+1} q_4 \quad (12)$$

In (11), q_1 equals $APL_{1-way}/[2N(N-1)]$ and q_2 equals $APL_{2-way}/[2N(N-1)]$. The q_1 stands for failure probability of 1-way paths because of one link failure and q_2 stands for failure probability of one of the two ways in 2-way paths when one link fails. In (12), q_3 and q_4 are equal to $(APL_{1-way}+1)/N^2$ and $(APL_{2-way}-1)/N^2$, respectively, and q5 which stands for failure probability of the source or destination switches in a 2-way path equals $4/N^2 - 1/N^4$.

B. Transpose Traffic

In the first and second transpose traffic patterns, a core (i, j) in the $N \times N$ mesh network sends the packets only to core $(N\text{-}1\text{-}j, N\text{-}1\text{-}i)$ and core (j, i), respectively. The *APL* parameter for these traffic models is computed using (13):

$$APL_{tp1} = \sum_{i=1}^{N^2-N} L(Path_i)/(N^2 - N) = 2/3.(N+1) \quad (13)$$

The *PR* and *PDP* parameters for these traffic models are computed similar to the uniform traffic model. But for XY-YX routing algorithm, *PDP* for one link failure is zero since there is no 1-way path in the transpose traffic models.

C. Hot-spot Traffic

In the hot-spot traffic, one or more cores called hot-spot receive more traffic (*E* extra packets) plus to the uniform traffic. The percentage of extra packets equals $E \times 100$. The derived formulas for the uniform traffic are also applicable to the hot-spot traffic, but *APL* used here is different and is computed by (14) when *M* hot-spot nodes exist in the mesh:

$$APL_{hs} = \frac{(N^2-1)\sum_{i=1}^{M} E_i \cdot APL_{rel_i} + N^2(N^2-1) \cdot APL_{uni}}{(N^2-1)(N^2+\sum_{i=1}^{M} E_i)} \quad (14)$$

In this equation, the *i*th APL_{rel} is the average path length of all the extra packets that enter the *i*th hot-spot node, which is computed based on its location, e.g. if each of four adjacent nodes of the mesh center are hot-spot nodes, then APL_{rel} equals $N^3/[2(N^2-1)]$ and $(N^3+N)/[2(N^2-1)]$ for even and odd mesh sizes, respectively.

To compute *PDP* for XY routing APL_{hs} is used, but for XY-YX routing we need to compute $APL_{1\text{-}way,hs}$ and $APL_{2\text{-}way,hs}$. The former can be computed similar to (14), but the latter can be computed using (6) when APL_{hs} and $APL_{1\text{-}way,hs}$ were computed.

IV. ANALYTICAL MODELS EXTENSION

A. Extension to Other Routing Algorithms

We can extend the reliability analysis to adaptive routing algorithms. These routing algorithms are still vulnerable to only one permanent failure when the source and destination nodes have the same row or column. In [5] three partially adaptive routing algorithms are introduced; west-first, north-last and negative-first. All of these algorithms, in average, use the fully adaptive paths in half of the paths and use one path in the next half. Since all the minimal routing algorithms suffer from the unreliability of 1-way paths, we can compute the minimum *PDP* for these algorithms based on PDP_{XY} and $PDP_{XY\text{-}YX}$, e.g. in west-first routing if x coordinate of destination is not less than x coordinate of source, the algorithm can use one of the total shortest paths between the source and destination otherwise, it can only use one shortest path between them. Thus, the minimum *PDP* in west-first routing algorithm for one link or switch failure is derived by (15) when the uniform traffic is used:

$$PDP_{min} = [\frac{1}{2} - \frac{1}{2(N+1)}] \cdot U + [\frac{1}{2} + \frac{1}{2(N+1)}] \cdot V \quad (15)$$

In (15) *U* stands for PDP_{XY} for one link or switch failure and *V* stands for $PDP_{XY\text{-}YX}$ according to (7) and (8). The coefficients are the average fractions of nodes in each half. However, for more failures these partially adaptive routing algorithms have less *PDP* than that of XY-YX routing. The extensions for the other routing algorithms with different traffic models can similarly be derived.

Since even the minimal fully adaptive routing algorithms have vulnerability against one failure, design of non-minimal adaptive routing algorithms is required to obtain more fault-tolerant routings. In the non-minimal and fault-tolerant routing algorithms although many failures cannot lead to a packet drop, but they can increase the latency at least for the packets traversing the 1-way paths. The latency is defined as the total time needed to convey a packet from a source to a destination and is the sum of the time needed to transmit a packet over the links, the time to send a packet from the core to the switch in the source node and from the switch to the core in the destination node, and the average waiting time (*AWT*) in the nodes multiplied by the number of nodes in the path. The average number of nodes and links in the path equal *APL+1* and *APL*, respectively. The packet length in flits (*Pkt_len*) minus one is added to the total latency. Therefore, if we suppose sending a packet from the core to the switch and vice versa each takes one clock cycle and we use the wormhole switching, we can compute the latency for a packet in cycles by (16):

$$Latency = AWT.(APL+1) + APL + Pkt\_len + 1 \quad (16)$$

On the other hand, in the non-minimal and fault-tolerant routing algorithms each link or switch failure in a 1-way path increases both the path length and the number of nodes in the path at least by two (Fig. 2). Thus, the minimum latency overhead because of one failure equals (*2.AWT+2*) cycles for the 1-way paths where the failure occurred.

B. Analysis of an Improvement Method

In [6] a decoupled router architecture with row-column switch is introduced in which for some permanent faults, the switch bypasses the incoming packets in the same direction. A similar work is presented in [7] in which the bypass is performed in a mode named through traffic or bypass mode using a wrapper. When a switch acts in the bypass mode, the local core and the cores that are accessible by turning in this switch are not reachable. Assuming XY routing algorithm and the uniform traffic, the maximum amount of *PDP* when a switch acts in the bypass mode is computed by (17):

$$PDP_{XY} = \frac{(N-1)^2 L_1 + 2(N^2-1)L_2}{N^2(N^2-1) \cdot APL_{uni}} = \frac{3N+1}{N^2(N+1)} \quad (17)$$

In (17), the 1st term in the numerator represents the number of turning paths in the switch multiplied by their average path length and the 2nd term represents the number of paths in which the local core acts as a source or destination multi-

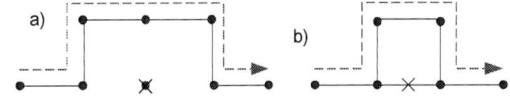

Figure 2. Effect of a) a switch and b) a link failure on the 1-way paths

-plied by their average path length. In average, L_1 and L_2 are equal to APL_{uni}. The simulation results show noticeable improvement when a switch acts in the bypass mode.

V. EVALUATION AND DISCUSSION

To verify the correctness of the analytical assessments, the appropriate simulations were conducted using C++ language. In all these experiments the buffer size is set to infinite, thus the packet injection rate and the packet length (except in Fig. 6) are not important. For simplicity the packet length is set to one flit except in Fig. 6 in which it is set to four flits. Fig. 3 represents the packet completion probabilities when one or two link failures occur in $N \times N$ meshes. Fig. 3a represents the results for XY and XY-YX routing algorithms with the uniform traffic, but Fig. 3b represents the results for the hot-spot traffic with the extra packets E equal to one when all of four adjacent nodes of the mesh center are hot-spot nodes. Fig. 4 shows the similar results for switch failures. In these figures the analytical results are exactly identical or very close to the simulation results. As depicted in Fig. 3b and Fig. 4b, when four hot-spot nodes are placed around the center, $PCPs$ become a little greater than the ones with the uniform traffic, but the effect of hot-spot nodes on $PCPs$ is too small.

To validate (17) Fig. 5 presents the results for XY routing with the uniform traffic when a switch fails. As shown, the analytical and simulation results are identical and using the bypass switches can greatly enhance the system reliability. In Fig. 6 the effect of some link failures on the average latency of the packets traversing the 1-way paths in a 5×5 NoC is presented when a non-minimal and fault-tolerant routing algorithm is used based on Fig. 2. The results are analytical and obtained with this assumption that at most one failure occurred in a 1-way path. The normal

latencies for average waiting times of 1, 2 and 3 when no failure occurs are 10, 13 and 16 clock cycles, respectively.

VI. CONCLUSION

In this paper using the probability models and defining some beneficial parameters, the inherent reliability of mesh-connected NoCs with different routing algorithms and traffic patterns is evaluated. This analysis can be groundwork for reliability assessment of more realistic NoCs to find the inherent reliability and design of more reliable ones. This can be performed by also considering parameters such as buffer size, packet size and latency to obtain more precise analytical reliability models.

REFERENCES

[1] L. Benini and G. D. Micheli, "Networks on Chips: A New SoC Paradigm", *IEEE Computer, Vol. 35, pp. 70-78*, Jan. 2002.

[2] M. Mondal, X. Wu, A. Aziz and Y. Massoud, "Reliability Analysis for On-chip Networks under RC Interconnect Delay Variation", *Proceedings of the 1st International Nano-Networks and Workshops, pp. 1-5*, Sep. 2006.

[3] T. Lehtonen, P. Liljeberg and J. Plosila, "Fault Tolerance Analysis of NoC Architectures", *Proceedings of IEEE International Symposium on Circuits and Systems (ISCAS), pp. 361-364*, May 2007.

[4] A. Dalirsani, M. Hosseinabady and Z. Navabi, "An Analytical Model for Reliability Evaluation of NoC Architectures", *Proceedings of IEEE International On-Line Testing Symposium, pp. 49-56*, Jul. 2007.

[5] C. J. Glass and L. M. Ni, "The Turn Model for Adaptive Routing", *Journal of ACM, Vol. 41, No. 5, pp. 874-902*, Sep. 1994.

[6] J. Kim et al., "A Gracefully Degrading and Energy-Efficient Modular Router Architecture for On-Chip Networks", *Proceedings of International Symposium on Computer Architecture, pp. 4-15*, Jun. 2006.

[7] D. Greenfield, A. Banerjee, J.-G. Lee and S. Moore, "Implications of Rent's Rule for NoC Design and Its Fault-Tolerance", *Proceedings of the 1st International Symposium on Networks-on-Chip (NOCS), pp. 283-294*, May 2007.

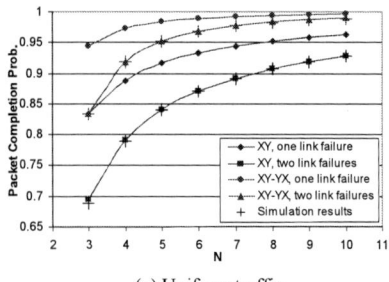
(a) Uniform traffic

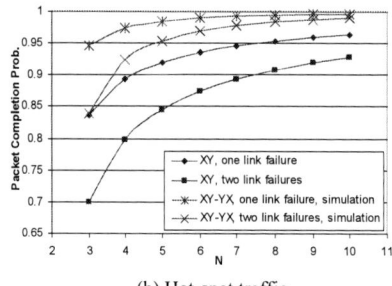
(b) Hot-spot traffic

Figure 3. The reliability as a function of mesh size and different number of link failures

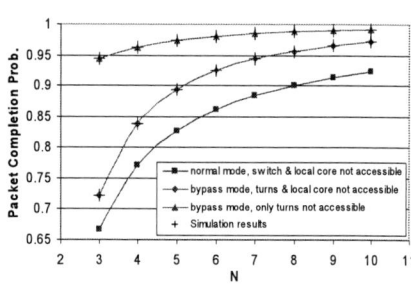
Figure 5. Effect of a bypass switch on reliability

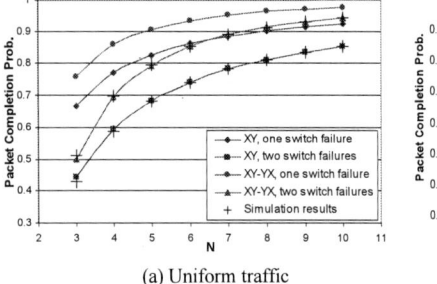
(a) Uniform traffic

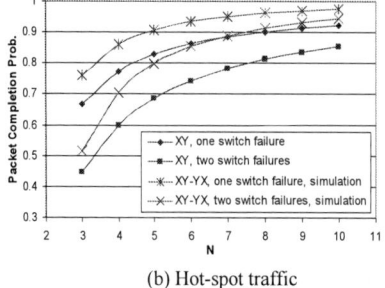
(b) Hot-spot traffic

Figure 4. The reliability as a function of mesh size and different number of switch failures

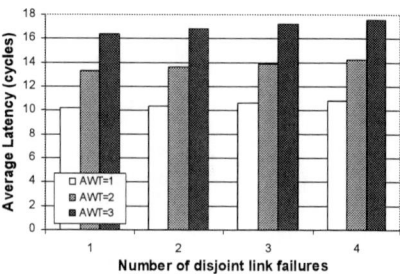
Figure 6. Effect of link failures and AWT on latency

978-1-4244-2541-9/08 $25.00 © 2008 IEEE

Implementation of W-CDMA Slot Synchronization on a Reconfigurable System-on-Chip

Fabio Garzia*, Claudio Brunelli*†, Carmelo Giliberto*, Roberto Airoldi‡, Jari Nurmi*

*Department of Computer Systems, Tampere University of Technology, Tampere, Finland

†Nokia Research Center, Tampere, Finland

‡ARCES, University of Bologna, Bologna, Italy

Abstract— This paper describes the implementation of the slot synchronization of a W-CDMA receiver on a reconfigurable system. The system includes a general-purpose processor core with floating-point capabilities and a reconfigurable array.
We mapped a 256-element correlation on the array and we evaluated its performance. The slot synchronization uses a large number of this correlations.
The stand-alone correlation produces a speed-up of 70X when mapped on the reconfigurable core in comparison with the software implementation on a general-purpose RISC core. The slot synchronization based on this implementation gives a speed-up of 33X against an area overhead of 4X.

I. INTRODUCTION

In the last years wireless communication research focused on multistandard systems [1]. These systems are able to access several wireless networks with different protocols from the same terminal. This approach is very appealing for the mobile phone developers, since a mobile phone can simultaneously access second and third generation cellular networks, as well as wireless local area networks (WLAN). The main advantage in this interworking is that the device can get the high data throughput typical of a WLAN and the mobility provided by the cellular networks. In other words higher data rate services with improved service quality and capacity can be provided to the mobile phone end-users.

This kind of multistandard systems can be implemented using a Software Defined Radio (SDR) approach. The main idea is that the receiver is based on a Digital Signal Processor (DSP). The signal received from the multiband antenna is band-pass filtered, amplified and then converted to a digital signal that will be processed by the DSP. This system can easily adapt to any network protocol only modifying the software running on the DSP.

Our department developed a programmable baseband receiver suitable for SDR systems. The details of the research work as well as the results obtained are in [2]. The main approach of that research work was to couple some application-specific coprocessor accelerators to a RISC processor. The radio technologies considered in that research work were *wideband code division multiple access* (W-CDMA [3]) and *orthogonal frequency division multiplexing* (OFDM [4]). The RISC processor has been developed in our group and details about its architecture can be found in [5] and [6].

The previous research work did not consider the usage of reconfigurable hardware to implement part of the receiver.

This paper evaluates the possibility to replace the dedicated accelerators with a reconfigurable array coprocessor developed in our group. Details about the basic architecture of the accelerator are given in [7]. The architecture has been further enhanced adding floating-point (FP) capabilities [8] and subword computation [9]. Also the general-purpose core processor used in this new system has been extended adding FP support. The architecture of the entire system with the RISC FP core and the reconfigurable array is described in [10]. The array is directly connected with two local memories. The first memory feeds at once one row of the array, the second one stores the results from the last row. It is possible to exchange the role of the two memories, so that the array can start processing the data from the second memory and store the results on the first.

In this paper we focused on the W-CDMA radio technology. The timing synchronization of the W-CDMA is composed of the cell search and the multipath delay estimation. The cell search includes slot timing synchronization, frame synchronization and scrambling code group identification. All these operations are described in detail in [11]. In particular, in the slot synchronization the application calculates several correlation over 256 complex elements. Due to the SIMD nature of the correlation kernel, it looks very convenient to implement it on our reconfigurable array. Since it is the first part of the W-CDMA receiver and it can get a significant speed-up from the implementation on the reconfigurable array, we decided to map the slot synchronization on our system. The general purpose processor core controls the global execution and performs some small operations, while the reconfigurable array calculates all the correlations.

This paper is organized as follows. In the next section we describe the correlation kernel from an algorithmic point of view. Details about the mapping and routing of the kernel on the reconfigurable array are in section III. In the last section we present our results related with the execution of the slot synchronization on our system in comparison with a pure software approach.

II. THE CORRELATION KERNEL

The correlation algorithm is performed on the sequence of input samples. The sequence of samples can be seen as a complex array of an indefinite number of elements. The correlation is performed every time on a window of 256 elements of this input array. The first window consider the first 256 samples received, then the following window is just shifted by one. In the beginning of the slot synchronization (the peak search) an unspecified number of correlations is calculated, until one of them is larger than a fixed threshold that identifies the peak.

The correlation on the 256-element window is implemented using a FIR algorithm. This means that we first multiply the 256-element array for an array of coefficients given *a priori*, then we add all these minterms together.

If R is the complex array corresponding to the current window, and C is the complex array of 256 coefficients, the formula to calculate the correlation is:

$$F = \sum_{i=0}^{255} F_i = \sum_{i=0}^{255} R_i \cdot C_i.$$

Each multiplication is a complex one. Defining as R_{Ri} and C_{Ri} the real part of element i, and as R_{Ii} and C_{Ii} the imaginary part, the real part of the product is

$$F_{Ri} = R_{Ri} \cdot C_{Ri} - R_{Ii} \cdot C_{Ii},$$

and the imaginary part

$$F_{Ii} = R_{Ii} \cdot C_{Ri} + R_{Ri} \cdot C_{Ii}.$$

A last consideration is needed to fully understand the implementation on the reconfigurable array. The application under study requires a sequence of correlations on a window of 256 elements. After each correlation the window is shifted by one element. This means that the second correlation is

$$F^2 = \sum_{i=0}^{255} R_{i+1} \cdot C_i,$$

and the n-th is

$$F^n = \sum_{i=0}^{255} R_{i+n-1} \cdot C_i.$$

III. MAPPING OF THE CORRELATION FUNCTIONALITY ON THE RECONFIGURABLE ARRAY

The implementation of the kernel described in Section II requires three contexts of configuration.
Two contexts are used for the actual calculation. One context is used to reset the accumulation every time the window is shifted.
In order to understand how these contexts work, we need first to describe how the inputs samples and the coefficients are organized in the local I/O memories.

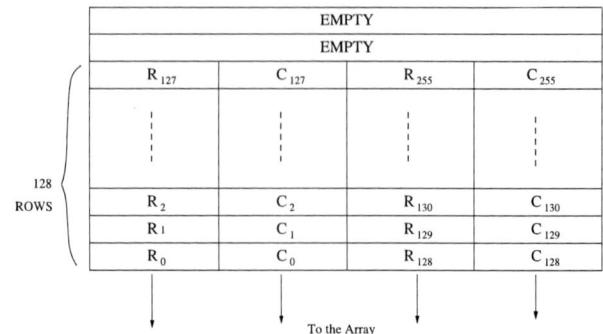

Fig. 1. Organization of the samples and coefficients in the input memory

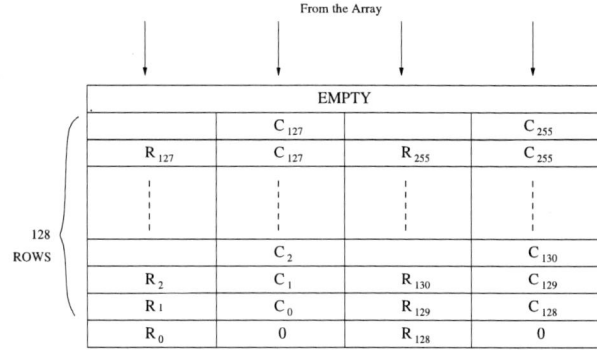

Fig. 2. Organization of samples and coefficients in the output memory after the processing of the first window.

A. Organization of the input samples and coefficients into the local I/O memories.

Fig. 1 depicts the organization of the input memory when the correlation calculation starts.

To exploit all the parallelism provided by the reconfigurable array we split the sequence of input samples and coefficients into two equal parts, 128 elements each. We feed the array with two samples and two coefficients per clock cycle. After 128 cycles we get the two partial results related with the two parts of the calculation. The general-purpose core fetches this two results and performs the last sum.
The interesting thing is that we can perform a new correlation on the shifted window with little overhead. During its processing the array sends to the local output memory the same input operands (samples and coefficients) delaying by one cycle the coefficient writing. After the processing of the first window, the output memory holds the input data as shown in Fig. 2.

As you can notice the coefficients are already correctly aligned with the samples of the new window: coefficient C_0 has to be multiplied by the sample R_1.
Some small adjustments are required. The sample R_0 should be removed from the memory, and this is done writing a 0 in that location. The sample R_{128} should be removed from its position and put to the top of the first column, while a new sample (R_{256}) should be added on top of the second column. These four single writings are performed by the

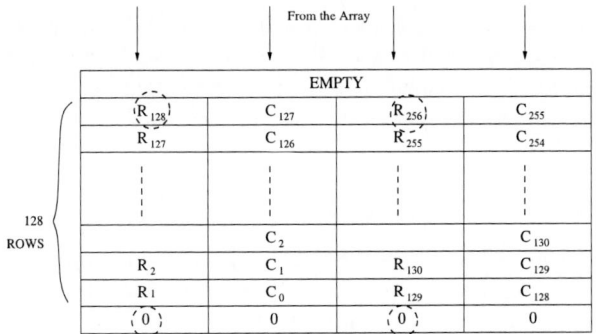

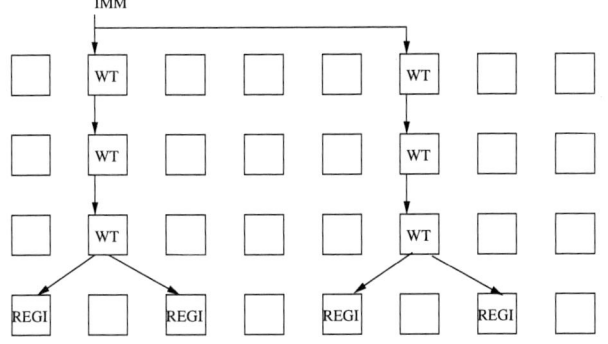

Fig. 3. Organization of samples and coefficients in the memory after the writings performed by the general-purpose core (indicated by the circles).

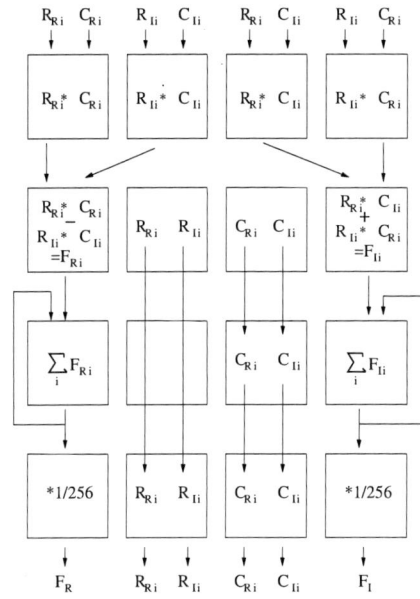

Fig. 4. First context: set of the immediate value in the PEs of the last row.

general-purpose core. The result in the memory organization is depicted in Fig. 3.

At this point we can perform the second 256-element correlation just swapping the I/O memories.

B. Mapping and routing on the Reconfigurable Array

Fig. 4 shows the first context used. Its purpose is only to put an immediate value in the odd PEs of the last row. This immediate will be used by the next context, that performs the actual calculation.

The first three PEs of the second and sixth columns are configured to write the input data into the output port with a one cycle delay (*write through* functionality, WT). In the last row the PEs configured with the *write immediate* functionality (REGI) sample the input data putting it into their immediate register.

The second context performs two complex multiplications in parallel and accumulates the result in one of its PEs. Mapping and routing are depicted in Fig. 5, while Fig. 6 shows in detail the functionality of each cell considering only one half of the array (one complex multiplication). We use the same notation as in Section II. In the first row we perform four FP products as required in the complex product (FPMUL). In the second row we get the real part and the imaginary part of the result using a subtraction (FPSUB) and a sum (FPADD) between the minterms. The central PEs just move ahead the input operands (WT). In the third row two PEs perform an

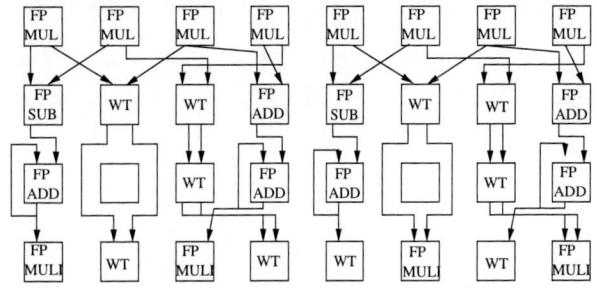

Fig. 5. Second context: two complex products and accumulation of results; final normalization.

Fig. 6. Functional description of the second context in one half of the array.

accumulation of the two parts of the result. The accumulation is realized with a floating-point sum (FPSUM) and a loop connection, that sends the output result back into the input register of the cell. The third PE performs an additional write-through of the coefficients. The samples are instead sent directly to the last row using an interleaved interconnection. This provides a one cycle delay between the storage of the samples and coefficients in the output memory, getting the memory organization depicted in Fig. 2. In the last row the final result of the processing (the correlation on 128 elements) is multiplied by the immediate 1/256 (FPMULI). This is the value stored by the first context and reproduces a normalization of the final result over the 256 elements.

The final context is used to reset the content of the accumulation register between two consecutive correlations. The PEs of the first and fifth columns are set for a write-through (WT) and a 0 is injected in the array in order to reset all the operand registers of those two columns (Fig. 7).

978-1-4244-2541-9/08 $25.00 © 2008 IEEE

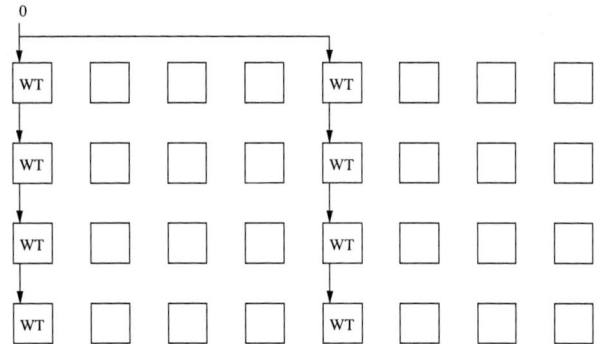

Fig. 7. Third context: reset of the accumulator register.

IV. IMPLEMENTATION RESULTS

The performance of our implementation is evaluated in comparison with a pure software approach. We evaluate separately the performance of the 256-element correlation and the performance of the slot synchronization using the kernel mapped on the reconfigurable array.

Table I shows the synthesis data for the reconfigurable array and the RISC core with floating-point capabilities on an Altera EP2S180 FPGA device.

TABLE I

SYNTHESIS RESULTS ON AN ALTERA EP2S180 FPGA DEVICE

	RISC FP Core	Reconfigurable Array
Logic Utilization	15%	47%
Combinational Adapt. LUTs	20285	70024
Adaptive Logic Modules	13181	44111
Dedicated Logic Registers	4466	8620
DSP Blocks	24	384

Considering the stand-alone correlation kernel the general-purpose core requires 17427 clock cycles, while the reconfigurable one only 251 cycles. In this calculation we are assuming that the array of samples and coefficients are already loaded into the local memory of the reconfigurable device, and the array is already configured. Thus the speed-up for the single correlation kernel is 70X. The implementation is convenient compared with the software one, since the area overhead is only 3X.

Some other figures can be obtained considering the whole slot synchronization that requires the computation of a large number of correlations. In this case we must use also the general-purpose core to perform the global control and some minor operations. In terms of area the overhead is between the entire system without the reconfigurable array and the system with it. Since the peripherals have a very low impact the area overhead is 4X (GP processor versus GP processor plus reconfigurable array). On the other hand the memory footprint of the two implementations is similar and is equal to $12KB$. Considering the performance the software implementation requires 93327993 clock cycles, while the reconfigurable one only 2779168 cycles. The speed-up is equal to 33X.

In this case we consider also the overhead related with the configuration and all the data transfers. Nevertheless the high speed-up is guaranteed by the massive usage of the local memories, that reduces significantly the number of transfers back and forth from the system memory to the array.

V. CONCLUSION

This paper focused on the implementation of the slot synchronization, part of a software W-CDMA software, on a system based on a general-purpose processor core with FP capabilities and a reconfigurable array. In particular we mapped a 256-element correlation on the reconfigurable array and used this kernel in the execution of the whole slot synchronization.

The single correlation kernel guaranteed a speed-up of 70X. The slot synchronization gave a speed-up of 33X, if compared with a pure software implementation, against an area overhead of 4X.

ACKNOWLEDGMENT

This work was partially supported by Nokia Foundation, which is gratefully acknowledged.

REFERENCES

[1] L. Harju and J. Nurmi, "Hardware platform for software-defined wcdma/ofdm baseband receiver implementation," *Computers & Digital Techniques, IET,* vol. 1, no. 5, pp. 640–652, September 2007.

[2] L. Harju, "Programmable receiver architectures for multimode mobile terminals." Ph.D. dissertation, Tampere University of Technology (TUT), Department of Information Technology, Institute of Digital and Computer Systems (IDCS), Tampere, Finland, August 2006, 160 pages, TUT Publication 604, ISSN: 1459-2045, ISBN: 952-15-1618-6.

[3] E. Dahlman, P. Beming, J. Knutsson, F. Ovesjo, M. Persson, and C. Roobol, "Wcdma-the radio interface for future mobile multimedia communications," *Vehicular Technology, IEEE Transactions on,* vol. 47, no. 4, pp. 1105–1118, November 1998.

[4] X. Wang, "Ofdm and its application to 4g," in *Proc. International Conference on Wireless and Optical Communications 14th Annual WOCC 2005,* 22–23 April 2005, p. 69.

[5] J. Kylliinen, T. Ahonen, and J. Nurmi, "General-purpose embedded processor cores - the COFFEE RISC example," in *Processor Design: System-on-Chip Computing for ASICs and FPGAs,* J. Nurmi, Ed. Kluwer Academic Publishers / Springer Publishers, June 2007, ch. 5, pp. 83–100, ISBN-10: 1402055293, ISBN-13: 978-1-4020-5529-4.

[6] "Coffee core project website," consulted 21 July 2008, URL: http://coffee.tut.fi.

[7] C. Brunelli, F. Cinelli, D. Rossi, and J. Nurmi, "A vhdl model and implementation of a coarse-grain reconfigurable coprocessor for a risc core," in *Proceedings of the 2006 International Conference on Ph.D. Research in Microelectronics and Electronics (PRIME '06).* IEEE, 11-15 June 2006, pp. 229–232, ISBN: 1-4244-0156-9 (printed), 1-4244-0157-7 (PDF).

[8] C. Brunelli, F. Garzia, and J. Nurmi, "A coarse-grain reconfigurable machine with floating-point arithmetic capabilities." in *Proceedings of the 2nd International Workshop on Reconfigurable Communication-centric Systems-on-Chip (ReCoSoC '06).* Univ. Montpellier II, July 2006, pp. 1–7, ISBN: 2-9517461-2-1.

[9] ——, "A coarse-grain reconfigurable architecture for multimedia applications featuring subword computation capabilities," *Real-Time Image Processing,* vol. 3, no. 1-2, pp. 21–32, 2008.

[10] F. Garzia, C. Brunelli, D. Rossi, and J. Nurmi, "Implementation of a floating-point matrix-vector multiplication on a reconfigurable architecture," in *Proceedings of the 15th Reconfigurable Architecture Workshop (RAW '08).* IEEE, 14-15 April 2008, pp. 1–6.

[11] Y.-P. Wang and T. Ottosson, "Cell search in w-cdma," *IEEE J. Select. Areas Commun.,* vol. 18, no. 8, pp. 1470–1482, Aug. 2000.

FlexPath NP - A Network Processor Architecture with Flexible Processing Paths

Michael Meitinger, Rainer Ohlendorf, Thomas Wild, Andreas Herkersdorf

Institute for Integrated Systems
Technische Universität München
Munich, Germany
Michael.Meitinger@tum.de

Abstract—In this paper we present a FlexPath Network Processor implementation, a platform with flexible, reconfigurable processing paths for packet processing. The path decision is made in hardware based on a packet's network application. Packets may be processed by a CPU or even completely in hardware. With our demonstrator the performance of different processing paths is shown for a scenario with simple IPv4 forwarding traffic mixed with IPSec packets. We show that flexible path selection significantly improves the system's performance. The specific FlexPath concepts are also applicable to other NP architectures.

I. INTRODUCTION

Rising bandwidths and increasingly ambitious processing requirements for packets demand more and more processing power that can not be reached with General Purpose CPUs. Hardware support for packet processing specific tasks in Network Processors (NP) helps to reach the needed throughput while preserving flexibility. This is important for future protocol adoptions and a high level of adaptability to different operational environments.

NPs can be divided in more or less two big classes. On the one hand, there are processor-centric architectures with a CPU-cluster as central element. All packets are processed by the cluster, often with additional hardware that can be accessed for special tasks. Well known commercial examples are the Intel IXP [1] or the AMCC nP series [2]. These architectures are quite flexible, since the function is mainly achieved in software. In such Symmetric Multiprocessor Systems (SMP) architectures the packet is stored in a central memory, memory bandwidth and the system interconnect are therefore critical points.

Another problem of parallel working processor-centric NPs is the distribution of packets to the single processing elements (PE). To keep the packet order per flow, all packets of one flow are usually processed by the same PE. This is usually done by hashing the IP 5-tuple, using this value as a processor index (see [3], [4], [5]). There are several attempts to switch flows in overload situations of single PEs in order

to optimize the system load, since highly active flows can produce a very high load to single CPUs. A more balanced load situation would be possible with spraying, i.e. packets are processed by the next free PE, ignoring a flow-PE-assignment. This may lead to packet reordering (packets of one flow are sent out not in the original order). Since packet reordering can have a significant degradation in the network performance for TCP (see [6]), spraying is not common.

NPs with a processing pipeline, on the other hand, try to address the memory and interconnect problem. Since the data travels through the pipeline there is no central storage that may cause a bottleneck. Of course a more complicated programming and a lack in flexibility are the price. The pipeline may consist of pipelined processors, e.g. [7]. There are already first products available with a full hardware pipeline built up on a FPGA [8]. Another hardware dominant approach is the Inline Security Processor SafeXcel-5160 by SafeNet [9]. This application-specific Processor fully offloads data plane processing to a dedicated hardware pipeline. The function of the embedded CPU is limited to connection setup and teardown and other management tasks.

Our FlexPath concept, which was already described in [10], is based on a processor-centric SMP-architecture amended with additional hardware in the NP's ingress and egress paths. Packets are processed on different processing paths. In contrast to related work, the path decision depends on the network application that has to be performed. While keeping flexibility in the CPUs, a full hardware pipeline path for simple and straightforward IP processing (e.g. IPv4 forwarding) boosts performance and decreases latency for the bulk of packets. The packet sequence is preserved for all flows within our NP. In contrast to previous publications we now have all FlexPath specific hardware support entities implemented and integrated with two PowerPCs in a single FPGA.

The rest of the paper is structured as follows: in section 2 we present our FlexPath concept as well as the FPGA implementation. In section 3, we introduce the demonstration setup and measurement results, followed by a conclusion and an outlook.

We would like to thank the German Research Foundation (DFG) for co-funding this work within the SPP1148.

978-1-4244-2541-9/08 $25.00 © 2008 IEEE

II. FLEXPATH NP

A. Idea and concept

The basic idea of FlexPath is a Network Processor (NP) with flexible and reconfigurable processing paths as shown in Fig. 1. Flexible means that the way of the packet through the system is not fixed, but may vary for each packet depending mainly on the applications to be performed. Reconfigurable means that the path decision, the so called rule base, is not fixed during runtime, but may be changed for various reasons.

This path decision which is made in hardware is the main difference to most common Network Processors, where the packet is always sent to a CPU for processing and handed over to other CPUs or accelerators, if needed.

The FlexPath NP combines the dedicated assignment of packets to CPUs – which is the usual way of packet distribution – with the possibility of spraying. With spraying all packets are distributed evenly among all or at least a part of all available CPUs. Packets are just served by the next free CPU without any preference. This creates a very well balanced load, since all CPUs will process more or less the same amount of packets. The overall system performance is high. Nevertheless, there are some situations where spraying creates problems, e.g. when processing stateful flow connections. Stateful applications usually use a central database. Consistent updates are needed, which is hard to guarantee, when many CPUs try to work in parallel on these databases. That is why it is practical that all packets of a single flow are processed by the same CPU. Another problem was already mentioned before: spraying may lead to packet reordering when packets of the same flow go to different processing elements which have a statistical delay characteristic. The packet order with flow-granularity is guaranteed by our system by re-sorting packets if needed. This is done by the Path Control that is introduced later.

We combine both mechanisms with a two stage packet assignment consisting of the Path Dispatcher and the Packet Distributor. The Path Dispatcher is one of the main modules of the concept, containing the already mentioned rule base and performing a path decision based on the packet data. The decision is done in hardware and is not limited to a CPU assignment. We are also able to send the packets directly to a hardware accelerator (e.g. Crypto Core) or to the egress path and process it with the help of additional modules completely in hardware (AutoRoute). If packets should be processed by a CPU, the packet is delivered to the Packet Distributor together with a path ID, indicating the path decision. The Packet Distributor consists of a configurable Multi-Processor Interrupt Controller along with queues for the different traffic types. With configuration registers inside the Interrupt Controller each path ID can be assigned to each CPU. This can be either a 1-to-1 (dedicated) or a 1-to-n assignment (spraying).

The AutoRoute path is a pure hardware path where the processing elements operate as a pipeline at the NP's aggregate line speed. This enables a high performance path for simple processing (like IP forwarding) and reduces the latency compared to a processor path, which was already described in [11].

Examples for dedicated traffic may be stateful flows, like IPSec packets, but also control packets that should be processed by the Control Plane or high priority packets that are sent to a reserved CPU. Spraying is used to distribute stateless flows on one or more CPU pools, e.g. differentiated by application or priority.

B. Architecture and Demonstrator

To demonstrate our concept we have developed a Multi-Processor System-on-Chip implemented on a single Xilinx FPGA.

The block diagram of our system is depicted in Fig. 2. As interconnect we are using the CoreConnect Processor Local Bus (PLB) with two PowerPC 405 CPUs connected, that are used for packet processing. One CPU runs always the Data Plane stack, whereas the second CPU can be configured to perform Control and/or Data Plane functions. For measurements, both CPUs have been configured as Data Plane CPUs. The stack mainly supports IPv4 forwarding and contains a full IPSec implementation, limited to manual keying. In the following we describe in short the main modules of our system:

Memory Management: The Memory Management – also called Buffer Manager – is an autonomously working DMA engine, which stores incoming packets in the main memory. A packet tag, called packet descriptor, is created and handed over to the Path Dispatcher. The packet descriptor contains pointers to the packet data in the memory. When receiving a packet descriptor at the transmit side (e.g. from the Data Plane CPU), the packet is read out again and sent to the Gigabit Ethernet Media Access Control (GEMAC).

Pre-Processor: The Pre-Processor parses the incoming packets, performs integrity checks (e.g. IPv4 header checksum) and extracts a set of important header fields depending on the protocol stack of the incoming packet. All this information is passed on to the Path Dispatcher for processing path selection. The Pre-Processor may also initiate a hardware next-hop lookup.

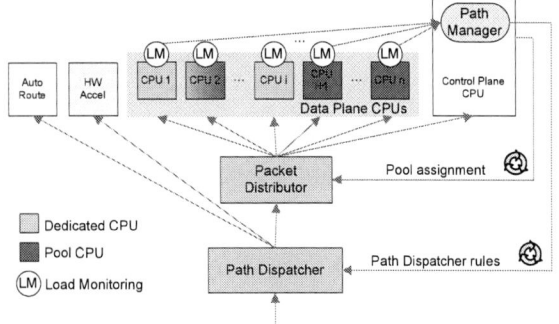

Figure 1: Packet Flow within the FlexPath NP

978-1-4244-2541-9/08 $25.00 © 2008 IEEE

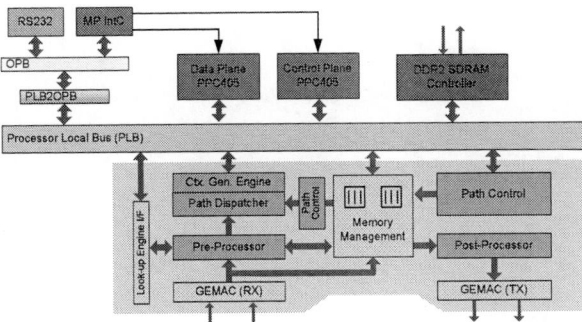

Figure 2: Block diagram of FlexPath NP architecture

Path Dispatcher: The Path Dispatcher decides where a packet should be sent to for processing. Using the extracted fields from the Pre-Processor, it evaluates a runtime reconfigurable rule base for every incoming packet (see [12]). After traversing the rule base, it is decided if a packet should be sent to a CPU or a hardware accelerator (e.g. for decryption). For very simple packets (e.g. only IP forwarding) we also can process the packet completely in hardware (AutoRoute) and send it directly to the Path Control and Post-Processor. The Path Dispatcher rule base can also be used for load balancing in a heterogeneous multi-processor system.

Context Generation Engine: Depending on the outcome of the packet classification, the Context Generation Engine either stores the extracted header fields from the Pre-Processor for later usage within a CPU or hardware accelerator. If the packet is eligible for AutoRoute, the correct instructions for the Post-Processor have to be generated and stored in memory.

Path Control: Since packets of one flow might be processed by different paths with different latencies, packets may be reordered. As shown in [6], packet reordering can have a significant performance penalty on network throughput. Therefore we propose the Path Control to re-sequence the packets at the egress side on a per-flow basis (see [13]). This creates additional freedom in processor assignment

Post-Processor: The Post-Processor is a hardware unit in the egress path that can perform basic packet manipulations

Table 1: Resource usage on a Xilinx Virtex-4 FX60

Module	Slices	BlockRAMs
Memory Mgmt.	3285	5
Path Dispatcher	1446	14
Ctxt Gen. Eng.	1978	4
Pre-Processor	682	1
Post-Processor	1615	3
MP-Interrupt C.	187	0
Path Control	1288	14
Others / Glue L.	5525	36
Σ	16006 (63%)	77 (33%)

Figure 3: Floorplan on a Xilinx Virtex-4 FX 60 FPGA

(e.g. data replace/insert, IP checksum update). It is programmed by specific assembler like instructions sent along with the packet (see [14]).

The benefits of HW offloading in a FlexPath NP compared to standard NPs has already been investigated in detail by simulations presented in [11].

We have built up our demonstrator on a Xilinx ML410 prototyping board with a Virtex-4 FX 60 FPGA. The board contains an external DDR2-SDRAM and two Gigabit Ethernet PHYs. The operating frequency of our system is 100 MHz. Besides the two PowerPCs we are consuming about 16,000 slices out of the 25,280 available on the Virtex-4 (63%). Additionally, we are using 77 (33%) built-in SRAM-blocks with 2 kbit each. Fig. 3 shows the floorplan of our system. Most of the modules are arranged around the central PLB. The most slice consuming modules are the Memory Management, Context Generation Engine, Post-Processor and the Path Dispatcher (cf. table 1).

III. FLEXPATH DEMONSTRATOR

In the last section we have presented our FPGA demonstrator with a working FlexPath NP on a single chip. In the next step now we would like to show the FlexPath benefits with a simple demonstration scenario, which is presented in the following. With the help of the demonstrator we mainly show three things:

- Pre-classification of packets in hardware – as it is done by the Path Dispatcher – can be helpful to enhance the performance of a NP.

- In our FlexPath NP we can easily combine the packet distribution techniques of dedicated assignment and spraying with benefits in the processor load.

- With additional hardware support by Pre-Processor, Path Dispatcher and Post-Processor the system performance is enhanced when using the AutoRoute path.

Of course we will not be able to demonstrate all of the benefits and aspects combined with FlexPath in this setup, but at least some major points should be extracted.

A. Demonstration Setup

For demonstration purposes we built up a setup with two TCP connections processed by our NP. The first connection consists of a simple Internet Mix (IMIX) as specified in [15] with 100 Mbps and a packet rate of 34.5 kpps. For this stateless flow we only have to perform IPv4 forwarding. For the second connection, our NP should be the beginning of an IPSec tunnel with AES encryption. The incoming packets with a uniform packet size of 512 bytes have to be encrypted, extended with additional tunnel and IP headers before being forwarded. Since IPSec is as stateful application, all packets of this flow must be processed by the same processing instance. We will increase the data rate and investigate its effect on the NP's performance and behavior. A schematic overview can be seen in Fig. 4.

The Path Dispatcher can easily identify these two connections by their IP addresses and choose different paths for each address. Of course, a packet with the same IP address does not necessarily belong to the IPSec connection. That is why in a second step, we have to perform a Security Policy Database (SPD) check in software, where the full IP 5-tuple is checked against the database. The database is implemented in the IPSec stack and tells whether a packet must be encrypted, forwarded or discarded.

In our configuration the Data Plane CPU 1 can process IP forwarding as well as IPSec packets. CPU 2 is also used as Data Plane CPU but only with forwarding capabilities. For our scenario we thus have different possible paths for processing, shown in table 2. Out of all possibilities, we have identified five configurations that we want to examine:

1. All packets are processed by CPU 1. This is the simplest case without usage of the FlexPath-specific capabilities.

2. Again, all packets are processed by CPU 1. But now by using different interrupt queues, there are different software entry points for IPv4 forwarding and IPSec.

3. The forwarding traffic (IMIX) is assigned to CPU 2.

Table 2: Processing capabilities of different paths

	CPU 1	CPU 2	AutoRoute
IPv4 Forwarding	X	X	X
IPSec	X		

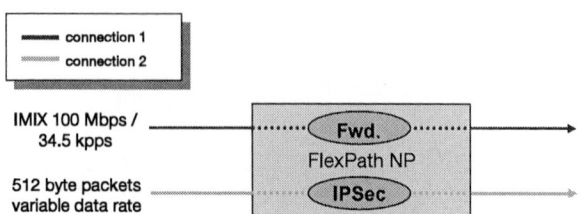

Figure 4: Demonstration setup

4. The forwarding packets are sprayed between CPU 1 and 2.

5. We use AutoRoute for forwarding traffic.

B. Results

For all scenarios we have measured the load of both CPUs as well as the output packet rate for the IMIX and for the IPSec packets. The results are presented over increasing IPSec data rates.

1) Scenarios 1 & 2: one Data Plane CPU

When comparing scenario 1 and scenario 2 in Fig. 5 we see quite a big difference in the CPU load in the case of no IPSec traffic. Here, only the IMIX traffic has to be forwarded, yielding a CPU load of 70% in case 1, but only 37% in scenario 2. The difference can be explained by the fact, that in scenario 1 CPU 1 has to perform the SPD check for all packets – in scenario 2 only on the small sub-set of packets pre-classified by the Path Dispatcher to be IPSec.

When increasing the IPSec traffic data rate, the CPU load increases dramatically (please take care of the broken y-axis for the packet rate in the diagram). An additional traffic of 1.6 Mbps (0.4 kpps) in scenario 2 already doubles the load of the CPU, which already shows that the encryption is very computational intensive. At the same time, the output packet rate of the IMIX traffic decreases. Without IPSec packets all IMIX packets can be served. With an IPSec data rate of 400 kbps we already have a loss rate of 8% for the IMIX packets – even though the CPU load is below 80%. This can be explained with the long processing latency of IPSec packets of more than 1 ms. During this time, the system is blocked.

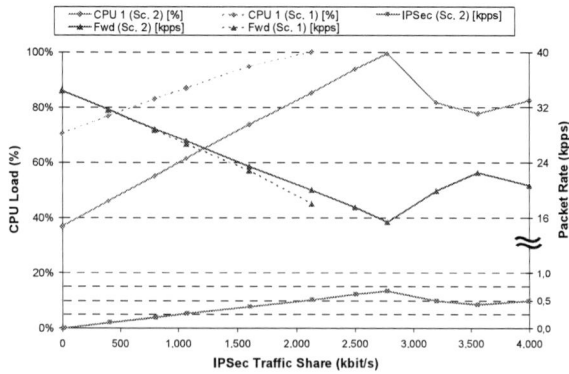

Figure 5: Measurement results scenario 1 and 2

978-1-4244-2541-9/08 $25.00 © 2008 IEEE

As soon as all input queues are filled, incoming packets are discarded at the system input. This leads to a packet loss even though there is enough processing power left between two IPSec packets.

The CPU load reaches its maximum of almost 100% with an IPSec data rate of 2.8 Mbps (scenario 2). Up to this point, there is no packet loss for IPSec packets. Since the CPU reaches its limit, also IPSec packets will be discarded. Since the processing effort per IPSec packet is high, each lost packet reduces the CPU load significantly, in turn increasing the forwarding rate. For higher input data rates the CPU load as well as the forwarding rate vary depending on the IPSec data rate and thus packet inter-arrival time.

2) Scenarios 3 & 4: two Data Plane CPUs

In a next step, we make use of the second CPU in the system. In scenario 3 all IPSec packets still are served by CPU 1, but all IMIX packets are assigned to CPU 2 for processing. In Fig. 6 you can see a constant processing load for CPU 2 of about 30% independent from the IPSec data rate. All IMIX packets pass the cluster. The load of CPU 1 starts from zero and increases with the IPSec data rate. In the spraying scenario (scenario 4) the IMIX packets will be distributed on both CPUs. Without IPSec packets, one would expect the same load for both CPUs. Nevertheless we have a load of 26% (CPU 1) and 10% (CPU 2). The reason for this lies in the implementation of the spraying mechanism. Our Interrupt Controller will notify both CPUs with an interrupt. The CPU that reacts first by reading out the Interrupt Service Register will get the packet. As long as both CPUs are idle, this will always be CPU 1, since it has the higher bus priority. When increasing the IPSec data rate, CPU 1 will mainly process IPSec packets, CPU 2 more and more IMIX packets. That is why the curves for the CPU load in scenario 4 are approaching the curves for scenario 3.

In both cases we identify a surprising decline of the IMIX output data rate when reaching 100% load for the IPSec processing CPU, which means a packet loss for the forwarding packets. The second CPU still has plenty of resources free. The explanation of this effect is Head-of-Line blocking in the ingress pipeline of our NP: CPU 1 is busy, processing an IPSec packet. In the meantime, another IPSec packet arrives, waiting for processing. At this time, the ingress pipeline is stuck. Packets – IPSec as well as forwarding packets – will be lost, even though there is enough processing power for the forwarding packets available.

3) Scenario 5: one Data Plane CPU plus AutoRoute

In our last scenario, we made use of our full hardware AutoRoute path for forwarding packets. Only the potential IPSec packets are now processed by a CPU. The results in Fig. 7 are quite similar to scenario 3. The load of CPU 1 is increasing linearly with the IPSec data rate and all IMIX packets can be processed. When the CPU load reaches 100% we see Head-of-Line blocking again. The main difference is the fact, that we get the same result now with only one CPU busy, which frees processing power for additional tasks (e.g.

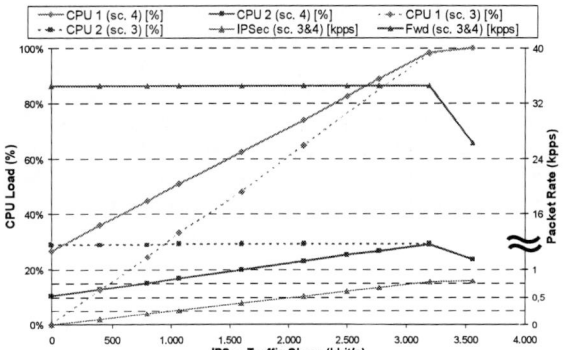

Figure 6: Measurement results scenario 3 and 4

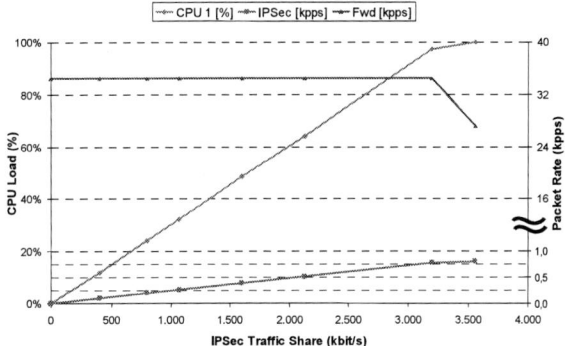

Figure 7: Measurement results scenario 5

higher IPSec data rates with two connections on two CPUs). In this scenario the AutoRoute path is not fully loaded and could handle much more than the 100 Mbps forwarding.

IV. CONCLUSION

In this paper we have presented a FlexPath network processor demonstrator on a Xilinx Virtex-4 FPGA. The main idea of FlexPath is the usage of configurable, flexible processing paths. Packets are classified in hardware and may be sent to different processing instances, depending on the network application. The path decision is based on a rule base that may be reconfigured during run-time. For simple processing tasks like IP forwarding, we are able to fully offload the complete processing to hardware.

With our measurement setup we have shown different scenarios with different path decisions for an example traffic constellation. We have demonstrated that the system behavior and performance is strongly dependent on the path decision and that the usage of FlexPath has a positive effect on the CPU loads. This is even true for the single CPU scenario (compare scenarios 1 and 2), without the need of full FlexPath support (like hardware next-hop look-up and Post-Processor). Simply by using the hardware pre-classification we can reduce the processor load by up to 33% in this case. We have also shown that we can easily combine two distribution mechanisms – dedicated assignment and spraying – within our system, depending on the current needs of the single flow. Whereas IPSec is a stateful application, with all packets of one flow being processed by the same CPU, IP forwarding is a stateless task that can be sprayed easily among many CPUs, producing a very well balanced load on the processors. We have also demonstrated the benefit of our hardware modules enabling a full hardware processing path for forwarding packets. This offloads the CPU, releasing processing resources for other tasks.

Because of Head-of-Line blocking a single blocked path can lead to a full system block – even if there are processing resources left for other paths. With the knowledge of a path decision in a FlexPath NP, we envision a selective discard of packets of that path in the Packet Distributor. This would limit the loss of packets to blocked paths, while others can go on processing as long as there is free processing capacity.

V. OUTLOOK

Although we demonstrated very well the basic benefits of our system, all scenarios have been static, which means there was no run-time reconfiguration of the rule base. In a next step, we now want to add this dynamic, in order to be able to optimize our system based on the current load situation.

This includes establishing an adequate monitoring possibility for the whole system, including CPU loads, queue fill levels etc. Out of this information, we want to develop an appropriate control algorithm optimizing the system performance and throughput while considering other requirements, like Quality-of-Service (QoS). For QoS it is not always desirable to have an even load distribution. Sometimes it is required to save processing resources for high priority tasks.

Another point is the update of the rule base in case of connection setup and teardown. For example, when a new IPSec tunnel is created the dedicated assignment of the new flow to an available CPU is needed. In overload situations (e.g. many IPSec flows on one CPU, creating a load near 100%) we also might think of a controlled migration of single, stateful flows to another CPU (including the migration of all state information).

We also want to extend our system with more CPUs and hardware accelerators, like crypto-cores. The idea here is that e.g. encrypted packets can be sent directly to the crypto core for decryption before being processed by the CPU. This would optimize the flow of packets through a heterogeneous MPSoC.

Although our FPGA demonstrator is surely not competitive with commercial NPs regarding packet throughput, we should mention, that the hardware performance in the FPGA is already exciting. All FlexPath-specific hardware modules are running at line-speed with a 32 bit data path at 100 MHz, equivalent to a packet data rate of 3.2 Gbps. In [11] we already described limitations in the memory and interconnect subsystem, limiting the possible data rate of the whole system to 400 Mbps. Nevertheless, transferring our architecture to an ASIC platform, with a more powerful interconnect, more CPUs and a clock frequency above 300 MHz, we should be able to reach data rates of 10 Gbps and more. Combining the FlexPath concept with a commercial NP architecture surely would be another interesting point. Since we are not limited to the PowerPCs, the combination of the FlexPath concepts with commercial NP architectures is conceivable. Our ingress and egress path hardware processing entities would serve as Pre-/Post-Processor for a cluster of Intel Microengines [1] or provide different entry points into the processing pipeline of a Xelerated NP [7].

REFERENCES

[1] Intel® IXP2800 Product Brief, http://download.intel.com/design/network/ProdBrf/27905403.pdf

[2] AMCC nP7300 Product Brief, https://www.amcc.com/MyAMCC/retrieveDocument/SNP/nP7300_060822.pdfNP 3

[3] G. Dittmann et al., "Network Processor Load Balancing for High-Speed Links", SPECTS 2002, July 2002

[4] W. Shi et al., "Load Balancing for Parallel Forwarding", IEEE transactions on Networking, August 2005

[5] X. Zhang et al., "An Efficient Packet Scheduler for Modern Network Processors: Guarantee Load Balancing an Packet Ordering", Workshop on High Performance Routing and Switching, June 2006

[6] S. Govind et al., "Packet Reordering in Network Processors", IPDPS 2007, May 2007

[7] Xelerated™ X10q Network Processor Product Brief, http://www.xelerated.com/uploads/files/62.PDF

[8] Ethernity ENET3000 Product Brief, http://www.ethernitynet.com/MEALX-ProductBrief.htm

[9] SafeNet SafeXcel 5160 Inline Security Processor, http://www.safenet-inc.com/Library/3/safexcel-5160.pdf

[10] R. Ohlendorf et al., "FlexPath NP - A Network Processor Concept with Application-Driven Flexible Processing Paths", CODES+ISSS'2005, USA, September 19-21, 2005

[11] R. Ohlendorf et al., "Simulated and measured performance evaluation of RISC-based SoC platforms in network processing applications", Journal of Systems Architecture; Vol. 53, No. 10, Oktober, 2007, pp 703-718

[12] R. Ohlendorf et al., "A Packet Classification Technique for On-Chip Processing Path Selection", Workshop on Application Specific Processors (WASP'07) Salzburg, Austria, October 2007

[13] M. Meitinger et al., "A Hardware Packet Resequencer Unit for Network Processors", International Conference on Architecture of Computing Systems (ARCS 2008), Dresden, Germany, February 2008

[14] M. Meitinger et al., "A Programmable Stream Processing Engine for Packet Manipulation in Network Processors", IEEE Computer Society Annual Symposium on VLSI, 2007 (ISVLSI '07), Porto Alegre, Brazil, May 2007

[15] Agilent, Mixed Packet Size Throughput, http://advanced.comms.agilent.com/n2x/docs/insight/2001-08/TestingTips/1MxdPktSzThroughput.pdf

978-1-4244-2541-9/08 $25.00 © 2008 IEEE

Micronmesh for Fault-Tolerant GALS Multiprocessors on FPGA

Heikki Kariniemi
Department of Computer Systems
Tampere University of Technology
Tampere, Finland
Email: heikki.kariniemi@tut.fi

Jari Nurmi
Department of Computer Systems
Tampere University of Technology
Tampere, Finland
Email: jari.nurmi@tut.fi

Abstract—System-on-Chip (SoC) circuits have evolved to single chip Multiprocessor systems. Due to increasing variance of process parameters, which produces synchronization problems on large SoCs, a Globally-Asynchronous Locally-Synchronous (GALS) design style must have been mobilized. In addition, the large VLSI circuits are also becoming more susceptible to transient and intermittent faults which can corrupt their operation. This paper presents a new Micronmesh Network-on-Chip (NoC) which is targeted to fault-tolerant communication of GALS Multiprocessor SoCs (MPSoC). It is fully synthesizable with current design tools and it can be used for prototyping MPSoCs on FPGA circuits. The Micronmesh incorporates a new improved Fault-Diagnosis-And-Repair (FDAR) system which is able to diagnose and repair also buffer memories in addition to wire connections while Fault-Tolerant DOR (FTDOR) routing is used for routing packets to their destinations around defected parts. Owing to the FDAR system and the FTDOR Micronmesh degrades gracefully as permanent faults appear and it is able to recover from transient and intermittent faults. The fault-tolerance of the Micronmesh is also improved by switch-to-switch (S2S) level retransmissions which reduce the number of end-to-end (E2E) level retransmissions that produce considerably higher latencies. These methods targeted at improving the fault-tolerance are also becoming necessary for improving the manufacturability of the circuits in the future.

I. Introduction

SoC circuits have evolved to MPSoCs where several processors communicate over NoCs. The increasing variance of the process parameters produces large clock skews and makes the synchronization of the sub-blocks of the MPSoCs more difficult [1]. Therefore, the sub-blocks must communicate asynchronously and the NoCs must provide methods for the implementation of GALS systems which consist of multiple clock domains with distinct clocks. Because the variance of process parameters will also increase the number of manufacturing defects, the NoCs' fault-tolerance will be crucial in the future as permanent

This research was supported by the Academy of Finland under grant 122361

manufacturing defects will further reduce the chip yields and increase manufacturing costs [1]. MPSoCs contain redundant resources for both computation and communication which can be used for replacing the faulty ones, and therefore, the chip yields and the manufacturing costs can be improved considerably, if it is possible to allow a certain part of these resources be defected without losing the chip [1, 2]. This can be achieved by making the NoCs more fault-tolerant. Fault-tolerant systems must be able to detect, locate, and isolate faults. Furthermore, they must be able to recover from faults. Therefore, the NoCs of the MPSoCs must supply methods for both fault detection and repair. Permanently faulty parts of the NoCs must be isolated from the other parts of the system so that they could not disrupt their operation, which can be considered a self-repairing. In addition to the permanent faults circuits' operation can be corrupted by transient and intermittent faults caused by, for example, crosstalk between wires and high-energy particles. These faults produce bit errors and corrupt packets, which degrades NoC's performance, because corrupted packets must be retransmitted. The growth of communication latencies and performance degradation can be reduced by S2S level retransmissions and error correcting codes [3, 4, 5, 6] which both reduce the number of E2E retransmissions. Transient and intermittent faults can also produce false state transitions in control logic which can corrupt the operation of the whole MPSoCs if appropriate methods for online fault detection and recovery are not usable on the chip.

Due to the increasing number of transistors, wires, functional blocks, and sub-systems the testing of the MPSoCs has become a more complex and time consuming task which must make use of different built-in test methods on the chip. It is possible to test the NoCs of the MPSoCs efficiently at full operation speed with functional level tests including switch logic [7, 8] and communication links [3, 6, 9, 10]. The functional level testing of the NoCs can be accomplished by software exploiting the same fault detection hardware which is used for online diagnosis of the NoCs operation. It can also be used online for testing faulty blocks in order to find out if their operation was corrupted by transient faults so that the faultless blocks could be returned into use. In the testing of the MPSoCs the NoCs can also be exploited as a Test Access Method (TAM) like in [9].

The fault-tolerance of the Micronmesh is partly implemented with an improved distributed FDAR system which is targeted for online detection and repair of faults like its previous version presented in [11]. It uses Cyclic Redundancy Checks (CRC) for detecting bit errors occurring on wire connections between switch ports and in buffer memory segments. If connections between some particular port blocks produce a lot of bit errors, which indicates that they may be permanently defected, the ports can use the FDAR system for disabling just these particular connections and continue their operation normally with other port blocks. Therefore, only such port blocks which do not have faultless functioning connections to other ports left are practically disabled. If buffer memory segments produce a lot of bit errors, the new improved FDAR system disables just the faulty segments while port blocks can continue their operation. Finally, the port block is disabled, if all of its buffer memory segments are disabled. The FDAR system controls also the duration of the arbitration between port blocks. If the duration exceeds a certain threshold, the blocked port, which experiences the large arbitration delay, disables the blocking ports. After this both ports can continue normally their operation with other ports, although they are not connected to each other any longer. These simple methods make it possible to detect, locate, and isolate faulty resources efficiently while faultless resources can be kept in use. The local processors of the Micronmesh nodes control the operation of the FDAR system and they are able to return the disabled ports into use again. This is necessary, because ports may be disabled due to transient faults, although they would not be permanently defected. Owing to the FDAR system the Micronmesh is able to degrade gracefully, if its nodes are defected.

After the FDAR system has isolated faulty parts the Micronmesh recovers from faults by rerouting packets around faulty disabled parts by the FTDOR [11] which is a software-based fault-tolerant routing algorithm. Processors are running software which reroutes blocked packets in two phases via intermediate nodes to the destination nodes. Basically the FTDOR resembles the e-sft [12] routing which is one of the first software-based fault-tolerant routing algorithms. The e-sft was designed for torus networks and it is not usable without modifications in Micronmesh. Both of the algorithms implement oblivious rerouting of packets by software controlled functions, which enables simple switch hardware (HW). The FTDOR selects intermediate nodes randomly, but it can be changed easily to take into account also the location of the faulty parts of the network and to implement shortest path routing. Simple partially adaptive turn model routing algorithms [13] could also be used in Micronmesh. These simple algorithms do not use additional virtual channels or physical channels which are usually required and exploited by adaptive routing algorithms [14]. However, the software-based rerouting functions should also be used with turn-model routing, since the turn model algorithms are not able to reroute packets around all faults, which is their disadvantage.

This paper is organized as follows. Section 2 presents the Micronmesh topology and its special modification named full 2D-mesh configuration. Section 3 presents the switch node of the Micronmesh which is named Micronswitch. Section 4 presents different methods which implement the fault-tolerance of the Micronmesh. The performance and the costs of the Micronmesh are presented in Section 5, and finally, Section 6 concludes this paper.

II. MICRONMESH TOPOLOGY

The Micronmesh has a two-dimensional (2D) mesh topology which seems to be the most commonly researched and used topology in NoCs up to now [4, 5, 10, 11]. The 2D-mesh is popular, because its regular topology allows regular and dense VLSI circuit layouts with short communication links. The 2D-mesh topology is also scalable for different systems sizes, because its size grows linearly as the system size increases while the length and the wire delay of the communication links remains unchanged. Due to short and fast communication links it provides also a good performance. The Micronmesh can be used with a special configuration named full 2D-mesh configuration where heterogeneous blocks like e.g. DSP accelerators, memory sub-systems, IO-interface blocks, spare components etc. can be connected to the edges of the 2D-mesh.

In the 2D-mesh topology every node contains a processor and a local switch. Owing to direct connections between the processors and the switches the on-line diagnosis performed by the FDAR system is able to respond very quickly in fault situations. As a consequence of this, reconfigurations can also be done with low latencies. The short response time reduces the Mean Time to Repair (MTTR), which improves the availability of the system. This is an advantage and one of the main reasons for choosing the 2D-mesh topology. The short MTTR makes it also easier to prevent faults from producing more faults in the system. Furthermore, the direct connections between processors and switches simplify the diagnosis and testing of the operation of the NoCs.

III. MICRONSWITCH ARCHITECTURE

Micronswitch is the switch of the Micronmesh. It has a distributed architecture where routing, arbitration, switching, and the FDAR system's functions are distributed to port blocks named Micronmux. The Micronswitch is depicted in schematic Figure 1 which shows also connections between neighboring switch nodes. Cut-lines illustrate edges of neighboring Micronmesh nodes. They are also edges of different clock domains, because every Micronmesh node forms a distinct clock domain.

The Micronswitch consists of two different sub-blocks which are Micronmuxes (MM), and a local processor port (µP IF). The local processor port contains also two Micronmuxes and it is connected to a system bus interface

978-1-4244-2541-9/08 $25.00 © 2008 IEEE

which the local processor uses for communication with other nodes and for diagnosis, reconfiguration, and testing of its local Micronswitch. Processors can lock and reset such Micronmuxes of their local Micronswitches which are locked by all of the Micronmuxes connected to them. Processors can also lock and reset the Micronmuxes of the neighboring Microswitches, if they have locked all of the Micronmuxes of the local Micronswitch. Processor nodes' local memories contain also packet buffers.

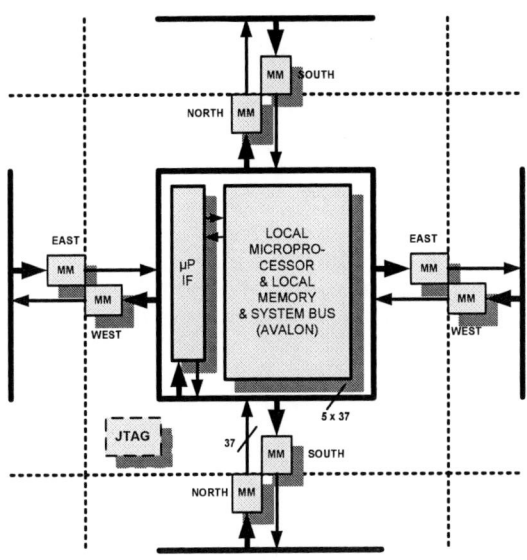

Figure 1. The Micronswitch architecture and its connections to its neighboring nodes.

The Micronmuxes (MM) are combined blocks of input and output ports and they perform functions of both the ports. They can be placed at the four sides of the Micronmesh nodes like schematic Figure 1 illustrates, which makes distinct communication links unnecessary. Because Micronmux can be used as an asynchronous bridge between two clock domains, this arrangement enables the implementation of the GALS systems. The following two sub-sections present the operation and structure of the Micronmux in more detail.

A. Combined port block Micronmux

Usually switch nodes are connected to each other by communication links. Because there are port blocks at both ends of the links, packets are usually also written twice into buffers and read twice out of them as they are transferred across the links. Furthermore, due to large clock skew, links must implement asynchronous self-timed communication the operation speed of which is determined by wire delay. The elimination of the long asynchronous communication links and the reduction of communication latencies belong to the primary targets of the Micronmux design depicted in Figure

2. These targets were achieved by combining the functions of the input and output ports in one block where communication is performed across an asynchronous dual-port buffer memory instead of self-timed asynchronous communication link. Owing to this arrangement, packets must be written into and read out of the buffer memory only once as they traverse the Micronmuxes. In a typical switch, input ports perform routing and output ports multiplex packets coming from different input ports to the output links. In the Micronmux the rx-port (RX-PORT, RX-CONTROLLER) performs the output port functions and the tx-port (TX-PORT, TX-CONTROLLER) performs the input port functions. The Micronmuxes are like asynchronous bridges the rx-ports of which are clocked by the local clocks and the tx-ports of which are clocked by the clocks of the neighboring nodes. Although all of the Micronmesh nodes would be clocked at the same clock rate, the phases of the clocks could differ considerably due to clock skew, and asynchronous communication should be used between nodes anyway.

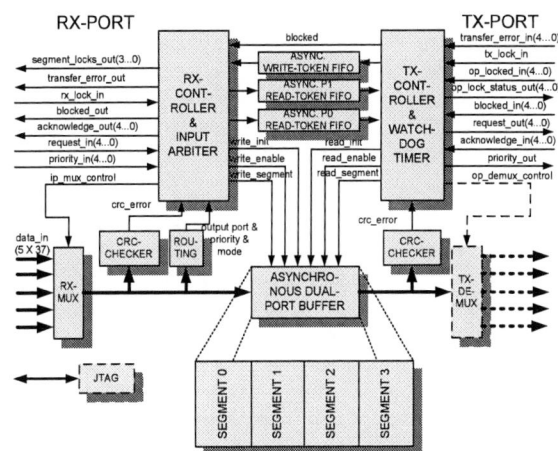

Figure 2. The Micronmux architecture.

The main sub-blocks of the Micronmux are a rx-controller (RX-CONTROLLER), a tx-controller (TX-CONTROLLER), an input multiplexer (RX-MUX), and an asynchronous dual-port buffer memory (ASYNCHRONOUS DUAL-PORT BUFFER). It contains also other blocks for CRC checking (CRC-CHECKER) and for routing function (ROUTING). The rx-controller and the tx-controller use read and write-tokens for managing the reading and the writing of the segmented buffer memory. Three small asynchronous FIFOs (ASYNC. P0 READ-TOKEN FIFO, ASYNC. P1 READ-TOKEN FIFO, ASYNC. WRITE-TOKEN FIFO) are used for transferring low-priority (P0) and high-priority (P1) read-tokens from the rx-controller to the tx-controller and write-tokens to opposite direction. It would also be possible to insert a demultiplexer (TX-DEMUX) at the output of the

Micronmux like Figure 2 illustrates for connecting to other Micronmuxes with distinct connections.

The Micronmux is actually a small switch which can be used in different network topologies as an independent switch node, although it is used like a port block in the Micronswitch. The number of its rx-ports and tx-ports, the number of its buffer memory segments, and its routing decision function can be changed so as to adapt it for different topologies. Therefore, the Micronmux can also be considered a general-purpose switching element which is configurable for different purposes.

B. Flow control and buffering architecture

The Micronmesh is targeted to MPSoCs which use message-passing communication. Messages are transferred by small fixed sized packets of four 37 bits wide words through the Micronmesh. Each word consists of 32 data bits and 5 CRC bits. In Micronmesh small packets of fixed size are used, because they enable efficient resource allocation of both bandwidth and buffer memories. Larger packets of eight words could also be used, but short packets block other traffic shorter times. The tx-controller of the transmitting Micronmux requests the transfer by asserting a *request*-signal and the rx-controller of the receiving Micronmux starts the transfer by asserting an *acknowledge*-signal after it has finished arbitration and granted the request. Since packet transfer takes always four clock cycles and arbitration takes one clock cycle, the total transfer time is five clock cycles.

Received packets are stored into dual-port buffer memory which is divided into four segments of four words which is the packet size. The writing and reading of the buffer memory is controlled by write-tokens and read-tokens. As the Micronmuxes are reset the tx-controller writes four write-tokens to the WRITE-TOKEN FIFO so as to enable writing to the buffer memory segments. Every write-token contains a segment number and one bit for indicating if a bit error was detected from the packet as it was transmitted. All of the write-tokens contain different segment numbers which are also used as the two most significant bits of the memory address. The rx-controller, which controls also writing to the buffer memory in addition to the arbitration and the input multiplexer, reads from the WRITE-TOKEN FIFO write-tokens the segment numbers of which determine which segments are written next. If the WRITE-TOKEN FIFO is empty, the dual-port buffer is full and the rx-controller is not able to receive more packets. After the reception of the whole packet the rx-controller writes a read-token either to the P0 READ-TOKEN FIFO or to the P1 READ-TOKEN FIFO according to packets' priority. Distinct READ-TOKEN FIFOs make it possible to read high-priority P1 packets out of the buffer memory always before low-priority P0 packets, which reduces their routing latencies. Every read-token contains a segment number which points the segment that contains the packet. The read-tokens carry also error bits and the packets' output port numbers which are computed by the FTDOR routing algorithm (ROUTING) as the packets are received. The tx-controller reads read-tokens from the READ-TOKEN FIFOs and starts arbitrations with the output Micronmuxes determined by the output port numbers if the packets are correct. After packets' transmissions and possible retransmissions are finished they make write-tokens of the segments and write them to the WRITE-TOKEN FIFO.

Although the previously described flow control mode implements *store-and-forward* (SAF) routing [14], *virtual-cut-through* (VCT) routing [14] is also possible. In VCT routing the rx-controller writes read-tokens to the READ-TOKEN FIFOs immediately after the first word of the packets is received so that the tx-controller could start routing packets forward as soon as possible, which reduces routing latency. Because the tx-controller does not write write-tokens to the WRITE-TOKEN FIFO before the whole packets are transmitted and possibly retransmitted to the next Micronmux, the segments are always empty before the next packet is written into them.

Owing to the token-based buffer segment allocation, all packets can be stored into the same memory which has a dense small logic and allows simple online diagnosis and testing. For example, in Æthereal the combined GS-BE switches, with so called distributed programming architecture, have distinct buffers for high-priority GS (Guaranteed Service) packets and low-priority BE (Best Effort) packets [15]. This architecture requires more buffer FIFOs and control logic the testing of which requires also more complex HW structures. Switches with virtual channel flow control have also a large number of buffers [5, 10, 14] which makes online testing and diagnosis more complex. In Micronswitches SRAM-cell based buffer memories have considerably less clocked registers than flip-flop based FIFOs of the same size would have, and therefore, they consume also less dynamic power. Furthermore, distinct retransmission FIFOs are not needed like in [10], because packets can be read and retransmitted several times from the dual-port buffer, which saves HW-resources.

IV. FAULT-TOLERANCE OF THE MICRONMESH

The fault-tolerance of the Micronmesh is based on five methods which are presented in the following sub-sections.

A. CRC checks for detecting bit errors

Connections used for packet transfers between Micronmuxes are potential sources of the bit errors, because in addition to permanent faults, packets can be corrupted by transient faults like e.g crosstalk which is caused by coupling effects between wires. Therefore, every 37 bits wide data word contains four data bytes and five bits wide CRC for detecting bit errors. The CRC is computed only once for every word before its transmission to the network and it is transferred from end to end through the whole routing path.

Buffer memories are also potential sources of transient faults, because high-energy alpha particles can charge or discharge memory cells and change their bit values as they hit silicon substrate. Therefore, the CRC checks are also used for detecting bit errors from the dual-port buffers' output words. If the packet was corrupted already in the buffer memory, the tx-controller does not retransmit it, although the receiving Micronmux would assert *transfer_error*-signal. The corrupted packets are removed by the receiving Micronmux, if SAF routing is used. Packet headers carry sequence numbers so that packets' destinations could detect packet losses and would request E2E level packet retransmissions from the source nodes.

B. Switch-to-switch (S2S) level packet retransmissions

The S2S level packet retransmissions are used for improving the reliability of communication between Micronmuxes and for reducing communication latencies. If Micronmuxes detect CRC errors as they receive packets, they request packet retransmissions by asserting *transfer_error*-signal. The tx-controllers retransmit packets only if their CRCs are correct at the dual-port buffer's output and the destination Micronmuxes assert *transfer_error*-signals after detecting CRC errors. The number of retransmissions must be limited, because otherwise permanently defected connections between Micronmuxes could congest the NoC by producing a large number of retransmissions. Currently Micronmuxes retransmit packets only once, but the number of retransmissions can be increased. If SAF routing is used and if packets are not received successfully despite retransmissions, the rx-controllers mark the read-tokens by an error bit so that the tx-controllers do not transmit the packets forward, but just return the segments by write-segment tokens to the rx-controllers through the WRITE-TOKEN FIFOs.

The S2S level packet retransmissions do not make the E2E retransmissions unnecessary, because if S2S retransmissions can not be accomplished successfully, processors must request E2E retransmission of lost packets. However, it is not possible to use the E2E level packet retransmission alone either, because they produce much higher communication latencies than S2S level retransmissions [3, 6].

C. The FDAR system

The FDAR system implements the fault repair method of the Micronmesh. Its operation is based on a simple functional fault model which assumes that hardware resources are faulty, if they produce a lot of bit errors or block packets permanently. According to this fault model the Micronmuxes are faulty, if their memory segments or their connections to other Micronmuxes produce a large number of bit errors. Therefore, as the tx-controllers transmit packets to the next Micronmux, they check the CRCs, and if bit errors are detected at the buffer memory's output, they mark the write-segment token by an error bit. The tx-controllers write the write-segment tokens into WRITE-TOKEN FIFO after the content of the corresponding segment is transmitted and possibly retransmitted. The rx-controllers use the error bits for counting bit errors produced by different buffer segments. They disable such segments which have produced a larger number of bit errors than a threshold value by

removing their write-segment tokens. This is called segment locking. After locking, the segment's lock bit is asserted and the segment can not be used any longer. The threshold value is determined by the VLSI technology, MPSoC application, and by other such things which affect the error probability. Local processors can check the segment status from the lock bits anytime and reset the Micronmuxes so as to return them all into use again. If VCT routing is used, Micronmuxes do not remove corrupted packets, and higher thresholds must be used so that faulty packets would not lock segments unnecessarily. Alternatively, the error counters could be reset or decreased by one after every correct packet, if their values are greater than zero. It would still be possible to find permanent static faults by using this procedure, but dynamic faults could necessarily not be detected any longer. The alternative procedures could also be used with the SAF routing.

The tx-controller counts the number of retransmissions and disables such connections between Micronmuxes which produce a lot of bit errors if the number of retransmissions exceeds a certain threshold value. Alternative procedures are also usable with retransmission counters as with error counters. The disabling of Micronmuxes is also called port locking. The port locking does not affect Micronmuxes' operation and they can still continue operation with other Micronmuxes normally, although Micronmuxes, which performed locking, do not transmit packets to them any more. The local processors of Micronmesh nodes can also monitor and lock the Micronmuxes, and return them into use by reseting and unlocking. The Micronmuxes can be reset or locked anytime without disrupting the operation of the other Micronmuxes, because rx-controllers and tx-controller operate independently without handshaking during transfers.

According to the fault model, the Micronmuxes are also faulty, if they are not able to route packets forward within a certain limited time. Therefore, the FDAR system controls also the duration of arbitration to detect permanent packet blockings. The tx-controllers have watchdog timers for measuring the arbitration times. They start their timers as they are requesting transfers and the requested target Micronmuxes have not asserted *blocked*-signals to indicate that they are also waiting and blocked themselves. If the target Micronmuxes are not blocked, they should be able to acknowledge the requests after a waiting time of at most few tens of clock cycles. If they do not acknowledge the requests before timers trigger timeouts, the tx-controllers will lock them when operating in a diagnosis mode. Like previously the locking of the target Micronmuxes does not disable them. Transient faults on *request* or *acknowledge*-signals can not corrupt the operation of the rx-controllers and tx-controllers either, because their state machines control the duration of the transfer and the handshaking signals do not affect their state transitions after the transfer request is acknowledged.

D. The FTDOR routing

The FTDOR routing [11] is a software-based fault-tolerant routing algorithm. It is able to route packets in a partly defected NoC by routing them via intermediate nodes around faulty parts in two phases and it implements part of the fault recovery method of the Micronmesh. In the first phase it routes packets to the intermediate nodes from which they are routed to the final destination. Packet headers can carry two node addresses for this purpose and appropriate control data for controlling the routing. Rerouting is performed by local processors of the faulty Micronmesh nodes where packets were blocked. If the local processor port is also locked, the FDAR system removes packets. Packet sources can also use oblivious routing and choose faultless intermediate nodes for rerouting packets. The number of reroutings is limited to two so as to prevent so called livelock. The FTDOR is a deadlock free algorithm, because the DOR is a deadlock-free routing algorithm [14]. Additionally, since corrupted packets are removed and in 2D-meshes the FTDOR prohibits routing packets from y-dimension back to x-dimension, faults can not produce deadlocks. The rerouting function is software, and therefore, it is possible to choose routing paths in a clever manner, which would otherwise require more complex and expensive switch hardware. Because packet rerouting increases routing latencies, it is primarily intended to be used only in such exceptional situations where the operation of the MPSoC could be completely corrupted by faults.

The operation of the FTDOR is illustrated in Figure 3 in a 4×4 full 2D-mesh configuration where nodes are addressed by their xy-coordinates (x, y) and the FTDOR routes packets at first to the direction of x-axis and then to the direction of y-axis. Black arrows illustrate how packets can be routed around a faulty connection between nodes (4, 3) and (4, 4) to node (4, 5). Packets would be permanently blocked at (4, 3) if they could not be rerouted (RR) from it to node (3, 4) which reroutes them again to (4, 5). Grey arrows show how information of blocking at node (4, 3) is sent backwards so that the Micronmuxes along the routing path would not lock more Micronmuxes and trigger more packet reroutings.

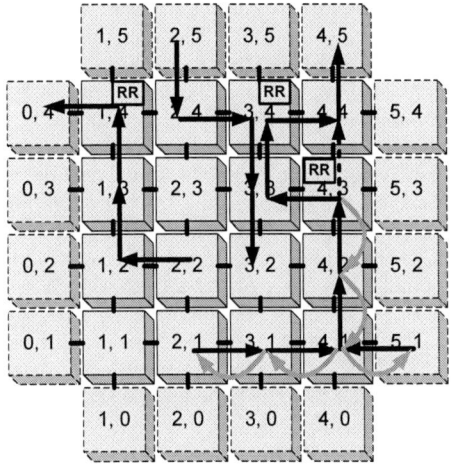

Figure 3. The operation of the FTDOR routing and the FDAR system in a 4×4 full 2D-mesh configuration

Rerouting is also needed for routing packets in the full 2D-mesh configuration between the edge nodes and the 2D-mesh nodes in the middle. Communication between the edge nodes on the top and on the bottom of the full 2D-mesh configuration can be performed without rerouting, because the FTDOR routes packets first at the direction of the x-axis after which it can route them directly to the destination nodes by routing at the direction of the y-axis. Black arrows from node (2, 5) to node (3, 2) illustrate a simple example of this. Routing at the opposite direction could also be performed without intermediate nodes. Communication from the nodes in the middle of the full 2D-mesh configuration to the nodes of the leftmost and the rightmost columns requires rerouting via one intermediate node which must be on the same row as the destination node like black arrows from node (2, 2) to node (0, 4) illustrate. Communication at the opposite direction does not require packets' rerouting, but communication between the leftmost and the rightmost columns requires. Packets' rerouting could also be avoided, if the nodes of the leftmost and the rightmost edge columns would communicate only with the nodes which are on the same rows with them.

E. Online functional level testing

The MPSoCs can perform software-based functional level testing online to test the operation of connections between Micronswitches. The processors can, for example, test connections by sending loopback test packets to any other processor node which sends them back. It is basically also possible to do manufacturing tests to the switches of the NoCs at the functional level like, for example, in [7].

V. PERFORMANCE AND COSTS

The Micronswitch was synthesized to Altera's Stratix III EP3SL150F1152 FPGA [16] which has a sufficient capacity for prototyping small Multiprocessor systems. One Micronswitch consumes about 3% of logic resources, which are combinational ALUTs (2952 of 113600) and logic registers (1643 of 113600), and less than 1% of block memory bits (3944 of 5630976). An operation speed of over 200 MHz was achieved.

A 4×4 2D-mesh (2DM) and a 4×4 full 2D-mesh configurations (F2DMC) were simulated with uniformly distributed traffic pattern, with different error probabilities, and with different number of locked ports and buffer segments to study how their performance changes. All of the simulation results are averages of the result of three simulations and they present received average maximum throughputs through processor interfaces in percentages. During simulations traffic sources injected new packets to the network always it was able to receive them and processor interfaces removed corrupted packets from the received packet streams. Probabilities P_{serr} and P_{terr} specify probabilities that bit errors corrupt packet words in buffer segments and on wire connections between Micronmuxes respectively. Simulations were also performed with different number of segments (#Seg) so as to study how segment locking affects the throughputs.

The first simulations were performed with the SAF and the VCT routing with both 2DM and F2DMC. The leftmost sub-column of Table 1 presents the average maximum throughputs when bit errors do not corrupt packets. The F2DMC produced clearly lower throughputs with both SAF and VCT routing than the 2DM, because it is connected to twice as many processors as 2DM with the same amount of routing resources. The F2DMC produced also quite similar performance with SAF and VCT routing whereas the 2DM produced clearly higher throughputs with VCT. The second sub-column from the left presents how much the average maximum throughputs degrade as the words are corrupted by bit errors with probability $P_{terr} = 0.005$ and corrupted packets are retransmitted once. The third sub-column presents how much the maximum throughputs degrade as the words are corrupted also with probability $P_{serr} = 0.005$ in buffer segments. The SAF and VCT routing produced practically equal throughputs as the bit errors corrupted words, but VCT routing seems to suffer more of the segment faults. This is because, if the VCT routing is used in Micronmesh, corrupted packets are removed in the network interfaces, which increases the traffic load of both the network and the network interfaces. The two rightmost sub-columns present how much the throughputs degrade as the number of usable segments is decreased from four to three and from three to two in all of the Micronmuxes. Based on these results it can be concluded that the locking of only a few buffer segments would not affect the throughputs very much.

TABLE I.   THE PERFORMANCE OF THE MICRONMESH WITH DIFFERENT CONFIGURATIONS AND ERROR PROBABILITIES.

Micron-mesh config.	Maximum throughput (%)				
	#Seg= 4 $P_{terr}=$ 0.0 $P_{serr}=$ 0.0	#Seg= 4 $P_{terr}=$ 0.005 $P_{serr}=$ 0.0	#Seg= 4 $P_{terr}=$ 0.005 $P_{serr}=$ 0.005	#Seg= 3 $P_{terr}=$ 0.0 $P_{serr}=$ 0.0	#Seg= 2 $P_{terr}=$ 0.0 $P_{serr}=$ 0.0
4×4 2DM (SAF)	44.1	41.9	40.0	33.7	21.4
4×4 F2DMC (SAF)	20.9	20.0	19.0	15.6	9.4
4×4 2DM (VCT)	47.8	42.6	39.4	38.2	25.1
4×4 F2DMC (VCT)	22.1	19.1	18.0	17.8	11.1

The next simulations were performed to study the throughputs with different error probabilities. Simulation results achieved by the SAF routing are only presented, because the VCT routing produces basically quite similar results. The normalized maximum throughputs produced by the 2DM and the F2DMC with different error probabilities are presented in Figure 4. The throughputs are normalized to the result of the leftmost sub-column of Table 1. The throughput of the faultless 2DM does not considerably

degrade before the error probabilities exceed 10^{-3}, which can be tolerated, because the true error probabilities can be assumed to be smaller in faultless networks. The performance of the F2DMC does not degrade as much as that of the 2DM as the error probability is increased.

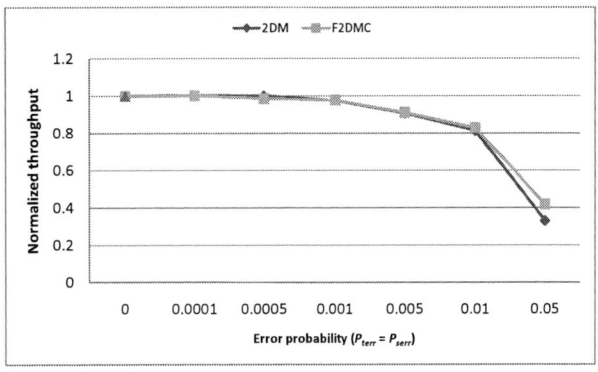

Figure 4. Normalized average maximum throughput (y-axis) of faultless 2DM and F2DMC with different error probabilities (x-axis) and with $P_{terr} = P_{serr}$.

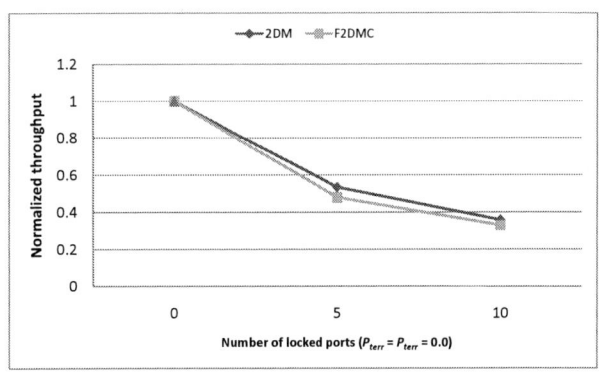

Figure 5. Normalized average maximum throughputs (y-axis) of 2DM and F2DMC with different numbers of faulty ports (x-axis) and with $P_{terr} = P_{serr} = 0.0$.

The effect of Micronmux locking on throughput was studied by simulations with such 2DM and F2DMC, where randomly chosen WEST, SOUTH, EAST, or NORTH Micronmuxes of randomly chosen Micronswitches were locked. The results in Figure 5 are also normalized to the results of the leftmost sub-column of Table 1. They show that the throughputs degrade quickly as the number of locked Micronmuxes grows from 0 to 5 and from 5 to 10. The throughput of the F2DMC seems to be slightly more sensitive to the number of faulty Micronmuxes than that of the 2DM. This is because the F2DMC connects larger number of processors than the 2DM with the same number

978-1-4244-2541-9/08 $25.00 © 2008 IEEE   101

of routing resources and because the traffic load of individual Micronmuxes grows slightly more, if the F2DMC is partly defected and its Micronmuxes are locked. However, the throughput depends also on the traffic pattern and the performance of the faulty networks could be improved by shortening the average communication distances between processes and by mapping processes to processors again in such a way that the packets could be routed around faulty parts without rerouting.

As was earlier mentioned both of the Micronmeshes were also simulated with three and two segments in every Micronmux so as to study the effect of segment locking on their throughputs. Simulation results in Table 1 show that the average throughput of the 2DM would fall off by 23.6%, if every Micronmux would have only three segments in use, and by 51.5%, if only two segments would be in use. Simulations with the F2DMC produced similar results. Since simulations with three segments produced higher throughput than simulations with only five locked Micronmuxes, it can be concluded that the locking of Micronmuxes degrades the throughputs clearly more than the locking of a few buffer segments. This motivates the augmentation of the FDAR system with the segment locks, because the segment locks improve the management of faulty hardware resources. As a consequent of this, they improve also the graceful degradation of the Micronmesh.

The usage of the segment locks improves also the performance of the Micronmesh, because the number of corrupted and lost packets would be higher, if segment locks would not be used. As a consequence of the increasing number of lost packets the number of E2E retransmissions would also grow, which would reduce system's performance. Furthermore, Micronmuxes could also be locked, if packets would be corrupted by their faulty buffer segments and if VCT routing would be used in the Micronmesh.

VI. CONCLUSIONS

This paper presents a new fault-tolerant Micronmesh NoC for MPSoCs. Its fault tolerance is implemented with an improved FDAR system which is able to detect, locate, and repair faults. Micronmesh can recover from faults owing to the FDAR system and the software-based FTDOR routing algorithm. Micronmesh nodes are also able to perform S2S level retransmission of corrupted packets. Simulation and synthesis results show that Micronmesh is usable for prototyping MPSoCs on FPGA circuits. Furthermore, it is usable for prototyping GALS systems on the FPGAs. Based on the simulation results the usage of the FDAR system augmented with segment locks can be justified, because it allows Micronmesh to degrade gracefully and reduces performance degradation produced by faults.

At his moment the research is continued by developing the Micronswitch Interface (MSI). A simple message passing protocol named Micron Message Passing Protocol (MMPP) is also under research. It will implement a sub-set of the functions of the Message Passing Interface (MPI) protocol

which is widely used in supercomputers. The MSI and the MMPP will be primarily developed for delivering fault-tolerant and fast communication in MPSoC platforms.

VII. AKCNOWLEDGEMENTS

This research was supported by the Academy of Finland under grant 122361.

REFERENCES

[1] G. Martin, H. Chang, Winning the SoC revolution, Kluwer Academic Publishers, Massachussetts, USA, 2003. Ch. 11.

[2] T. Dumitras, S. Kerner, R. Marculescu, "Towards on-chip fault-tolerant communication," Proceedings of the Asia and South Pacific Design Automation Conference (ASP-DAC 2003), Kitakyushu, Japan, Jan. 21-24, 2003, pp. 225-232.

[3] C. Grecu, A. Ivanov, R. Saleh, E.S. Sogomonyan, P.P. Pande, "On-line fault detection and location for NOC interconnects," Proceedings of the 12th On-Line Testing Symposium (IOLTS 2006), Lake of Como, Italy, July 10-12, 2006, pp. 6.

[4] A.P. Frantz, M. Cassel, F.L. Kastensmidt, E. Cota, L. Carro, "Crosstalk- and SEU-aware networks on chips," IEEE Design & Test of Computers, Vol. 24, July-Aug. 2007, pp. 340-350.

[5] J. Kim, C. Nicopoulos, D. Park, V. Narayan, M.S. Yousif, C.R. Das, "A Gracefully degrading and energy-efficient modular router architecture for on-chip networks," Proceedings of the 33rd International Symposium on Computer Architecture, Boston, MA, USA, June 17-21, 2006, pp. 4-15.

[6] S. Murali, T. Theocharides, N. Vijaykrishnan, M.J. Irwin, G. De Micheli, Analysis of error recovery schemes for networks on chips, IEEE Design & Test, Vol. 22, Sept.-Oct. 2005, pp. 434-442.

[7] R. Jaan, R. Ubar, V. Govind, "Test configurations for diagnosing faulty links," Proceedings of 12th IEEE European Test Symposium (ETS'07), Freiburg, Germany, May 20-24, 2007, pp. 29-34.

[8] A. Alaghi, N. Karimi, M. Sedghi, Z. Navabi, "Online NoC switch fault detection and diagnosis using a high level fault model," Proceedings of the 22nd IEEE International Symposium on Defect and Fault-Tolerance in VLSI Systems, Rome, Italy, Sept. 26-28, 2007, pp. 21-29.

[9] A. Grecu, A. Ivanov, R. Saleh, P.P. Pande, "Testing network-on-chip communication fabric," IEEE Transactions on Computer-Aided Design of Integrated Circuits and Systems, Vol. 26, Dec. 2007, pp. 2201-2214.

[10] D. Park, C. Nicopoulos, J. Kim, N. Vijaykrishnan, C.R. Das, "Exploring fault-tolerant network-on-chip architectures," Proceedings of International Conference on Dependable Systems and Networks (DSN'06), Philadelphia, PA, USA, June 25-28, 2006, pp. 93-104.

[11] H. Kariniemi, J. Nurmi, "Fault-tolerant 2D-mesh network-on-chip for multiprocessor systems-on-chip," Proceedings of Design and Diagnosis of Electronic Systems (DDECS'06), Prague, Czech's Republic, Apr. 18-21, 2006, pp 184-189.

[12] Y.J. Suh, B.V. Dao, J. Duato, S. Yalamanchili, "Software-based rerouting for fault-tolerant pipelined communication," IEEE Transactions on Parallel and Distributed Systems, Vol. 11, March 2000, pp. 193-211.

[13] C.J. Glass, L.M. Ni, "The turn model for adaptive routing," Proceedings of the 19th International Symposium on Computer Architecture, Queensland, Australia, May 19-21, 1992, pp. 278-287.

[14] J. Duato, S. Yalamanchili, L. Ni, Interconnection networks: an engineering approach, Morgan Kaufmann Publishers, USA, 2003.

[15] K. Goossens, J. Dielissen, A. Radulescu, "Æthereal network on chip: concepts, architectures, and implementations," IEEE Design & Test of Computers, Vol. 22, Sept.-Oct. 2005, pp. 414-412.

[16] Altera Corp., "Stratix III Device Handbook", Vol. 1-2, 2007, http://www.altera.com/literature/hb/stx3/stratix3_handbook.pdf

Configuring Smart Objects over Cognitive Radio

K. Nikunen, H. Heusala, J. Komulainen
Department of Electrical Engineering
University of Oulu
Oulu, Finland
First name. Last name@ee.oulu.fi

Abstract— the most promising application of Field Programmable Gate Arrays (FPGA) is reconfigurable computing, which gives basis for prototyping a swarm intelligence and the behavior of smart objects. With different configurations, the functionality of an FPGA can change different modular units of Smart Objects. With a swarm of these FPGA-based objects, it is possible to take advantage of a new kind of data processing method: Cognitive Computing. The use of cognitive radio makes this technology more dependable. These Smart Objects have to be developed to a level where they communicate safely before any bigger evolution can happen. So, how do we reconfigure these Small Smart Objects safely over a wireless network? This paper introduces an implement prototype which enables the safer reconfiguration of hardware logic over a wireless radio link by using frequency hopping and by searching for available noiseless channels.

I. INTRODUCTION

There are plenty of designer-friendly intellectual-property (IP)-radios on the market that are capable of sending several bytes of data in one data packet. These radios often have properties that enable better use of the existing bandwidth. The circuit used in this paper is Nordic Semiconductor's nRF24L01, which has inner counters to register lost and re-sent data packets. Successful transmission is reported to the transmitter via acknowledgement. The number of sent packets is easy to count. The radio's Serial Peripheral Interface (SPI) can also display numerical information on the channel quality, which helps determining whether the channel should be changed. These properties enable use of cognitive radio on reconfiguration process and other important data exchange.

This paper presents an implementation scenario of the cognitive radio that seeks the best channels available for the communication link through which a configuration file or other data can be sent to another device or a Smart Object [1]. It is possible to use frequency hopping during a transmission, depending on the channel quality and the security demands. Basic reconfiguration functionality is tested and confirmed where configuration data is sent straight-forward to another wireless object [3] and it is reconfigured successfully.

II. BASIC RECONFIGURATION PROCESS

The FPGA device used in this paper is Altera EP1C6. The configuration file needed to configure this device limited to a maximum of 1048576 bits, 131072 bytes. The configuration file specifies the digital hardware structure of the FPGA of a smart object. When transmitting a configuration file all the bytes must be in order with bit-level robustness. If any of the bits are harmed, the final configured system will be inoperative, in the way of miss-guided signal lines inside FPGA or possibly even short-circuit which can lead to destruction of the physical implementation. Normally though, a misconfigured FPGA will not be bootable. When this happens, the FPGA needs physical manipulation and a new configuration file from a computer. Radio has limited ways of detecting defected data packets.

Figure 1. A Smart Object with display.

The configuration device used in this paper, EPCS4, has 4 Mbits of data storage, which can simultaneously hold up to four configuration files. The configuration file that configures the FPGA always starts from the memory address 00000h. Bigger FPGAs and configuration devices are also available for bigger requirements. Any of the possible stored configuration files can be sent to a target object, which then

978-1-4244-2541-9/08 $25.00 © 2008 IEEE

reconfigures itself with the new configuration. The object can also choose which configuration it reconfigures itself with, assuming it carries several configurations. The chosen configuration has to be copied from the "non-zero" location to the "zero"-location. The original configuration, if needed later, has to be moved to a different location before the new configuration process begins. With the EPCS4 and other configuration devices, all the extra space can be used for the application memory. If one configuration is stored, which is the minimum during reconfiguration process; there will be approximately 3 Mbits of free space left. This configuration file size applies only for the EP1C6, used in this work.

Theoretically, a configured object can erase the whole memory temporarily to create more available memory space, but it needs to reload the configuration file from another location before any new reconfiguration process or reboot can take place. In case of a power shortage, the object would need a manual reconfiguration. With the configuration file stored to "zero"-location, the object will always reconfigure itself from there and is able to recover from errors.

A prototype of a wireless smart object has been built with the aforementioned properties. Figure 1 illustrates the dimensions of the experimental prototype. One side is approximately 4.3 cm and the height is around 1 cm. Technology makes it possible for the object to be assembled in a much smaller, integrated packet.

III. BUILT-IN PROPERTIES OF THE RADIO

Objects communicate through a radio link. Radio used in this work can operate in 126 different channels. A radio specifies what data is sent from transmitter to receiver. During a transmission process, objects have to know which object is the transmitter and which one is the receiver. The receiver can send data to the transmitter when answering to an acknowledgment which is sent from the receiver to a set transmitter. This enables two-way data communication with a single setup of radio properties. A transmitter can send data to 5 different smart objects in one channel. Different objects in one channel are separated with individual addresses.

The radio has built-in transmission automation, which confirms that a transmission has been sent by receiving acknowledgement from the receiver, which functions as a hand shake of a successful transmission. It is possible to add data to be sent with the acknowledgement for the original transmitter.

The receiver decides if received packet is correct by calculating two different cyclic redundancy check (CRC) sums from the data and comparing them to the CRC sums received, along with the data. A failed CRC sum comparison results in a discarded packet. All the packets also have a packet id (PID) which ensures that the same packet will not received twice. CRC and PID are not visible for the user and both can be corrupted during transmission which normally leads to a discarded packet, but both still enable possibility of accidental reception and approval of corrupted packet.

The radio enables a 32 -byte transmission per one packet. In the basic reconfiguration process data can be sent in full packages of 32 bytes. This requires 4096 transmissions for reconfiguration process to complete.

With cognitive radio property in use, 16 bytes out of the total available are reserved for actual data. The remaining 16 bytes are divided among Object ID, Channel Number, Configuration Page Number, Configuration Command and non-used frame. Figure 2 displays structure of one data Frame. Nine bytes are free for possible future expansions.

The Process of transmitting a new configuration file requires the transmitting and storing of 131072 bytes. It is executed one memory page at a time; 16 bytes of configuration data is read from the memory of configuration device at once and stored to buffer register. These 16 bytes are then sent in a single packet, which means a total of 8192 iterations of transmission.

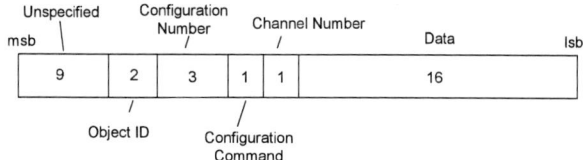

Figure 2. A transceiver vector, total of 32 bytes.

IV. USE OF COGNITIVE RADIO

Before the transmission can start, the transmitter has to find out the most appropriate channels. The process is done as follows: The transmitting radio is the master in the process. It uses another object as a slave. The master sets itself to operate as the receiver and starts listening to the available channels.

A master object can read the carrier level, after a certain amount of time, after listening to the channel. The value is registered into the CD-register included in the radio's SPI-interface. This information is stored for later use. After that the master object changes to the next possible channel and repeat the previous steps.

After all the channels have been gone through, the stored information is processed. All the channels that had a certain carrier level detected are used. The channel quality level is represented as an eight-bit binary number where a high value means a jammed channel. The number of used channels can be limited to fewer if needed. The list of channels chosen by this algorithm is used during transmission of the configuration. The list of used channels is stored only in the object that transmits the configuration file. The process lists all the used channels in the order they are accepted. From this created list, the channels are then used in the order of skipping 2 consecutive channels according to the (1). In (1) k represents the original channel indexing and m new, mixed indexing.

$$m(k) = (3*k \bmod 126) + \mathrm{floor}(k/42),\ k=0,1,2,..125. \quad (1)$$

The transmission algorithm is based on transmitter-based commands. Each packet carries data and channel information in Channel number and Configuration Number. If the channel needs to be changed, a new channel will be designated in Channel Number. The next transmitted packet will be on the new designated channel within a specified time window.

Based on the number of robust channels, the transmitter decides the number of frequency hopping channels used for the transmission. The transmission can also be done using only a single channel, or a limited number of channels. While the transmission is taking place, the receiver keeps count of the received packets by following Configuration Number, as described in Figure 2. Configuration Number consists of three bytes. It can express all the packets needed to finish the configuration until the configuration file exceeds 286 Mbytes. This ensures that bigger FPGA devices and memory devices can be used if needed. Object ID always points to the object that sent the data. This field limits the number of smart objects in one swarm to the amount of 65536. Length of an Object ID -field can be increased on later versions. The communication uses the same protocol both ways.

The frequency hopping procedure works as follows: The reconfiguration process starts from the channel zero. After the transmitter generates list of channels it will use, it takes one Channel Number at a time and copies it to transmission payload and then uses that channel for the next transmission. The receiver sets the radio frequency to match the next channel number through SPI, after receiving new Channel Number on data package.

If any of the data packets are dropped for some reason or another, Configuration Number, when received, skips one or more values. In this case, a re-start command, which has been included in the acknowledgement, is sent to the transmitter. The restart command is given in Configuration Command -field. The transmitting object responds to this request by restarting the transmission from the location needed and resending the missing values.

If transmitter cannot send data to the receiver in certain channel and does not get acknowledgment within set time window, it returns to the previous operative channel. In this case, also receiver returns to the previous operative channel after set period of time. Simultaneously, transmitter discards channel where it could not send data through. After that transmission continues normally.

After each successful transmission, the radio checks how many times it had to resend packet before it was sent successfully. If this number exceeds the set value, the logic can optionally discard the frequency hopping frequency that was previously used, and continue with the remaining frequencies. If there are no frequencies left, the transmitter scans through the channels again to find appropriate channels.

V. APPLICATION

The future aspects of wireless reconfigurable Smart Objects have been described in these papers [1, 3]. Use of cognitive radio makes reconfiguration process more secure. The discussed properties can also be used when utilizing over-the-air-programming (OTAP) for wireless sensor network platforms (WSN). They have some similarities with Smart Objects but a main difference is that their main purpose is the collection of environmental data [6, 7, 8]. Improvements targeting to the configuration transmission issues have been dealt also with in [5] along with the common issues listed in [4].

Frequency hopping procedure with bad channel removal works also with streaming applications. Possible problem is if all the channels are revealed jammed. This leads to the situation when new channel search procedure is needed. This leads to possible stop for continuous data flow.

VI. CONCLUSION

The processes described in this paper enable a more secure wireless transmission of a configuration file for the FPGA. These functional methods help prevent channel noise, reduce the possibility of data packets missing, and reduce possibility of corrupted data packets to be accepted. More likely, a corrupted packet is noticed and the full re-transmission of a Configuration File is not needed in case of missed package.

With this implement, the communication uses channels that are best suited for the transmission. The list of used channels is decreased after sensing that the quality of the channel used has weakened to a certain level. This allows for communication adjustment to always match the environment.

The time required to complete a configuration process is longer when using a cognitive radio, rather than a straightforward, basic reconfiguration. However, the difference decreases when the channel used is noisy because the implement begins rejecting bad channels. This way the configuration process cannot possibly be halted in case of an occupied or disturbed radio transmission channel.

VII. ACKNOWLEDGMENTS

We would like to thank University Program of Altera and Nordic Semiconductor for the donation of devices and evaluation boards used to implement the prototypes described in this paper.

REFERENCES

[1] H. Heusala. (2007) "Technology trends and Design Aspects of Data Processing Cores of Future Small Smart Objects". DIPSO 2007 proceedings.

[2] Storrs Hall J. (2005) Nanofuture – What's next for Nanotechnology. New York, Prometheus Books.

[3] K. Nikunen, H Heusala, J Komulainen (2008) "Configuration method of Wireless Smart Object". University of Oulu, Department of Electrical Engineering. DIPSO 2008 proceedings

[4] Qiang Wang; Yaoyao Zhu; Liang Cheng; "Reprogramming wireless sensor networks: challenges and approaches". Network, IEEE May-june 2006 Volume 20, issue: 3, pp. 48-55

[5] Hagedorn, A.; Starobinski, D.; Trachtenberg, A.; "Rateless Deluge: Over-the-Air Programming of Wireless Sensor Networks Using Random Linear Codes". Information Processing in Sensor Networks, 2008. IPSN '08. Digital Object Identifier 10.1109/IPSN.2008.9. pp. 457 – 466

[6] Hinkelmann H, Reinhardt A, Varyani S, Glesner M; "A Reconfigurable Prototyping Platform for Smart Sensor Networks".

Programmable Logic, 2008 4th Southern Conference Page(s):125 – 130

[7] Portilla J, Riesgo T, de Castro A; "A Reconfigurable FPGA-Based Architecture for Modular Nodes in wireless Sensor Networks," in Proc. 3rd Southern Conference on Programmable Logic, 2007, pp. 203-206.

[8] Prabal Dutta; Grimmer, M.; Arora, A.; Bibyk, S.; Culler, D. "Design of a Wireless Sensor Network Platform for Detecting Rare, Random, and Ephemeral Events"; Information Proc. in Sensor Networks, 2005.

Real-Time Execution Monitoring on Multi-Processor System-on-Chip

Kalle Holma, Tero Arpinen, Erno Salminen, Marko Hännikäinen, and Timo D. Hämäläinen
Tampere University of Technology
Department of Computer Systems
P.O. Box 553, FI-33101 Tampere, Finland.
Email: kalle.holma@tut.fi

Abstract— In system-level design, design space exploration (DSE) produces large amounts of data when exploring myriad of alternatives for application mapping and the underlying platform. Visualization of the essential execution data makes the right design decisions essentially easier. This paper presents Execution Monitor, a versatile monitoring tool implemented in Java, for multi-processor systems-on-chip (MPSoCs). It allows monitoring both the application and the underlying platform in real-time, and also viewing the previously recorded execution trace. Execution Monitor can be used both during the simulation and prototyping. Moreover, the designer can rapidly evaluate in run-time the performance of multiple application mappings via intuitive drag-and-drop mechanism. The case study shows that the visualization of the monitored execution data significantly eases optimizing the performance of the video codec after addition of new application functionality.

I. INTRODUCTION

System-level design needs automated tools to obtain high performance, minimize human errors, and to save design time. Design space exploration (DSE) produces large amounts of data when exploring thousands of alternatives for application mapping, scheduling, as well as for hardware (HW) and software (SW) platform. Extracting the most important issues out of that data may be extremely difficult [1], but this can be relieved by data visualization. For example, observing anomalies in the system behavior is much easier by looking at graphs drawn in real-time compared to textual log file interpretation afterwards.

Visualization is a very effective tool for the system designer as it provides immediate feedback, especially, when rapidly prototyping different system configurations [2], [3]. Visualization tool should not be tightly integrated with a single simulation environment, but it should be able to visualize data from various sources, for instance from execution on FPGA board or from system-level simulators. Moreover, the tool should provide detailed execution information both from the application and the underlying platform and it should be easily extendable to visualize application dependent aspects when necessary.

The main contribution of this paper is a novel tool, Execution Monitor, for visualizing the application execution on a multi-processor system-on-chip (MPSoC) in real-time. It is a generic, modular, and an interactive tool that provides detailed execution information to the system designer in an easily digestable form. In addition, the designer can rapidly evaluate the performance of various application mappings dynamically. We have previously presented automatic back-annotation of the performance information to high abstraction level design environment from FPGA execution [4]. In addition, we have monitored the application execution on FPGA platforms, such as presented in [5]. This paper extends the monitoring to virtual platform simulation as well as for monitoring the execution from collected execution traces.

The rest of the paper is organized as follows. Section II discusses the related work. Section III describes the overall Koski design flow [6] in which Execution Monitor is utilized and Section IV the tool itself. Section V presents an example design case using Execution Monitor, and finally, Section VI concludes the paper.

II. RELATED WORK IN MONITORING

Visualization has not yet become an active research area in system-level design. Pimentel pleads for the development of generic methods and techniques to provide scalable and interactive run-time visualization of system-level computer architecture simulations for DSE [1]. Here, we respond to his request by presenting the Execution Monitor.

For pure SW domain, monitoring the parallel SW execution has been an active research area [7]. Many of the visualization aspects widely used for analyzing the parallel SW in multicomputer domain can also be adapted to real-time SoC domain. In [8], Heath and Etheridge introduce a tool called ParaGraph for message-passing systems. It can visualize e.g. processor utilization, inter-processor communication, and task activities which all are important properties also in the SoC domain. Another monitoring approach for similar multicomputer systems is presented in [9].

In real-time systems, the monitoring process must not interfere with the application it is monitoring. This can be achieved with e.g. additional HW block which is responsible in monitoring the rest of the system, as seen in [10].

There are also approaches for low-level architectural visualizations as presented in [2], [11] but they are intended for visualizing the application execution on instruction level on a single pipelined processor rather than on a whole SoC. Thus, they lack the necessary detail for analyzing a parallel application on an MPSoC.

978-1-4244-2541-9/08 $25.00 © 2008 IEEE

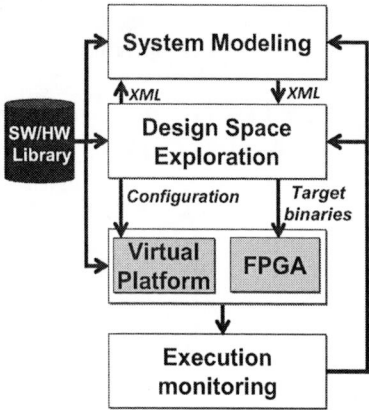

Fig. 1.   Koski design flow.

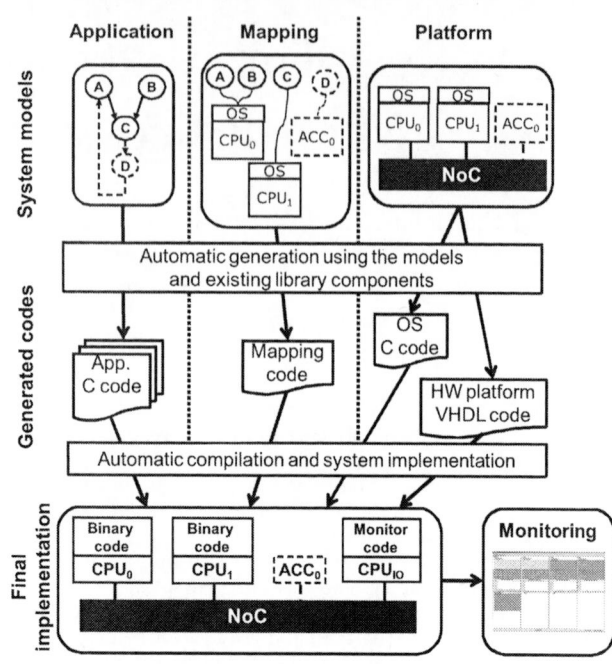

Fig. 2.   Extending the system functionality using Koski. The additions are shown in dashed lines.

For SoCs there are only few tools intended for visualizing architectural values such as processing element (PE) and network-on-chip (NoC) utilization either during the simulation [3], [12] or in run-time [13]. However, they lack the visualization for the application execution running on the architecture. Furthermore, Hassan et al. present an approach for mapping task graphs to multiprocessor SoCs [14]. Their approach is intended only to visualize the task scheduling and energy distribution during simulation.

To summarize, no single tool has been presented yet that:

- visualizes the results both in real-time and afterwards
- visualizes the execution both from application and underlying platform perspectives
- supports multiple execution data sources via generic interface

The presented Execution Monitor meets these requirements and, in addition, it enables the system designer to rapidly evaluate the performance of multiple mappings by moving the tasks between threads or processing elements (PEs). The information collected during the monitoring process can also be utilized for run-time resource management [15].

III. KOSKI DESIGN FLOW

Fig. 1 presents a simplified view of Koski design flow [6] which is a fully automated complete framework for designing and constructing MPSoCs. The system is first modeled in a Unified Modeling Language (UML) 2.0 design environment. Second, the model is automatically transformed into abstracted model which is stored in an intermediate eXtensible Markup Language (XML) System Model (XSM) file format between the tools in Koski.

Next, automated DSE optimizes various system parameters, such as resource allocation, task mapping, scheduling, and NoC configuration. The goal of the DSE is to minimize a case-dependent, user-defined cost function, e.g. $runtime \cdot area$, in a heuristic fashion. The system is executed on a virtual platform or on an FPGA board, and the execution can be visualized using the Execution Monitor. Finally, the optimized system from automated DSE and the execution data from virtual

platform or FPGA can be automatically back-annotated to the system modeling environment for further investigation.

Fig. 2 illustrates the implementation flow in Koski when new functionality is added to the system. The system is modeled from three perspectives: application, HW and SW platform, and mapping between them. The example system is originally composed of three tasks (A-C) which are mapped to two processors (CPU_0 and CPU_1). Adding a new task (D) to the application model and mapping it to a separate HW accelerator (ACC_0) causes changes only to the system models. Because the whole design flow from the system modeling to the implementation is fully automated, the system with extended functionality can be monitored without altering the monitoring code or Execution Monitor.

A. Virtual Platform Monitoring

Virtual prototypes are used to obtain early estimates on the behavior and performance of the system. Here, the system components are modeled in a simulator instead of a physical prototype. Simulator approach gives more freedom in exploring the parameters, allows full visibility, and allows SW and HW development to proceed concurrently. Design abstraction with appropriate models and automated model transformations are key enablers in designing complex MPSoCs.

A system-level simulation engine called Transaction Generator (TG) [16] is utilized for executing the application model on a virtual platform. It is implemented in SystemC. TG models the application execution on SW platform and PEs, but the network model is external. It is accessed through Open Core Protocol (OCP) [17] Transaction Level (TL)2 interface. Thus, any network model having an OCP TL2 interface can

978-1-4244-2541-9/08 $25.00 © 2008 IEEE

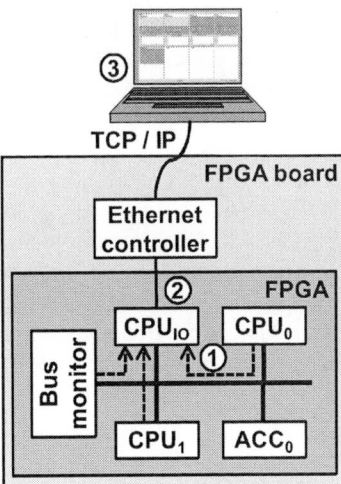

Fig. 3. Monitoring procedure on FPGA.

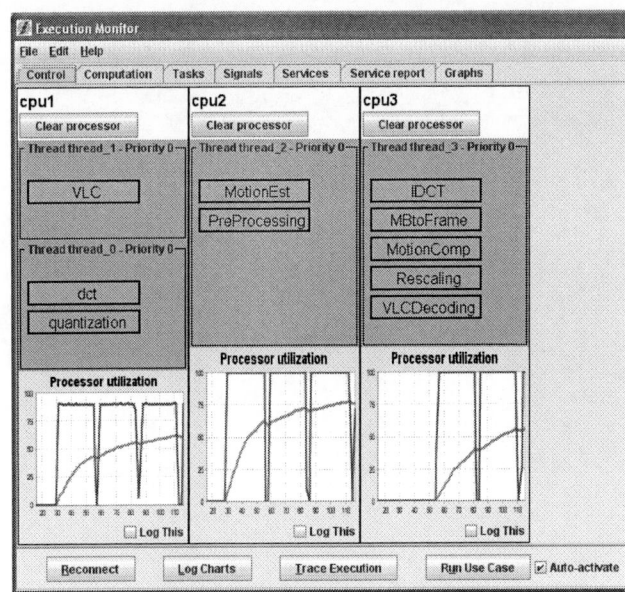

Fig. 4. Control view in Execution Monitor.

be used with TG. The model of computation (MoC) in TG as well as XSM are being adopted and standardized by Open Core Protocol International Partnership (OCP-IP) NoC benchmarking workgroup [18].

The statistics both from the application and the platform are collected during the simulation, and printed to a log file. An intermediate Tool Command Language (TCL) script reads the log file and transforms the execution data to the XML format supported by Execution Monitor.

B. FPGA Monitoring

Models implicitly contain some simplifications and, hence, produce estimation errors. Fortunately, novel automated design tools enable rapid prototyping that gives very accurate information. In Koski, the high-level models can be automatically transformed into implementation on FPGA.

Traditionally, the visibility inside FPGA is very limited. For example, the synthesizable logic analyzers are intended to capture cycle-accurate and bit-accurate trace of internal signals. Although these tools are very useful in low-level HW debugging, they are unsuitable for MPSoC monitoring because they are tightly memory-bound. Thus, only few signals can be monitored simultaneously and the trace time is short.

Our monitoring procedure on FPGA is presented in Figure 3 and consists of three phases:

1) Each processor (*CPU*) sends (dashed lines) at one second intervals execution data of itself and the tasks mapped to it to the dedicated I/O processor (CPU_{IO}). The execution information from HW accelerators (*ACC*) is collected in the SW platform layer of the processor which calls the accelerated functionality. In addition, a separate bus monitoring block measures the bus related execution data.

2) The I/O processor has a monitoring task mapped to it. The monitoring task gathers the execution information from all PEs together and transforms the information to the XML format supported by Execution Monitor.

The sent XML data amount depends on the platform and application and is in the order of few kB/s. The information is regularly sent to a workstation via TCP protocol on Ethernet. Note that this way the memory requirements are low, and the length of the monitored execution is not limited. Moreover, the time scales are different in virtual platform and FPGA prototypes. The former captures statistics in nanoseconds to milliseconds granularity whereas the latter in milliseconds to seconds.

3) Finally, the execution information is visualized on a workstation.

IV. EXECUTION MONITOR

Execution Monitor is a visualization tool implemented in Java that supports any source that provides execution information in a generic XML format. The information in XML format is passed to Execution Monitor through TCP socket.

The monitoring process consists of monitoring and visualization phases. The monitoring phase is a lightweight process that occurs either on FPGA or on virtual platform. On FPGA, the monitoring is implemented in C code and it occurs in real-time. On virtual platform the monitoring code is implemented in SystemC. The visualization takes place in a workstation, and can spend more resources, because it is separated from the actual monitoring phase.

As can be seen from Fig. 4, Execution Monitor contains several views to the system in different tabs: *Control, Computation, Tasks, Signals, Services, Service report*, and *Graphs*. The data is visualized in three forms: graphs, tables, and flow chart. The different views are presented next according to their visualization form.

978-1-4244-2541-9/08 $25.00 © 2008 IEEE   109

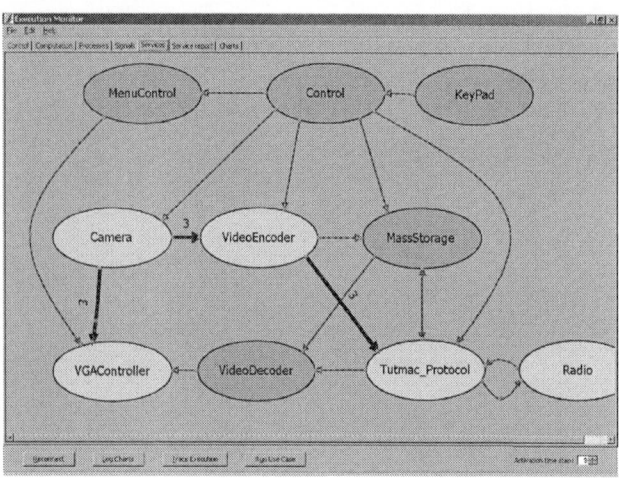

Fig. 5. Services flow view in Execution Monitor.

A. Graph Views

Fig. 4 shows the *Control* view to a system consisting of ten tasks mapped to three processors. Each processor column consists of the current task mapping on top and an optional graph on the bottom. The graph can present e.g. processor utilization, as shown in the figure, size of dynamically allocated memory or amount of sent and received data from other PEs. The time from the beginning of the execution is shown in the x-axis. In this case, the time is presented in milliseconds. Here, *cpu1* has three tasks mapped to two threads whereas the tasks in *cpu2* and *cpu3* are mapped to one thread. Moreover, the processor utilization charts on the bottom show two lines: the darker line is measured from a shorter time frame and the lighter is the average over the whole execution time. Visualizing minimum and maximum values is also possible but not shown here.

It is possible to explore different task mappings in the *Control* view in run-time just by dragging and dropping a task between the threads or PEs. It is a very intuitive way for exploring the mapping, and the results can be seen immediately. Although this method is not intended for exploring large amounts of different mappings it can guide the designer towards the right design decisions, e.g. by providing a good initial mapping for automated DSE. A task can be dynamically remapped also from a processor to an appropriate HW accelerator, or vice versa.

In the *Graphs* view, application specific graphs can be shown. Examples of these include the frame rate of a video codec and the radio throughput of a wireless communication protocol. It is also possible to visualize any selected value from the tabular views together with its minimum, average, and maximum values.

B. Flow View

If the designer has grouped the application tasks into services according to their functionality, their interactions are presented in the *Services* view. It is a high level view to the application under execution. The view consists of services and the connections between them. The inter-service communication signal counts are also depicted beside each active connection. Moreover, the currently active services are shown with a different color.

An example *Services* view of a wireless video player system having ten services is presented in Fig. 5. Services *VideoEncoder* and *VideoDecoder* correspond to the tasks shown in Fig. 4. The tasks of *VideoEncoder* are mapped both to *cpu1* and *cpu2* whereas the tasks of *VideoDecoder* are all mapped to *cpu3*. *Camera*, *VGAController*, *VideoEncoder*, *Tutmac_Protocol*, and *Radio* services are currently active and the connection lines between them are bold indicating ongoing data transmission.

C. Tabular Views

The *Tasks* view is illustrated in Fig. 6. It consists of a single table in which each task is shown in its own row. In the table, multiple statistics for each task are shown: e.g. execution counts, signal queues, response times, as well as data amounts and cycle counts spent for intra-PE and inter-PE communication. The tasks can be sorted according to any column. Here, they are sorted according to their mapping to the PEs. In addition, every value can be easily illustrated in the *Charts* tab just by selecting the value with a mouse button.

Similarly to the *Control* view, the *Computation* view is divided into different sections presenting the PEs in the platform. Each section representing one PE consists of a table which lists the computation properties of the tasks mapped to the corresponding PE. Available information includes the execution counts, average and total execution cycles, and the execution time proportions for each task.

The *Service report* view is similar to the computation view with the exception that the view is divided into different sections according to the services instead of PEs.

Signals view depicts the inter-task communication amounts in a matrix. From this view the designer can observe how much and how often the tasks communicate with each other.

D. Summary

All the monitored properties in Execution Monitor are summarized in Table I. The properties are grouped into three groups: application, mapping, and platform. Further, they are divided into eight categories: application specific, service, task communication, task general, mapping, SW platform, PE, and network. All the values presented in a table form, e.g. task execution cycles, can be visualized also in a graph.

V. CASE STUDY: ADDING NEW FUNCTIONALITY TO VIDEO SYSTEM

This section presents a case study illustrating the applicability of Execution Monitor to system design. We used a scenario where new functionality representing web client was added to an existing video codec system in Fig 4 and the system was optimized based on the monitored information. All the functionality was modeled by its workload and simulated in SystemC using TG. The workload model of the video codec

978-1-4244-2541-9/08 $25.00 © 2008 IEEE

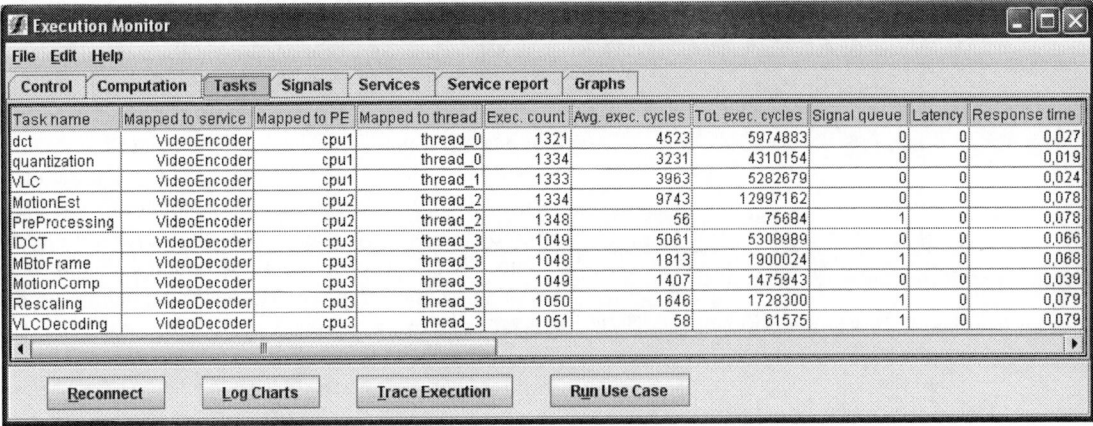

Task name	Mapped to service	Mapped to PE	Mapped to thread	Exec. count	Avg. exec. cycles	Tot. exec. cycles	Signal queue	Latency	Response time
dct	VideoEncoder	cpu1	thread_0	1321	4523	5974883	0	0	0,027
quantization	VideoEncoder	cpu1	thread_0	1334	3231	4310154	0	0	0,019
VLC	VideoEncoder	cpu1	thread_1	1333	3963	5282679	0	0	0,024
MotionEst	VideoEncoder	cpu2	thread_2	1334	9743	12997162	0	0	0,078
PreProcessing	VideoEncoder	cpu2	thread_2	1348	56	75684	1	0	0,078
IDCT	VideoDecoder	cpu3	thread_3	1049	5061	5308989	0	0	0,066
MBtoFrame	VideoDecoder	cpu3	thread_3	1048	1813	1900024	1	0	0,068
MotionComp	VideoDecoder	cpu3	thread_3	1049	1407	1475943	0	0	0,039
Rescaling	VideoDecoder	cpu3	thread_3	1050	1646	1728300	1	0	0,079
VLCDecoding	VideoDecoder	cpu3	thread_3	1051	58	61575	1	0	0,079

Fig. 6. Tasks view in Execution Monitor.

TABLE I

SUMMARY OF MONITORED PROPERTIES.

	Category	Values
Application	Application specific	E.g. frame rate, radio throughput
	Service	Service interaction graph, avg./tot. execution cycles, communication
	Task communication	Signals in/out, avg./tot. communication cycles, communication % of execution time, intra/inter-PE communication bytes and cycles, communication cycles/byte
	Task general	Execution count, avg./tot. execution cycles, execution % of thread/service total, signal queue, execution latency, response time
	Mapping	Task to thread/PE/service
Platform	SW platform	Thread priority, thread avg./tot. computation cycles, computation load, dynamically allocated memory
	PE	Utilization, allocated memory, power, inter-PE communication bytes, SW platform load, avg./tot. execution cycles
	Network	Utilization, efficiency, power, address cycles, data cycles

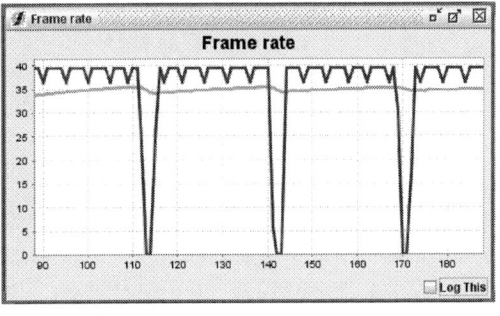

Fig. 7. Application-specific graph showing initial frames per second (FPS). The darker line is a short time average and the lighter the average over the whole execution. The system is able to meet the 35 FPS requirement.

Task name	Mapped to PE	Signal queue
VLC	cpu1	233
WebClient	cpu1	104
dct	cpu1	427
quantization	cpu1	304
MotionEstimation	cpu2	0
PreProcessing	cpu2	0
IDCT	cpu3	1
MBtoFrame	cpu3	0
MotionCompensation	cpu3	0
Rescaling	cpu3	1
VLCDecoding	cpu3	1

Fig. 8. Signal queues for extended application. *cpu1* becomes the bottleneck and the signals accumulate to it.

was profiled from real FPGA execution trace whereas the model of the web client was only an early estimate of its behavior. Hence, it was modeled using a single task. This exemplifies the use of mixed-level models in Koski.

The performance requirement of the video codec was set to 35 frames per second (FPS). Thus, an external event representing the camera triggered at 35 Hz frequency. The HW platform consisted of three processors connected through a shared bus. The operating frequencies of the processors were 150 MHz, 120 MHz, and 120 MHz. The frequency of the bus was 100 MHz. Monitoring confirmed that the initial system met the FPS requirement, as presented in Fig. 7.

Next, functionality for the web client was added to run in parallel with the video codec. The web client was mapped to *cpu1* because it was observed that the utilization of *cpu1* was the lowest in the original system (see Fig. 4). It was observed that the performance of the video codec was now only 14 FPS. In addition, *cpu1* was fully utilized whereas the utilizations of the other two processors decreased. Thus, *cpu1* clearly became the bottleneck and could not forward the data fast enough to the other processors.

This could also be observed from the task signal queues in Fig. 8 where the tasks are sorted according to the mapping.

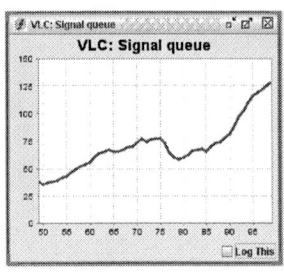 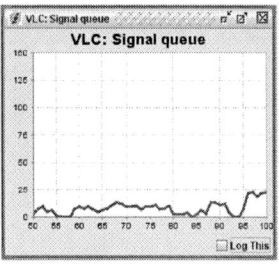

(a) Before mapping exploration.   (b) After mapping exploration.

Fig. 9. Signal queues for task *VLC* before and after mapping exploration.

The environment (model of the camera) produced raw frames so fast that they started accumulating at the *cpu1*. Further, the queues kept increasing as the execution advanced. It should be noted that in the simulation the signal queues have no upper bounds. A signal may carry arbitrary data types, for example integers, arrays, lists etc., and their size is not limited either.

We then tried remapping the application because the workload of the processors was clearly imbalanced. The mapping was done manually so that all the encoder tasks were mapped to *cpu1*, the decoder tasks to *cpu2*, and the web client functionality was mapped alone to *cpu3*. However, this improved the FPS only to 22.

Because the manual mapping did not result in the required performance, the next phase was automatic exploration of the task mapping. The result mapping was a non-obvious because the tasks of the encoder and decoder were distributed among all the processors. Hence, it is unlikely that we had ended to it with manual mapping.

The system became more balanced and the video codec performance increased to 30 FPS, but it did still not meet the required 35 FPS. These observations would have been impossible to make without having the information both from the application and the platform perspectives. *Cpu1* was still the bottleneck and the signal queues of the tasks mapped to it kept increasing. However, they were not increasing as fast as with the unoptimized mapping, as presented in Fig. 9. Fig. 9(a) illustrates the queue before the mapping exploration and Fig. 9(b) after the exploration. The signal queues are shown for the time frame of 50 to 100 ms, and the scale of the y-axis is 0-150 signals.

Finally, we performed automated exploration for the operating frequencies of the processors. The result of the exploration was that the frequency of *cpu1* was increased 40 MHz to 190 MHz, and the frequencies of the other two processors were increased 20 MHz to 140 MHz. The monitored frame rate showed that the FPS requirement was met, and the tasks could process all the signals which they received.

VI. CONCLUSIONS

Design automation, execution monitoring, and its visualization are very effective methods in MPSoC design. This paper presented a versatile monitoring concept for MPSoC that can be used both during the simulation and prototyping. It allows monitoring both the HW and SW and also the previously recorded execution trace. Moreover, the designer can rapidly evaluate in run-time the performance of multiple application mappings via intuitive drag-and-drop mechanism. The case study showed that solving the problems in the system performance would have been significantly more difficult without visualization. In addition, without automated DSE the system optimizing would have spent considerably more resources.

Future work will include more detailed monitoring of the SW platform performance. Moreover, extending the real-time related monitoring ascepts, e.g. observing the required maximum execution time for consecutive application tasks, will be an important part of the Execution Monitor development.

REFERENCES

[1] A. D. Pimentel, "A Case for Visualization-integrated System-level Design Space Exploration," in *Proc. SAMOS '05*, July 2005, pp. 455–464.

[2] C. Stolte, R. Bosch, P. Hanrahan, and M. Rosenblum, "Visualizing application behavior on superscalar processors," in *Proc. Info Vis*, 1999, pp. 10–17, 141.

[3] H. C. Kok, A. D. Pimentel, and L. O. Hertzberger, "Runtime Visualization of Computer Architecture Simulations," in *Proc. Workshop on Performance Analysis and its Impact on Design (in conjunction with the ISCA)*, June 1997, pp. 15–24.

[4] P. Kukkala, M. Hännikäinen, and T. D. Hämäläinen, "Performance Modeling and Reporting for the UML 2.0 Design of Embedded Systems," in *Proc. Int. Symp. on SoC*, Nov. 2005, pp. 50–53.

[5] T. Arpinen et al., "Configurable Multiprocessor Platform with RTOS for Distributed Execution of UML 2.0 Designed Applications," in *Proc. DATE*, Mar. 2006, pp. 1324–1329.

[6] T. Kangas et al., "UML-based Multi-Processor SoC Design Framework," *ACM TECS*, vol. 5, no. 2, pp. 281–320, 2006.

[7] N. Delgado, A. Q. Gates, and S. Roach, "A taxonomy and catalog of runtime software-fault monitoring tools," *IEEE Transactions on Software Engineering*, vol. 30, no. 12, pp. 859–872, 2004.

[8] M. Heath and J. Etheridge, "Visualizing the performance of parallel programs," *Software, IEEE*, vol. 8, no. 5, pp. 29–39, Sep 1991.

[9] D. Haban and D. Wybranietz, "A hybrid monitor for behavior and performance analysis of distributed systems," *IEEE Trans. Software Engineering*, vol. 16, no. 2, pp. 197–211, Feb 1990.

[10] J. Tsai, K.-Y. Fang, H.-Y. Chen, and Y.-D. Bi, "A noninterference monitoring and replay mechanism for real-time software testing and debugging," *IEEE Trans. Software Engineering*, vol. 16, no. 8, pp. 897–916, Aug 1990.

[11] P. Marwedel and B. Sirocic, "Multimedia components for the visualization of dynamic behavior in computer architectures," in *Proc. WCAE*. New York, NY, USA: ACM, 2003, p. 13.

[12] P. S. Coe, F. W. Howell, R. N. Ibbett, and L. M. Williams, "Technical note: a hierarchical computer architecture design and simulation environment," *ACM Trans. Model. Comput. Simul.*, vol. 8, no. 4, pp. 431–446, 1998.

[13] C. Ciordas, A. Hansson, K. Goossens, and T. Basten, "A monitoring-aware network-on-chip design flow," *Journal of Systems Architecture*, vol. 54, no. 3-4, pp. 397–410, Mar./Apr. 2008.

[14] M. Hassan, E. Okushi, and M. Imai, "A SystemC simulation modeling approach for allocating task precedence graphs to multiprocessors," in *Proc. ASICON '07*, Oct. 2007, pp. 1205–1208.

[15] A. Rasmus et al., "Flexible Management of Shared Resources on Multiprocessor System on Chip," in *Proc. SPIE Electronic Imaging*, Jan. 2008.

[16] T. Kangas et al., "Using a Communication Generator in SoC Architecture Exploration," in *Proc. Int. Symp. on SoC*, Nov. 2003, pp. 105–108.

[17] Open Core Protocol International Partnership (OCP-IP). (2008, May) OCP Specification 2.2. [Online]. Available: http://www.ocpip.org

[18] E. Salminen, C. Grecu, T. D. Hämäläinen, and A. Ivanov. (2008, May) OCP Network-on-Chip Benchmarks Specification Part 1: Application Modeling and Hardware Description 1.0. [Online]. Available: http://www.ocpip.org

Using Soft Processors for Component Design in SOC: A Case-Study of Timers

M. Ortiz, M. Brox, F. Quiles, A. Gersnoviez, C. Moreno, M. Montijano

Universidad de Córdoba. Departamento de Arquitectura de Computadores, Electrónica y Tecnología Electrónica
Edificio Leonardo Da Vinci. Campus Universitario de Rabanales, 14071-Córdoba, Spain
el1orlom@uco.es

Abstract—**System on Chip (SOC) could be considered as a very useful alternative in the design of real-time systems, especially due to the possibility of integrating several processors in just one FPGA. This strategy enables the use of soft processors to design the system's components, which have traditionally been developed by hardware. In this paper we study a HW/SW co-design of a timer pool for its use in SOC, which is constructed by a Picoblaze soft processor. Our approach offers a novel alternative among hardware and software timers that increases the overall system performance, and achieves a higher precision than software timers with a considerable reduction in cost and area occupied.**

I. INTRODUCTION

Timers play an important role in any scheduler for real-time systems. In this case, the scheduler must be carefully designed in order to have few overheads in the system, especially in timer management. Some authors implement the scheduler in hardware to address the overhead [1, 2, 3]. On the other hand, applications related to periodic data acquisition, motor control, signal generation, pulse counting and loop timeout use a great number of timers.

The use of soft processors can be an alternative in hardware-software co-design. This alternative is explored in [4] where three different scheduler implementations are investigated. A software implementation uses a processor to run the scheduler and the application tasks. A software-software implementation uses a processor to run the application tasks, and a co-processor to run the scheduler. In a hardware-software implementation, the scheduler is implemented directly in the hardware.

Different timer resolutions are required depending on the application. When a low precision is required, a software solution is a good option. On the other hand, if a high precision is demanded, a hardware implementation is necessary.

When a system is considered, it is analysed the number of hardware and software timers to implement depending on the cost and processing capacity respectively. Later on in this paper, we are going to perform a study of HW and SW timers showing a very interesting alternative, especially for SOC, by using a simple soft processor.

II. HARDWARE TIMERS

Hardware timers are timers implemented by hardware circuit logic. They are counters that work to a fixed frequency. They have a high precision and do not create an overhead for the processor. Implementing hardware timers is the most efficient way to achieve timers in a computing system. This is the most accurate option and the one that involves the lowest overhead in the system. However, the implementation cost and the area occupied on the FPGA is high.

Theoretically the number of HW timers could be unlimited. However the occupied size on the FPGA, the power consumption and programming complexity impose some limitations in SOC. Because of this, it is not usual to implement a high number of hardware timers in systems. For instance, a circuit with sixteen autoload timers of sixteen bits implemented into a Xilinx Spartan-xc3s200 consumes 496 Slices of the FPGA, approximately 12% of the available resources, without including the external interface logic.

III. SOFTWARE TIMERS

Hardware timers are limited in a system. Because of this, it is necessary to use software timers for applications that require an unlimited number of timers.

Software timers are a piece of code connected to the system clock interrupt. Each timer is represented by a data structure. A basic data structure is shown in Fig. 1.

```
struct timer {
        int used; /* TRUE if in use */
        TIME time; /* time left */
        TIME period; /* time to wait */
        int *event_timeout; /* set to TRUE at timeout */
} timers[MAX_TIMERS]; /* set of timers */
```

Fig. 1. Basic data structure of timer.

978-1-4244-2541-9/08 $25.00 © 2008 IEEE

The system reserves memory for a fixed number of timers which will depend on the available memory. This memory limits the number of timers that can be defined. However, the number of software timers and the precision of these timers will strongly influence the processor efficiency, as we will see later.

In order to connect to the periodic timer interrupt, a code, shown in Fig. 2, is used:

```
/ Interrupt handler
for all timers objects in use
{
    time is decreased;
    the count timer reaches zero
    {   /if true
        load de initial count;
        event_timeout signalling;
    }
}
```

Fig. 2. Pseudoce of periodic timer interrupt

All the systems present a clock that connects to an interrupt routine generating a basic timing. Software timers are implemented using the clock interrupt routine in order to decrease the associated variables for each time period. These variables are the software timers. Thus, the number of software timers can be considered unlimited if we only estimate the memory capacity.

These timers are a good solution for non real-time systems and all general operating systems use this technique. However, software timers should be used carefully in hard real-time systems. Because this code runs periodically, by introducing an overhead into the system the computing time and precision will limit the number of timers.

Certain hard real-time tasks demand precise timing of events but software timers driving for periodic interrupts only provide precision in the millisecond range. For instance, we obtain a computing time of 8.8 µs for a MicroBlaze processor [5] at 50 MHz and a number of sixteen timers. The overhead is low if the interrupt period is of 1ms. However, if the period decreases to 0.1ms, the overhead involves consumption of about 8.8 %. According to this idea if the number of timers increases to 100, the computing time would be 50 µs, which involves consumption of about 5 % for an interruption period of 1 ms.

Therefore, by using a MicroBlaze processor and software timers, the use of periodic interrupt timers lower than 1 ms is not reasonable.

IV. HARDWARE SOFTWARE CODESIGN OF TIMERS

The simplicity of integrating soft processors on FPGAs allows us to implement particular functions by using simple soft processors. The use of soft processors to carry out SOC components provides an efficient solution in terms of resources and cost. The soft processor solution is very attractive since its area size is low. In this paper, we present a timer pool by using a simple soft processor such as a PicoBlaze microcontroller processor [6] that occupies 192 logic cells, which represents just 5% of a Spartan-3 XC3S200 device.

This system of timers presents advantages and drawbacks compared with hardware and software timers. The precision of HW/SW timers will be always lower than the precision obtained by using HW timers and it will depend on the response time of the system of timers. The minimum period of timers and the time to access the external bus will be different, depending on the soft processor frequency and the HW/SW partitioning that the designer has made.

Timers are in charge of increasing the count, carrying out the autoload and attending to the main processor. An important point is the HW/SW partitioning of the system functions. According to this idea, all the non-timing crucial complex control functions will be executed by software. This decision is crucial since it affects the response time. For each particular case, a response time analysis of the system must be performed, considering it as a hard real-time system.

For our study we have decided to minimize the necessary hardware in order to implement the timers. Therefore the tasks related to the increase of timers, the autoload and the timer initiation by the main processor, are going to be programmed. On the other hand the control operations of the timers such as overflow clearing and timer stopping are implemented by hardware.

Fig. 3 shows a block diagram of our timer system with sixteen autoload timers of sixteen bits. The logic that performs the interface to the external bus has not been shown in order to simplify the diagram. The count value and timers load are stored in the 64-bytes scratchpad RAM PicoBlaze.

The interface to the main processor is composed by a command register, an overflow register and a gate register. The pre-scaler is written by the main processor and produces the periodic interrupt for Picoblaze. The IRQ to the main processor is active and it is requested when overflow and gate bits are set for a particular timer.

978-1-4244-2541-9/08 $25.00 © 2008 IEEE

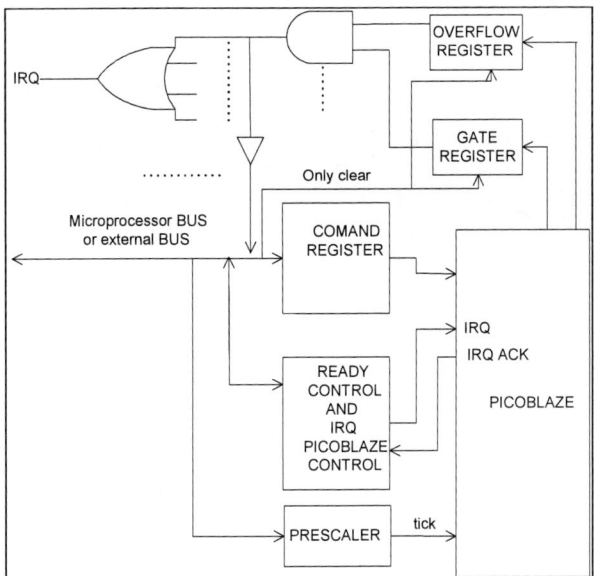

Fig. 3. Block diagram of HW/SW timers

The command register is a 32-bit register. Fig. 4 shows the meaning of the bits. The bits t_3 to t_0 represent one of the 16 timers. If a 16 or 8 bit external bus is used, the writing to the byte control timer (Byte 3) will start the timer with "initial count high and low" values.

bit 7 …….. bit 0

unused	Byte 4
gate x x x t_3 t_2 t_1 t_0	Byte 3
initial count high	Byte 2
initial count low	Byte 1

Fig. 4. Command register.

The overflow and gate registers are set by PicoBlaze and cleared from the external bus. It is not possible to use PicoBlaze in order to clear these bits because the response time of PicoBlaze is high in comparison with the external bus. The main processor finishes the bus cycle before the register is cleared by PicoBlaze, and a new false interrupt occurs. The timers are started by setting the gate bit in the command register and PicoBlaze sets the gate register. The timers are stopped by clearing the gate bit in the gate register by the main processor.

The logic ready and IRQ Picoblaze control are in charge of generating the interrupt for PicoBlaze and extending the external bus cycles, as we will see later.

A. Soft processor code

The main processor requests to Picoblaze to generate interrupts. Thus, the attention to the main processor has the highest priority. Fig. 5 shows the pseudocode, which is executed in PicoBlaze. The program is performed in assembler language.

```
/Main code
while TRUE
{
    wait tick;  /trigger
    for all timers
    {
        disable interrupt;  /critical section begin
        increment timerL;
        increment timerH;
        if  overflow
        {
            set bit overflow register;
            load initial count timerL;
            load initial count timerH;
        }
        enable interrupt;  /critical section end
    }
}
/Interrupt handler (ISR)
read command register;
if gate set  /initialization timer
{
    clear bit overflow register;
    load initial count timerL;
    load initial count timerH;
}
```

Fig. 5. Soft processor pseudocode.

There are two tasks: the ISR (Interrupt Service Routine) and the main code that controls the timers. The ISR code loads the initial count of the timers and sets the gate bit. The main code waits for the tick and the timer count is increased. (All the SW timers in Picoblaze are running continuously). The gate register controls the activation of a specific timer. If increasing this produces an overflow in the timer, the overflow bit is set and an interrupt is received in the main processor.

B. Response time analysis of HW/SW timers.

The worst-case execution times at 50 MHz are 14 µs for the main code and 760 ns for the ISR code. The access to the 64-byte scratchpad RAM (timers) is shared from the ISR and the main code; therefore it will be necessary to perform mutual exclusion. In our case mutual exclusion will involve a blocking time of the interrupt. The maximum block time is 600 ns at 50 MHz. We can conclude that the worst-case execution time of a request from the main processor is the ISR execution time plus the block time. In our case, this execution time is 1.360 µs.

The access time from the main processor is very low since the requests are registered. However a second access

978-1-4244-2541-9/08 $25.00 © 2008 IEEE   115

would produce a larger bus cycle since it is not possible to attend a request while another is being attended. The response time to the main processor request has no influence on the access time if they are performed punctually.

Since the purpose of our study is a hard real-time system, we must perform an analysis which allows us to find out a limit in the response time of the system. We are going to show how it is possible to find out a limit in the response time by performing the following approximations. We will consider that the tasks, the main code and the ISR code are periodic and the relative deadline of the task is equal to its period. The main code is obviously periodic but the ISR code is not. Although this involves a worst-case approximation, we are going to take the ISR as periodic. Then we can use a rate–monotonic algorithm and a very simple schedulability test, chosen because of their simplicity, despite neither being particularly accurate [7].

If we use a rate-monotonic algorithm for both ISR and main code tasks, the two tasks meet their deadlines if

$$\sum_{i=1}^{N}\left(\frac{C_i}{T_i}\right) < N\left(2^{1/N}-1\right) \qquad (1)$$

where N is the number of tasks, C is the worst–case computation time for the task and T is the task period. For our case:

$$\frac{C_{main}}{T_{main}} + \frac{C_{ISR}}{T_{ISR}} < 0.828 \qquad (2)$$

This equation is not ideal for our case, but it allows us to have an analytic equation for response time. With the purpose that the main code meets its deadline, the time that must be reserved between accesses can be determined. For instance, for a case of sixteen timers with a precision of 100 μs (T_{main}) at 50 MHz, a period between accesses of 1'725 μs must be guaranteed.

The times between accesses to the HW/SW timers system can be guaranteed in the main processor program. However we can also have a little counter which extends the bus cycles to the system, acting as an arbiter/scheduler. In this way the arbiter/scheduler is very simple.

This simple arbiter/scheduler is based on a very pessimistic approximation that reserves computing time of PicoBlaze for the increased timer operations. However this approach leads to a very simple arbiter/scheduler. This simple arbiter/scheduler is enough because the bus extensions are only produced when the timers are started.

V. CONCLUSIONS

The use of soft processors in FPGAs facilitates the development of HW/SW co-design of parts of the system. The use of little soft processors simplifies the control logic design and reduces the area size. In particular, the timer pool proposed uses a smaller FPGA area and provides slightly better accuracy than software-only timers. A well-known soft processor which has associated tools that minimize the development time has been used.

Following the proposed procedure, more timers can easily be implemented with other characteristics. Two PicoBlaze could be used in order to have thirty two timers using the same memory block for the program. If a higher number of timers is required, ram memory can be added to PicoBlaze in order to store timers together with the initial count.

Another advantage for the system on chip multiprocessor (MPSOC) is the possibility of obtaining coprocessors whose code is loaded in the execution time of the main processor.

The analysis of these systems, which have been performed with soft processors, must be studied as a hard real-time system and therefore it is possible to design arbiters/schedulers in this way.

VI. REFERENCES

[1] P. Koout, Ganesh, and B. Jacob."Hardware support for real-time operating system". Proceedings of the First International conference on Hardware/Software Codesign and System Synthesis (CODES-ISSS), Newport Beach, California, 2003.

[2] V. Mooney III. Hardware/software partitioning of operating systems. In Design, Automation and Test in Europe Conference (DATE'03), 2003, pp. 338–339.

[3] D. Andrews, D. Niehaus, and P. Ashenden. Programming models for hybrid CPU/FPGA chips. IEEE Computer, v..37(1), 2004, pp.118–120.

[4] M. Vetromille, L. Ost, C. Marcon, C. Reif, and F. Hessel. RTOS Scheduler Implementation in Hardware and Software for Real Time Applications. Proceedings of the Seventeenth IEEE International Workshop on Rapid System Prototyping (RSP'06), 2006

[5] Xilinx Company, "MicroBlaze User Guides", http://www.xilinx.com/products/design_resources/proc_central/microblaze.htm

[6] Xilinx Company, "PicoBlaze User Guides", http://www.xilinx.com/ipcenter/processor_central/picoblaze/picoblaze_user_resources.htm

[7] Liu and J. Layland," Scheduling Algorithms for Multiprogramming in a Hard Real-time Environment", Journal of the ACM, 20(1):46--61, Jan. 1973.

978-1-4244-2541-9/08 $25.00 © 2008 IEEE

Synthesis for Variable Pipelined Function Units

Yosi Ben-Asher   Nadav Rotem

Computer Sci. dep.

Haifa University, Haifa.

Email: yosi@cs.haifa.ac.il

Abstract—Usually, in high level hardware synthesis, all functional units of the same type have a fixed known "length" (number of stages) and the scheduler mainly determines when each unit is activated. We focus on scheduling techniques for the high-level synthesis of pipelined functional units where the number of stages of these operations is a free parameter of the synthesis. This problem is motivated by the ability to create pipelined functional units, such as multipliers, with different pipe lengths. These units have different characteristics in terms of parallelism level, frequency, latency, etc. In this paper presents the variable pipeline scheduler (VPS). The ability to synthesize variable pipelined units expands the known scheduling problem of high-level synthesis to include a 2D search for a minimal number of instances and their desired number of stages. The proposed search procedure is based on algorithms that find a local minima in a d-dimensional grid, thus avoiding the need to evaluate all possible points in the space. We have implemented a C language compiler for VPS. Our results demonstrate that using variable pipeline units can reduce the overall resource usage and improve the execution time.

I. Introduction

Embedded systems are dominated by extensive loop processing and low power budget. In many cases these loops can be compiled to hardware circuits that are significantly faster and more power efficient compared to their software versions. Efficiently compiling from high-level languages, such as C or Java, to gate-level may reduce time-to-market, ease verification, and lower the design costs. Such a tool must achieve good performance in terms of design size, execution time, and power consumption. These restrictions are the fundamental problems in high level synthesis (HLS) [8].

Arithmetic operations are usually thought of as atomic. However, they are often implemented as pipelined operations that take several cycles to complete. This design has several advantages. Most notably is the ability to operate at a high frequency. Variable pipeline stages means that the scheduler is able to decide the pipeline depth of each of the functional units. This is usually done by selecting the proper implementation from a library of modules. In addition to operating at a high frequency, pipelining allows execution in parallel which may benefit some applications. For example, assume that we have six independent multiplications. Instead of using six separate multipliers, we can use one multiplier with six pipeline stages. This will overlap the execution of these six operations. Thus, at $1/6$'th of the hardware costs, it is possible to implement all six operations increasing the execution time by one cycle. In this study, we consider scheduling for the case in which there is the freedom to select the desired number of pipeline stages of functional units. We have developed the variable pipeline scheduler (VPS), to address this scheduling problem, which has not addressed before in the context of hardware synthesis.

When scheduling real code, the problem is complex. The loop's body usually includes resources of different types, namely adders, subtracters, shifters, and memory ports. Clearly the delay, throughput, and other constraints of these operations can affect the configuration of the design. In this work, we aim to find the optimal hardware configuration for programs. It is possible to have different tradeoffs in design. One design may take less resources while another design may execute faster. We define a scoring formula to evaluate the grade of the design. Solving for either best performance on one hand or for the smallest design on the other hand is trivial. A more practical scorer needs to evaluate the design based on the different traits (Cycles to complete, Design Frequency, Size). We optimize the design to achieve a high grade on the scoring formula. The same methodology presented in this paper can achieve optimal results for various convex score functions.

In this study, we present the variable pipeline scheduler (VPS) which is made of a search procedure combined with a modified list scheduling algorithm. The VPS finds an optimized hardware configuration for a given loop which includes numerous variable pipelined operations.

II. The Effect of Using Pipelined Operations on the Synthesis of Memory References

Synthesis of memory operations affects scheduling and resource-selection as r parallel memory references can be synthesized concurrently using r memory ports or more serially using less memory ports. Though practically very few ports are available to each memory module, in this work we assume that memory can be synthesized with more than two memory ports. Since the number of memory ports is limited, they are regarded as a very expensive. The memory latency for each transaction is assumed to be relatively high. Following the approach in [4], memory operations can be synthesized with different values of Memory Delay Factor (MDF) determining how many clock cycles a memory operation takes to be completed. Using $MDF = 1$ is the minimum, indicating that the value is returned by the memory port one clock cycle after the memory address has been placed in a memory port. If the clock frequency is dominated by the memory latency then larger MDF values may allow operating at a higher clock frequency. However, larger values of the MDF can lead to

978-1-4244-2541-9/08 $25.00 © 2008 IEEE   117

no-op clock cycles wherein the scheduler cannot find useful operations to fill the MDF rows between the intialization of a memory operation and its completion. Consequently if the program is not heavily memory bound and we can schedule functional units and other calculations during the delay phase, then having a larger MDF value may benefit the HLS. Obtaining higher MDF values without increasing the resulting execution time is an important goal for the search procedure.

III. RELATED WORKS

There are many works in HLS that implement forms of pipeline synthesis. However, to the best of our knowledge, the synthesis of pipelines with a variable number of stages of functional units has not been addressed. Therefore, a short survey of the main types of pipelined synthesis is in order.

A basic concept common to all of these systems is to pipeline a DDG of operations by inserting registers along cuts in the DDG. A good representative of this type of pipelining synthesis is [23], who describes a system which implements a PCU (pipelined control unit) for vectorized code. Weinhardt used the idea of pipelining parallel operations to pipeline vector operations in a loop. The combinatorial parts of the computations that are not included in a dependency cycle are pipelined in Weinhardt's synthesis by adding registers in the appropriate cuts. Special care is given to optimize the location of the cuts and registers. Another approach to the pipelined synthesis of loops that was studied extensively in HLS is modulo scheduling (MS) [14] which is a well known compiler optimization. MS is a loop optimization in which some of the intra-dependencies of the loop's instructions are eliminated by overlapping successive parts of successive iterations. The main difficulty involved with MS synthesis is to find a pipelining of the loop's iterations such that resource and delay constraints are preserved. Perhaps the most general way to combine MS constraints is to model all constraints as linear inequalities and use an integer linear programming (ILP) solver to compute an optimized solution. Many constraints such as dependencies, number of registers, power consumptions, and initiation interval can be modeled this way. This approach was used in [13] which attempts to minimize three factors: number of functional units (including memory units), register storage, and wire length. To the best of our knowledge, the issue of determining the desired number of pipeline stages of basic operations and the number of instances in conjunction with combining pipelining and state exploration has not been addressed by previous works. As indicated before, previous works on pipelined synthesis target pipeline synthesis of loops iterations and do not partition basic operations to pipeline stages.

IV. SEARCHING FOR THE LOCAL MINIMA

Here we describe the search procedure used by the VPS to explore the HC (Hardware Configurations) space. Each point in the HC-space is a tuple with d coordinates. As in [4], evaluating each point and finding the global minima is an expensive operation that can not be completed in the short

times needed for interactive work. Note that the evaluating of a point in the HC space requires full scheduling for the 2D table and then computing a score for that point. The score of a given circuit accounts for both the expected execution time and the amount of hardware resources used. Instead of an exhaustive search, we consider an easier problem that does not require testing/evaluating all the points, namely finding a local minima. We assume that a local minima in the HC space represents a unique state wherein any change in one of the coordinates (increasing/decreasing a resource, delay of pipeline stages) does not improve the score. Intuitively, if we reach a local minima by increasing the number of resources it is unlikely that if we continue to increase the resources we will reach another minima. The reason for this is that if adding more resources does not help the scheduler to improve the score after the first minima, it is unlikely that adding even more resources will suddenly cause the scheduler to improve the score and reach a second minima. Consequently we believe that the HC space contains only very few local minima and that each of them is close to the global minima. We also rely on the assumption that each point in the space we are searching has a unique score, i.e., no two different configurations will receive the same score. For example, adding resources that cannot be used by the scheduler alter the score. Although the scheduler produces the same circuit (it cannot use the additional resources), the score decreases as more resources are used. Under these conditions it is well known that finding a local minima does not require the evaluation of all the points in the space. Consequently we can search for a local minima instead of searching for a global minima.

Searching for a local minima in a d-dimensional space is a well known problem [12], [7], [3]. Though these are known techniques, their use in the context of high-level synthesis is not known.

V. THE LIST SCHEDULING ALGORITHM AND THE OVERALL SYSTEM

We first describe the list scheduling (LS) variant used in this work and then the overall system that uses this LS to assign scores to the sampled points in the HC space. The goal of list scheduling [9] is to schedule a sequence of instructions with dependencies into a 2D table $[T = clock\_cycles \times resources]$ such that the following conditions are maintained. First, data dependencies and delays between the instructions are preserved. Secondly, the number of cycles (rows/stages) used is minimized. The input to the LS is a data dependency graph (DDG) G of operations where each edge $u \longrightarrow v$ is labeled by the real_latency between u and v. There are several features that are unique to the use of LS for VPS. Each row of the 2D table contains operations that will be executed in the same clock cycle. The resources of the 2D table include: multipliers, shifters, dividers, and memory ports. Other resources such as xor-operations or registers are not restricted, and the LS uses them as much as needed. Unlike the regular use of LS for synthesis, the delays on the edges of the DDG are delay factors, not real latencies (the MDF for memory operations, All the function units are pipelined excluding the MDF-delay which is non-pipelined. The LS looks for the first available place in the 2D table to schedule an operation based on its dependencies on other operations that are already scheduled. This can reduce the execution time of the resulting schedule compared to a more naive scheduling where an operation is scheduled on the next free row. When many different stages are assigned to use a single resource, a MUX with many entries is created. Large MUX circuits take up valuable mux-primitives on FPGAs and may not operate at a high frequency due to increased gate chain length. In order to prevent this problem, the scheduler attempts to balance the allocation of the shared resources. When multiple resources are available for scheduling simultaneously, the resource which was scheduled fewer times is selected. This helps to eliminate some of the big multiplexers we have witnessed. The number of resources in the resource table is usually small compared to the number of nodes in G. Hence the LS algorithm can be regarded as a linear time algorithm (in the number of nodes in G) that can be used multiple times to test different combinations.

The system we implemented is based on the LLVM[15] compiler toolkit and is called SystemRacer. The system is comprised of two parts. First, a compiler front-end parses C into the internal compiler representation. Secondly, a synthesis backend creates the hardware description language. In our case, Verilog. The synthesis unit contains the search algorithm which finds the best resources for the given high-level code using the above LS algorithm. We have extended the LLVM optimizations passes in order to expose parallelism. Each type of the LLVM bytecode has a resource reservation table allowing the scheduler to place a given instruction in the 2D table. The reservation tables are generated dynamically based on the current HC point that is evaluated. The search starts by determining the range of the eight dimensional OHC space including the number of arithmetic operations, memory ports, and delay factors. The LS is initially applied to determine an actual upper bound to the maximal number of resources that can be used for each basic block. This is done by running the LS with infinite resources and recording how many resources were actually used. This helps in reducing the search space for some of the parameters.

VI. EXPERIMENTAL EVALUATION

In this work, we report how the implemented search procedure was able to find optimized HC configurations for several benchmarks and what improvements were obtained. The pipelined 32-bit multipliers were synthesized using Xilinx's CoreGen [2] which is an automatic library that can synthesize multipliers with varying numbers of pipeline stages. The selected schedulings are compiled to Verilog and synthesized to FPGA (Virtex-4) using Xilinx's ISE(xst synthesis tool). We used CoreGen's results to estimate the number of LUTs needed for each pipelined multiplication. We have manually tested the frequency and size for each of the pipeline lengths we used. The size of other componentssuch as shifters, adders, and dividers was selected based on known numbers from Xilinx Specs. Some FPGAs, including the Virtex4 we have used, use built-in ASIC multipliers. However, for a fair comparison we

base our results only on LUT synthesized multipliers. This is in accordance with the fact that our synthesis ability is based on LUT synthesis of CoreGen. Namely, we believe that our results are also valid for ASIC circuits. For the kernel benchmarks, we have used constraints which may not necessarily be synthesizable, such as the case of the number of memory ports. However, these constraints demonstrate the design trade-offs and the massive reduction of resources. For the comparison to xPilot we synthesized the output of both programs using ISE.

We have tested our HC exploration system on several benchmarks with different characteristics. Importantly, all of these benchmarks use multipliers. They are profiled in the table below by the pipeline length of the multipliers, number of memory ports, memory delay factor, number of multiplication units, number of cycles the program needed to complete, the latency of the design and the size of the design. In each column there are two values $initial\_state / after\_optimization$: In the initial state of the exploration, where no pipeline is used. The design has as many ports as needed and no restriction on the MDF. The values after optimization represent the outcome of the proposed technique. It demonstrates improvements in the pipeline stages, MDF, and resource usage.

The following benchmarks were taken from MPEG and JPEG implementation, NVidia Cuda SDK[1], libOIL, LINPACK, FFTW, etc. As previously discussed, the latency of the design is directly related to the number of pipeline stages of the multipliers and the other pipelined operations. The size of the design is overwhelmingly affected by the number of functional units. In all of the test cases, the search algorithm improved the score of the designs. Naturally, by enlarging the pipeline depth and memory delay factor the number of cycles for completion was increased. However, when considering the benefits in frequency, design size, and resource usage, the overall design has improved. In the case of Cft1st, the number of cycles increased by 37%, however, the size was cut in half and the frequency of the design was almost three times as high. Figure 1 is a graphical representation of the improvements obtained due to the proposed approach. The bars in figure 1 indicate improvement ratio factors. Based on the assumption that the clock frequency is dominated by either the memory latency or by the multiplication latency, the time is computed by $max((\#cycles \cdot mem\_latency)/MDF, (\#cycles \cdot mem\_latency)/pm)$ hence the time improvement ratio is computed by $time.before/time.after$.

Prog.	Pipe dpth	Mem port	MDF	Muls units	Cycle	freq. (ns)	SizeK LUTs
Fdtc	1/5	62/9	1/3	68/3	16/91	11/3.6	75.5/6.9
Idct	1/2	8/4	1/3	6/2	13/18	11/6.7	6.6/4.7
Cft	1/5	16/7	1/3	8/2	24/33	11/3.8	9.9/4.9
Ddot	1/2	10/6	1/3	4/1	11/19	11/6.7	5.4/3.6
Dscal	1/5	8/3	1/3	7/1	8/23	11/3.8	9.7/3.6
Cpl	1/3	4/4	1/3	2/1	6/15	11/5.1	3.2/3.6
Conv	1/2	2/2	1/2	1/1	6/10	11/6.7	2.1/3.6
Wlsh	1/7	2/2	1/3	1/1	7/11	11/3.6	3.6/2.1
Mat	1/6	16/8	1/8	7/1	7/31	11/3.6	7.7/2.2
Fir	1/3	3/2	1/3	1/1	7/10	11/5.1	1.1/3.6
Mini	1/6	20/4	1/4	15/1	11/47	11/3.6	16.1/2.1
Mlt	1/1	0/0	1/2	4/2	7/8	11/11	4.3/2.1
Sqs	1/3	8/5	1/3	7/2	7/14	11/5.1	7.6/4.7

978-1-4244-2541-9/08 $25.00 © 2008 IEEE

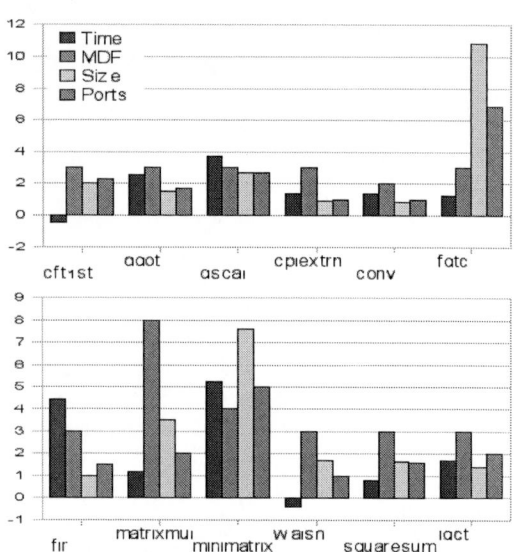

Fig. 1. Improvemnts in execution time, MDF, size, and ports usage.

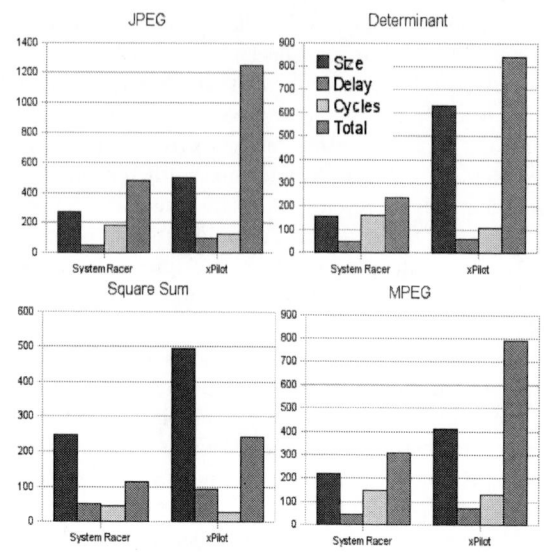

Fig. 2. Comparison with xPilot.

The performances of VPS were compared to xPilot [6] which is based on Integer Linear Programming techniques but does not include synthesis of variable pipelined units as is the case with System Racer. Figure 2 contains four charts comparing the reduction in resources ("size") obtained by System Racer versus the tested increase in the execution time after the synthesis ($cycles \cdot delay$). The results show that using System Racer's pipelined synthesis reduced the size by a factor of $2 - 3$ while the execution time remained about the same. The total columns indicate the cost of the delay, size, and frequency where lower scores are better. All of the benchmarks were synthesized to a Virtex4 FGPA. The frequeny and LUT-count numbers are taken from the reports of the place-and-route phase of the synthesis.

REFERENCES

[1] Nvidia cuda sdk: http://www.nvidia.com/object/cuda_showcase.html.
[2] Xilinx core generator: http://www.xilinx.com/ipcenter/coregen/updates.htm.
[3] D. Aldous. Minimization algorithms and random walk on the d-cube. *Annals of Probability*, 11(2):403–413, 1983.
[4] Y. Ben Asher and E. Schohat. Finding the best compromise in compiling compound loops to verilog. In *IEEE Computer Society Annual Symposium on VLSI*, 2008.
[5] R. Camposano. Path-based scheduling for synthesis. *IEEE Transactions on Computer-Aided Design*, 10(1):85–93, 1991.
[6] Deming Chen, Jason Cong, Yiping Fan, Guoling Han, Wei Jiang, and Zhiru Zhang. xpilot: A platform-based behavioral synthesis system. In *SRC TechCon'05, Portland*, 2005.
[7] C. Tovey D. C. Llewellyn and M. Trick. Local optimization on graphs. *Discrete Appl. Math.*, 23(2), 1989.
[8] Srinivas Devadas, Abhijit Ghosh, and Kurt Keutzer. *Logic Synthesis*. McGraw-Hill, 1994.
[9] D. Gajski, N. Dutt, A. We, and S. Lin. *High-Level Synthesis Introduction to Chip and System Design*. Kluwer Academic Publishers, 1994.
[10] Abhijit Ghosh, Sandeep K. Lodha, and Ranga Vemuri. Hierarchical scheduling in high level synthesis using resource sharing across nested loops. In *glsvlsi, p. 140, Ninth Great Lakes Symposium on VLSI*, 1999.
[11] C. T. Hwang, J.-H. Lee, and Y. C. Hsu. A formal approach to the scheduling problem in high level synthesis. *IEEE Transactions on Computer-Aided Design of Integrated Circuits and Systems*, 10(4):464–475, 1991.

[12] Eva Tardos J. Kleinberg. *Algorithm Design*. Addison-Wesley, 2006.
[13] M. Kudlur, K. Fan, and S. Mahlke. Streamroller:: automatic synthesis of prescribed throughput accelerator pipelines. In *CODES+ISSS '06: Proceedings of the 4th international conference on Hardware/software codesign and system synthesis*, pages 270–275, 2006.
[14] M. Lam. Software pipelining : an effective scheduling technique for vliw machines. In *Proceedings of the ACM SIGPLAN 1988 Conference on Programming Language Design and Implementation*, pages 318–328, 1988.
[15] C. Lattner and V. Adve. The llvm instruction set and compilation strategy. Technical Report UIUCDCS-R-2002-2292, Univ. of Illinois at Urbana-Champaign, 2002.
[16] J. Llosa. Swing modulo scheduling: A lifetime-sensitive approach. In *PACT '96: Proceedings of the 1996 Conference on Parallel Architectures and Compilation Techniques (PACT '96)*, page 0. IEEE Computer Society, 1996.
[17] Maheshwari and Sapatnekar. Efficient retiming of large circuits. *IEEETVLSIS: IEEE Transactions on Very Large Scale Integration (VLSI) Systems*, 6, 1998.
[18] N. Park and A. Parker. Sehwa: A program for synthesis of pipelines. In *25 years of DAC: Papers on Twenty-five years of electronic design automation*, pages 595–601, 1988.
[19] P. G. Paulin and J. P. Knight. Force-directed scheduling in automatic data path synthesis. In *DAC '87: Proceedings of the 24th ACM/IEEE conference on Design automation*, pages 195–202, 1987.
[20] B. Ramakrishna Rau. Iterative modulo scheduling: An algorithm for software pipelining loops. In *Proc. 27th Annual International Symposium on Microarchetecture*, pages 63–74, 1994.
[21] Mukund Sivaraman and Shail Aditya. Cycle-time aware architecture synthesis of custom hardware accelerators. In *CASES '02: Proceedings of the 2002 international conference on Compilers, architecture, and synthesis for embedded systems*, pages 35–42, 2002.
[22] R. Walker and S. Chaudhuri. High-level synthesis: Introduction to the scheduling problem. *IEEE Design & Test of Computers*, 12(2):60–69, 1995.
[23] M. Weinhardt. Compilation and pipeline synthesis for reconfigurable architectures - high performance by configware. *Reconfigurable Architecture Workshop*, 1997.
[24] M. Weinhardt and W. Luk. Pipeline vectorization. *IEEE Transactions on Computer-Aided Design of Integrated Circuits and Systems*, pages 234–248, 2001.

A Two-Phase Return-to-Zero (RZ) Asynchronous Transceiver Circuit for Pipe-Lined SoC Interconnects

Muhammad E. S. Elrabaa

Computer Engineering Department
King Fahd University of Petroleum and Minerals
Dhahran, Saudi Arabia
elrabaa@kfupm.edu.sa

Abstract—**A new delay-insensitive two-phase asynchronous handshaking protocol has been developed. The new protocol utilizes return to zero data format which simplifies communication circuits design significantly. Robust transceiver circuitry that implement this protocol have been developed and simulated using a 0.13μm, 1.2V technology to verify their performance.**

I. INTRODUCTION

Current SoCs not only feature multiple clock domains but also integrate a wide range of blocks (IPs) with various data communication needs and patterns. In addition, SoC designs usually have very short time-to-market demands. This requires efficient design flows that can achieve time closure of the whole SoC in short times. As a result of these requirements two main new design paradigms have emerged to satisfy the communication needs of these SoCs while enabling a reasonable timing closure of the complete SoC design; Network-on-Chips (NoCs) [1-3] and Globally Asynchronous Locally Synchronous (GALS) systems [4-6].

NoCs research aims at developing scalable interconnect architectures that can provide means for routing data between SoC IPs with minimum latency over shared interconnects. While research on GALS aims at developing circuits, methodologies and models for interconnecting synchronous blocks with separate clock domains using asynchronous interconnects. Hence NoCs can be viewed as a special case of GALS. In any case, both share the common problem of designing the point-to-point interconnect circuitry (repeaters, buffers, and pipeline stages) between routers and/or IP blocks. Hence developing high performance robust interconnect circuitry is essential for current and future SoCs.

GALS are categorized into three types based on their communication schemes [6]; pausible clocks; asynchronous,

and loosely synchronous. Pausible clock systems stop (or pause) the clock of the IP block during data transfer. This goes against the fundamental concept of decoupling 'computations' from 'communications' rendering this design style impractical. With each additional input channel, the percentage of idle time would increase even further. Loosely synchronous techniques would require some form of buffering (FIFOs) on the receiver and/or transmitter sides. Again coupling IP design with the communication (interconnect) design. This increases the SoC's design time significantly. Fully asynchronous interconnects offer the highest degree of robustness and decoupling of different SoC design activities. However, latency and throughput are major concerns. Due to handshaking, each datum transfer would require at least two round trips. Interconnect pipelining and repeaters can improve latency and throughput.

Many researchers have proposed new solutions to improve latency and throughput of asynchronous pipelines [7-11]. In [7,8] control pulses are used instead of traditional transition-coded control. This allows faster acknowledge at the expense of more complex circuit design to precisely control pulse widths and math the wire delays. Other researchers proposed a form of wave-pipelining called surfing interconnects [9-11] where they remove two way handshaking altogether. This adversely affects the robustness of circuits and increase the design time significantly. By trading off design time (complexity) for speed another important feature of asynchronous interconnects is sacrificed, flow control. Asynchronous handshaking not only ensure proper timing of valid data but it also allows receivers to control the flow of data, an essential feature in SoCs. Using FIFO buffers instead of handshaking as proposed in [11] would require flow control at higher levels of the protocol stack. Surfing interconnects resembles source synchronous communications with the request signal being used to strobe the data at the receiver and repeaters with adjustable delays as delay lines. Efficient

---

This work have been supported by King Fahd University of Petroleum and Minerals through grant # IN070367

source synchronous on-chip serial communication circuits have been proposed in [12,13] where the data and clock are re-timed at the receiver side instead of repeaters along the control line as in [11]. Again flow control would have to be handled at higher levels of the communication protocol stack, something that SoC IPs might not be designed for.

Another concern with asynchronous interconnects is the use of non-standard CMOS circuits. Hence developing robust asynchronous circuits that can be used as 'plug-and-play' hard macros is highly desirable. This can be achieved through the use of delay-insensitive design techniques.

In this work a robust pipelined asynchronous interconnect system is proposed. The proposed interconnect system combines a new handshaking protocol with an efficient delay-independent circuit implementation that keep the delay to a minimum. The new handshaking protocol is introduced in section II followed by the developed circuits that implement it in section III. Simulations results that verify the operation of these circuits are provided in section IV followed by conclusions in section V.

II. THE NEW HANDSHAKING PROTOCOL

In a typical asynchronous pipeline, Figure 1, data is transferred from one stage to the next via a sequence of handshaking signals. A stage would latch a datum when it receives a Request (**Req**) signal from the preceding stage while the next stage had already indicated that it had latched the previous datum (by de-asserting the Acknowledge signal). Traditionally, there have been two main handshaking protocols for asynchronous data exchange; four-phase handshaking and two-phase handshaking. When combined with dual-rail data encoding these protocols yield delay-insensitive (or at least Quasi-delay-insensitive) operation. The four-phase protocol, illustrated in Figure 2(a), uses a return-to-zero (**RZ**) data format requiring 4 steps (or *trips*) to complete a single datum transfer. The transmitter initiates a datum transfer by driving one of the pre-charged data lines low (or high depending on the pre-charged value). The receiver detects the difference between the data lines using a simple CMOS gate, generates the request, latches in the data if the acknowledge signal coming from the next stage is low and force its own acknowledge high. This signals the transmitter that the transfer is successful and it responds by pre-charging the data lines which is detected at the receiver as the request signal going down. The receiver now responds by lowering its acknowledge signal indicating to the transmitter that it is ready for a new data. Since data is level-encoded, conventional circuits can be used in the transmitter and receiver. The two-phase protocol is very similar except that it uses a non-return-to-zero (**NRZ**) data format (no pre-charging) requiring only two steps to complete a datum transfer as shown in Figure 2(b). For this protocol, data is transition encoded which require special circuitry to detect and handle the two possible transitions.

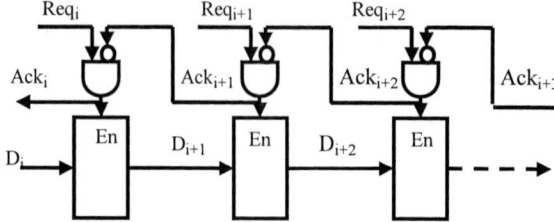

Figure 1. A typical asynchronous pipeline. For dual-rail or 1-in-n encoding, the Req signal is generated from at the receiver.

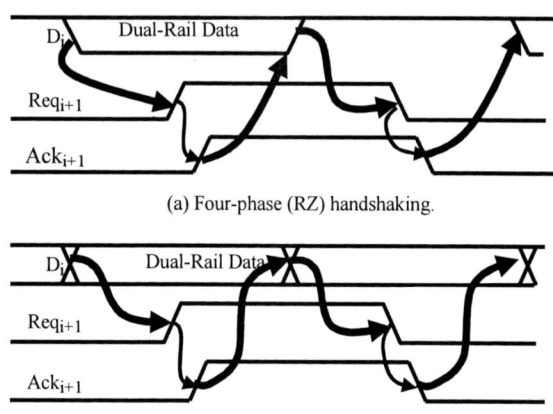

(a) Four-phase (RZ) handshaking.

(b) Two-phase (NRZ) handshaking.

Figure 2. Traditional asynchronous handshaking protocols with dual-rail data encoding for delay insensitive operation. Di is the data transfer initiated at the transmitter. Both Req and Ack signals are generated at receiver side. Thick arrow curves indicate trips from transmitter to receiver or vise versa.

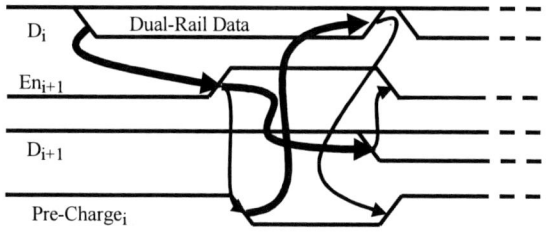

Figure 3. The proposed two-phase (RZ) handshaking. Di is the data transfer initiated at the ith stage. Each data segment is discharged from the transmitter side and pre-charged from the receiver side. Only two trips are required per datum transfer.

Noting that both request and acknowledge signals are generated at the receiver, a new handshaking protocol has been developed that combines the four-phase data level-encoding (i.e. RZ) with the two-phase data exchange steps (*trips*), as illustrated in Figure 3. When a new data initiated at stage *i* is received at stage *i+1*, an enable signal is generated at stage *i+1* (En_{i+1}). This enable signal would initiate the transfer of data to the next segment ($i+1^{th}$ segment) and at the same time activate a pre-charging signal (Pre-Charge$_i$) that would pre-charge the preceding data segment (i^{th} segment). This overlaps the transfer of data to the $i+1^{th}$ segment with the pre-charging of the ith segment. So

978-1-4244-2541-9/08 $25.00 © 2008 IEEE

within two *trips* the data is transferred, similar to conventional two-phase signaling. Because data lines are pre-charged between transfers, then simple level-sensitive circuits can be used, simplifying the circuit design significantly and enabling higher performance. Also, since a data line can only go down, there is no need for an actual data latch. The enable signal can be simply used to drive the data line low using a single NMOS switch, again simplifying the design and reducing the latency of the repeater. Hence each data segment in the pipeline is discharged from the transmitter side and charged from the receiver side. The developed protocol and circuits ensure delay-insensitive operation with no contention between the discharging and charging circuitry on the same data segment.

III. CIRCUITS DESCRIPTION

Figure 4 shows the circuit details of the repeater (transceiver) on one of the dual data lines. The circuit for the other line is similar with Di replaced by Di~ and Di+1 by Di+1~. It has four components as shown in Figure 4(a); a data driver circuit for the next data segment, an enable circuit to generate the control signal (En) for the data driver, a pre-charging driver for the preceding data segment, and a pre-charging control circuit that controls the pre-charging driver. The data driver circuit is a simple NMOS switch with a weak keeper to hold the data line low when the enable signal goes down. The enable circuit, Figure 4(b), has a behavior similar to a Muller-C element. It would assert the enable signal only when the input data (D_i) becomes low while both output data lines are high (indicating that previous data has been transferred). Only when the next data segment (D_{i+1}) is discharged the En signal is de-asserted.

En would remain low as long as the next segment is low. A weak keeper is added to hold the enable signal low when all data segments are in the pre-charged (high) state. The pre-charging driver for the preceding data segment is a simple PMOS transistor and a weak keeper to hold the data line high. The circuit that controls the pre-charging driver, Figure 4(c), would produce a low signal when both En and Di are low. When the preceding segment is charged (i.e. Di becomes high), the pre-charging signal goes high and the data line is held high by the weak keeper.

Unlike previous work, the widths of the En and Pre-Charge pulses are automatically set by the timing behavior of the data lines and need no special circuit sizing.

IV. CIRCUIT SIMULATIONS

Spice simulations using a 0.13μm, 1.2 V CMOS technology were used to verify the operation of the proposed circuits. Figure 5 shows the test pipeline consisting of three stages asynchronous transceivers, a data producer and a data consumer. Wire segments in between are modeled using lumped RC circuits that approximately represent 100μm wires. Transistors were simply sized to achieve 50 ps fall times and 100 ps rise times. No further optimization was carried out to illustrate the robustness of the circuits. Figure 6 shows the simulation waveforms of a single stage transceiver. It shows how the transceiver circuits achieve the appropriate sequence of events on input data, En and pre-charge signals, and output data. It shows also at this wire length, the throughput is ~ 5 Gbs.

To test the complete asynchronous pipeline the following scenario has been simulated; 1st the producer produces data at a constant rate (every 2 ns) while the consumer does not consume any data, Figure 7(a). The figure shows how the pipeline is filled after the injection of 4 data items (all data lines D1-4 are now low).

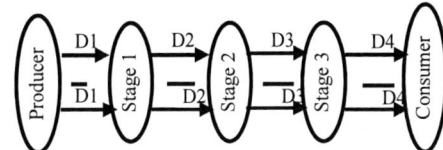

Figure 5. The setup used to test the new transceiver circuit. Wire segments between stages are 100μm long and represented as a lumped RC circuit.

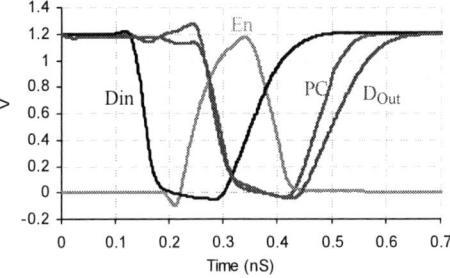

Figure 6. Signal waveforms of one of the transceiver stages. PC is the pre-charge control signal.

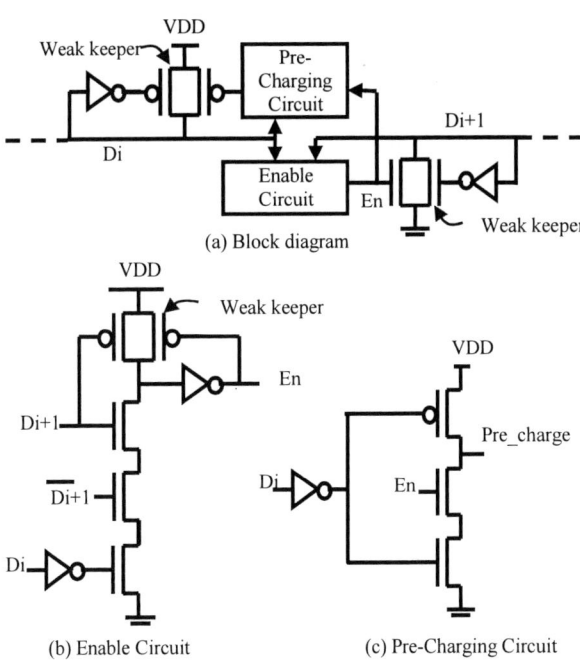

(a) Block diagram

(b) Enable Circuit   (c) Pre-Charging Circuit

Figure 4. Asynchronous transceiver Circuitry

978-1-4244-2541-9/08 $25.00 © 2008 IEEE   123

Next, the consumer starts consuming data items also at a rate of 2ns. Whenever the consumer consumes a data item (indicated by the pre-charging of D4), all the data in the pipeline move one step forward as evident from the consecutive pre-charging and discharging of the data lines in Figure 7(b). The producer continues to inject data at the same rate the consumer is consuming them, keeping the pipeline full while data move along the pipeline. The 2ns injection/consumption rate was used to have uncluttered waveform graphs that clearly show the movement of data along the pipeline. At this wire length, the injection rate could have been made as small as 200 ps.

V. CONCLUSIONS

A new two-phase asynchronous handshaking protocol that utilizes dual-rail RZ data encoding has been developed. Allowing simple circuit implementations that keep minimize latenc. Efficient robust circuit implementation of the protocol has been realized and tested using SPICE simulations. With almost no circuit optimization, the new transceiver circuit can achieve a throughput of 5Gbs with wire lengths of ~100μm. The robustness and delay-insensitivity of the developed circuitry would help decouple computations from communications in the SoC design process, significantly increasing the design productivity.

ACKNOWLEDGMENT

Facilities support by King Fahd University of Petroleum and Minerals is highly appreciated.

REFERENCES

[1] L. Benini and G. D. Micheli, "Networks on chips: A new SoC paradigm", *IEEE Computer*, Vol. 35, No 1, pp. 70 – 78, 2002.

[2] W. J. Dally and B. Towles, "Route packets, not wires: On chip interconnection networks", *Proc. 38thDesign Automation Conference*, pp. 684–689, June 2001.

[3] J. Henkel, W. Wolf, and S. Chakradhar, "On-chip networks: a scalable, communication-centric embedded system design paradigm", *Proc. 17th Int'l Conf. VLSI Design*, 2004, pp. 845-851.

[4] D. M. Chapiro, "Globally-Asynchronous Locally-Synchronous Systems", PhD thesis, Stanford University, October 1984.

[5] S. Moore, G. Taylor, R. Mullins, and P. Robinson, "Point to Point GALS Interconnect", *Proc. 8th Int. Symp. Async. Cir. & Sys.* (ASYNC'02), pages 69–75, 2002.

[6] P. Teehan, M. R. Greenstreet, and G. Lemieux, "A survey and Taxonomy of GALS Design Styles", IEEE Design and Test of Computers, pp. 418-428, September-October 2007.

[7] I. Sutherland and S. Fairbanks, "GasP: A Minimal FIFO Control", ASYNC'01, pp. 46–53, 2001.

[8] R. Ho, J. Gainsley, and R. Drost, "Long Wires and Asynchronous Control", ASYNC'04, pp. 240–249, 2004.

[9] B. D. Winters and M. R Greenstreet, "A Negative Overhead, self-timed pipeline", ASYNC'02, pp. 32–41, 2002.

[10] S. Yang, B. D. Winters, and M. R Greenstreet, "Energy Efficient Surfing", ASYNC'05, pp. 2–11, 2005.

[11] M. R Greenstreet and J. Ren, "Surfing Interconnect", ASYNC'06, pp. 1–9, 2005.

[12] M. E. S. Elrabaa, "A Digital Clock Re-Timing Circuit for On-Chip Source-Synchronous Serial Links", *Proc. IEEE Int. Conf. on Microelectronics*, pp. 206-209, 2006.

[13] M. E. S. Elrabaa, "Portable Clock Recovery Circuits (CRCs) For On-Chip and Off-Chip Serial Data Communication," AJSE Journal, pp. 109-117, Dec. 2007. http://www.kfupm.edu.sa/publications/ajse/

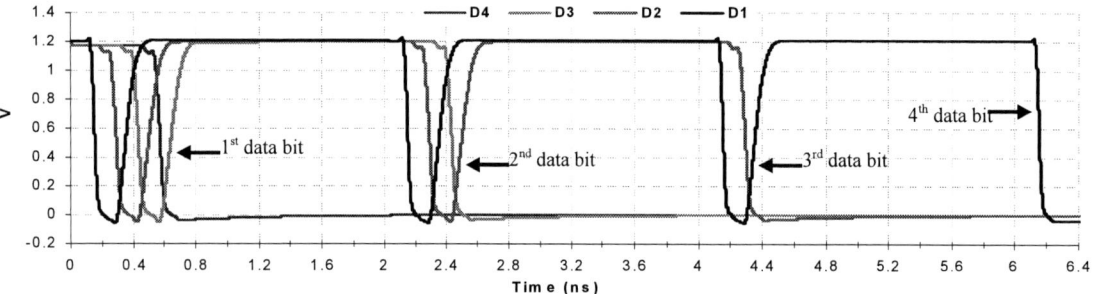

(a) Data is being injected by the producer (D1) at a rate of 2ns while the consumer is not consuming any data at all. After four data injections the pipeline is full and can not accept any new data (all data lines D1-D4 are low).

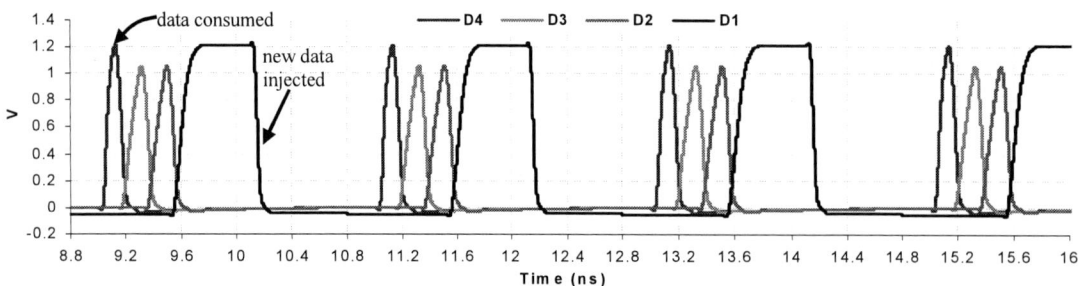

(b) Data is being injected by the producer (D1) and consumed by the consumer (D4) at the same rate of 2ns. The data is moving along the pipeline at a speed of ~200 ps/stage.

Figure 7. The data waveforms along the asynchronous pipeline stages for the two communication scenarios.

High Resolution Flash Time-to-Digital Converter with Sub-Picosecond Measurement Capabilities

Nikolaos Minas, David Kinniment, Gordon Russell and Alex Yakovlev

School of Electrical Electronic and Computer Engineering, Newcastle University

Newcastle upon Tyne, NE1 5UG

Email :{Nikolaos.Minas, David.Kinniment, G.Russell, Alex.Yakovlev} @ncl.ac.uk

Abstract- The paper presents a flash TDC implemented in a UMC 0.13um technology node. The maximum resolution of 0.6ps and a dynamic range of $\pm 17ps$, makes it ideal for measuring set-up and hold time violations and quantifying clock jitter. The method proposed has the effect of reducing the errors introduced by noise and process variations by a factor of two over present techniques. A novel method for overcoming the effects of process variability by counting the number of high outputs is also presented.

I. INTRODUCTION

With the recent advances in VLSI technology, timing issues have become a major concern in the design of high speed digital circuits. Investigation into the cause of timing problems cannot satisfactorily be undertaken by external equipment. This is due to the remoteness of external equipment from the source of the problem. On the other hand, on-chip measurement circuits such as Time-to-Digital Converters (TDCs) are ideal for measuring accurately the relationship between two or more physical events to establish whether they operate according to specifications. Time-to-Digital Converters are well suited for use in on-chip timing measurements systems because they can operate at high frequencies, offer low test time and are relative easy to integrate.

Time-to-Digital Converters have been used in a variety of application areas where accurate measurements of *Time Intervals (*TIs) are necessary, for example, particle physics, astronomy, telecommunication (evaluation of high speed data transfers), military (laser range finders), semiconductor industry (dynamic testing of ICs), and many more. Many of the techniques that today are considered for use in on-chip time measurements have their origins in the field of nuclear physics [1].

Depending on the application, various methods have been proposed for digitizing short TIs, with varied resolution and dynamic range. They are mainly based on fast digital logic

and include the use of fast counters [2], and various CMOS tapped delay line configurations [3,4,5]. The delay line method operates by comparing the STOP signal to the reference START signal. Buffers or inverters are utilised to create a single delay chain on the START signal. On the rising edge of the STOP signal all delay elements outputs are compared using an arbiter or a Flip-Flop. The resulting thermometer code will give an approximation of the time difference between the two signals. The main limitation of this method is that the resolution is equal to a single gate delay.

Resolution below one gate delay can be achieved by using the Vernier Delay Line (VDL) method. In this method each signal is delayed by independent delay lines. Here, the delay (t_1) of the delay element in the START delay chain is slightly greater than the delay (t_2) of the delay element in the STOP delay chain. The resolution of the VDL is determined by $t_R = t_1 - t_2$.

However, for high speed applications it is necessary to deskew the Stop signal due to the high latency introduced by the delay lines. Moreover, the resolution is very sensitive to error factors, such as process variation and noise.

An alternative solution is the Sampling-Offset TDC (SOTDC) [6]. This method rather than using delay lines, exploits the random offset of the sampling circuits used (Flip-Flops or arbiters) to perform time quantisation. This method relies on the assumption that each stage will have a transistor mismatch which will introduce some form of delay. Although this method can achieve resolution as low as 3ps, the offset generated from each stage will vary and each stage will have to be calibrated individually. Moreover, the calibration method requires external equipment and can be difficult to implement.

This paper presents a high resolution Flash TDC with sub-picoseconds resolution. This method relies on asymmetrical MUTEXes with increasing time offset to quantise a step interval rather than delay elements. This technique has the advantages of less area overheads, fast conversion times and high resolution. The effects of process variability are discussed in Section II, where an empirical study has been

978-1-4244-2541-9/08 $25.00 © 2008 IEEE

undertaken to demonstrate how the results obtained can remain unaffected by standard process variation errors. Section III describes the silicon implementation of the Flash TDC in 0.13um UMC technology node. The calibration method using Variable Delay Lines and the results obtained are discussed in Section IV. Conclusions and future work are presented in Section V.

II. EFFECTS OF PROCESS VARIABILITY

The timing between two signals can be compared by the means of a MUTEX circuit. The Signal waveform is compared to that of the Reference, and after resolving any metastability, the circuit indicates which of the two inputs occurred first. In a symmetrical MUTEX the accuracy of the result with respect to the time difference between the two input signals is limited only by noise. This indicates that resolution to an accuracy of 0.1ps could be achieved in a MUTEX circuit with a time constant, τ, of 100ps [7]. However, in reality imperfections in the fabrication stage may affect the transistor dimensions, which will affect the balance of the circuit with respect to the two inputs. To study the effect of process variability, the MUTEX circuit shown in **Error! Reference source not found.** was used to identify the time differences required between the Signal and the Reference waveform to bring the circuit back into balance, for a 10% variation in the dimensions of each transistor.

Using a large number of process variations in device geometries we performed 2000 calculations of the time offsets that showed that the distribution of the resulting offset was normal with standard deviation of 2.03ps. As a result each output in a TDC may not change at exactly the time required, but the probability for it changing at any particular time can be calculated from the cumulative error function with a standard deviation of 2.03ps. This calculation is demonstrated in **Error! Reference source not found.**, where a 10 MUTEX TDC is used as an example. Each stage is set to change state at 1ps intervals (from -4.5 to + 4.5ps). The probability of each MUTEX being high is calculated for an input time difference of 0ps. It can be observed that due to the random variation in the set point of each MUTEX, with standard deviation of 2.03ps, a MUTEX with 0ps input offset is 50% likely to be high and one with -1.5ps is still 84% likely to be high.

In this example the probability of a given number of high outputs can also be calculated. This is done by taking the probability of each individual pattern being the same as the corresponding bit in the required pattern and finding the product of these 10 bit level probabilities. For a 0ps time difference it is expected that there will be 5 high and 5 low outputs. The results showed that, 86% of the TDC outputs will give a count of 4, 5 or 6 high outputs, which is within one correct value of the expected results.

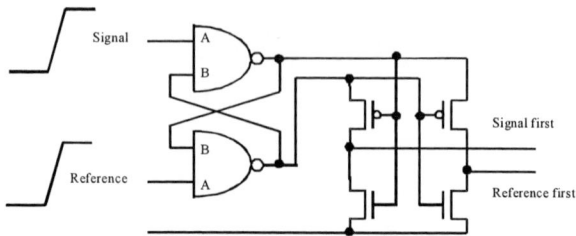

Figure 1.   MUTEX time measurement

TABLE I.   PROBABILITIES OF HIGH OUTPUTS

MUTEX	0	1	2	3	4	5	6	7	8	9
Offset, ps	-4.5	-3.5	-2.5	-1.5	-0.5	+0.5	1.5	2.5	3.5	4.5
Probability of a high output, %	98.8	96.0	89.4	77.3	59.9	40.1	22.7	10.6	4.0	1.2

Since a change in the count of one is equivalent to a 1ps change in the input, we can use these statistics to get the standard deviation of the error when we use the number of high outputs as a measure of time. Here this deviation is 1.1ps rather than 2.03ps. It can be said that with a spacing of 1ps, there are typically 4 MUTEXes contributing to the measurement because they are within the standard deviation of the set points (2ps + 2ps)/1ps = 4, and the effective accuracy is improved by $\sqrt{4}$, from 2ps to 1ps. In general, if there is a random variation in the offset with a standard deviation of σ, then the standard deviation of the measurement error due to this variation will be approximately $\sigma / \sqrt{\frac{2\sigma}{s}}$, or $\sqrt{0.5 \cdot s \cdot \sigma}$ where s is the time step between successive MUTEXes.

This shows that with sufficient MUTEXes, accuracies of much better than σ can be obtained by reducing the time step so that each measurement is the result of many individual MUTEX outputs. Unfortunately this may require a large number of MUTEXes, and the averaging effect is reduced at the extreme ends of the scale. This is further explained in the example of Figure 1. When the input time difference falls in the middle of the scale as in the top register, the uncertainty can be fully contained within the 16 bit output, and the correct results of 8 high and 8 low outputs will be obtained. However if the time difference is at the extreme ends, as in the middle register, the uncertainty presented will not be fully contained and an incorrect value will be obtained. By adding extra stages at the extreme ends of the TDC, as in the bottom register, it is possible to fully contain the uncertainty and thus to obtain the correct results. The number of extra bits that need to be added at each end is given by σ/s, and considerably increases the cost of the TDC as more accuracy is sought.

978-1-4244-2541-9/08 $25.00 © 2008 IEEE   126

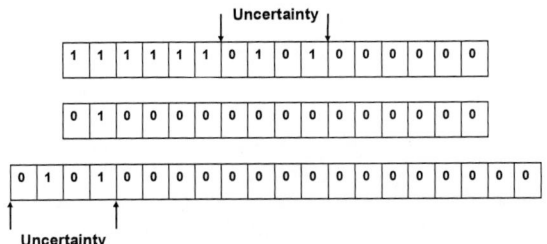

Figure 1. Effects of uncertainty

III. FLASH TDC ON-CHIP IMPLEMENTATION

To demonstrate the effectiveness of the proposed method, a flash time to digital converter was implemented in UMC 0.13um technology node. The TDC consists of two identical 32-stages arrays, with the inputs reversed on the second array so that the negative time difference between the two input signals can be measured. The proposed flash TDC although it relies on the VDL principle, it does not use independent delay lines to quantise a time interval. Instead, the quantisation is performed using asymmetric MUTEXes with progressive built-in offset, where one of the inputs is favoured over the other.

Due to the large input capacitance in the MUTEXes and to ensure fast rise-times, large buffers were used to drive the two input signals. The outputs from the MUTExes will settle on one state until the input time interval matches the offset and then it will settle on the other state. The outputs of the TDC are registered, after a delay of approximately 1ns, using an array of Flip-Flops. The outputs of those Flip-Flops are then connected to a series of cascading adder arrays to determine, in a 5-bit binary format, the time difference of the two input signals. In that way, even if a MUTEX doesn't change state at the time required due to process variability, the correct time difference of the input signals could still be determined by counting the number of high outputs as a measure of time.

Of course the very fact that MUTEXes are used means that the closer the time interval is to the offset, the longer the resolution time (onset of metastability). This would be a limiting factor for bursts of events. When a MUTEX is presented with two rising edges which have a very short time interval separating them, the time taken for the MUTEX to resolve which came first is given by

$$t_{res} = \tau \ln \left(\Delta + t_{offset} \right)$$

Here, Δ is the time interval between input rising edges. For a symmetrically constructed MUTEX, t_{offset} should be zero. The resolution/settling time of a symmetrical or balanced latch has been addressed by several authors, [8] and [9]. However, if there is asymmetry which results in the offset being non zero, then it is possible for the MUTEX to resolve in favour of the signal with the *later* edge.

If signal A is in advance of signal B when the signals are applied to a symmetrical MUTEX, the MUTEX would resolve in favour of A.

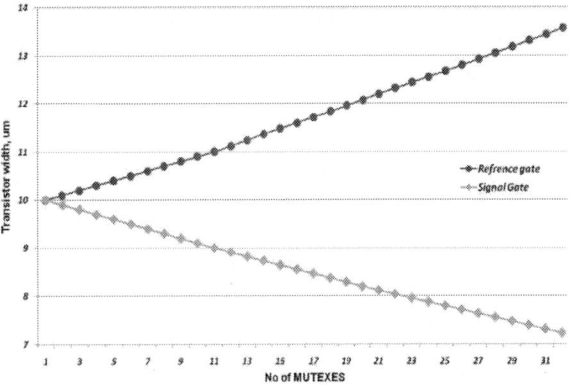

Figure 2. Transistor widths

However, when the same signals are submitted to an asymmetrical MUTEX which favours B the MUTEX will resolve in favour of B until the time interval (of A in advance of B) is greater than the offset.

The input offset of each stage was realised by altering the widths of the n-type transistor of input A in the Signal and Reference gate of the MUTEX circuit. A linear progressive offset was achieved by increasing while decreasing the width of the n-type transistor in the Reference and Signal gate respectively; the resulting widths are shown in Figure 2. The linearity of the input offset also depended upon the time constant of the MUTEX circuit, thus it was decided to increase the nominal widths of all the transistors in the circuit to 10μm.

The silicon implementation of the flash TDC is shown in Figure 3. The proposed design has been implemented in a single die with dimensions of 1525 x 1525um and has 48 I/O. It has been submitted through IMEC for manufacturing and will be available for testing by the end of June.

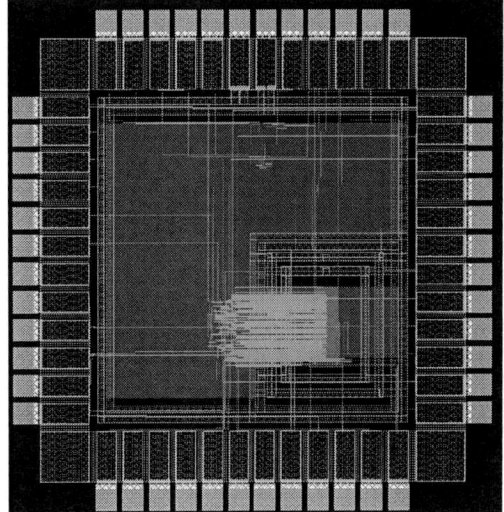

Figure 3. TDC on-chip implementation

IV. CALIBRATION

The calibration of the TDC is performed using two Variable Delay Lines to control the Start and Stop inputs of the TDC. Each Delay line is controlled by an 8-bit counter, which can be externally enabled, so that both negative and positive input time differences can be measured. The variable delay line is based on a current mirror structure and has been proposed by Maymandi-Nejad and Sanchdev [10]. Its main advantage compared with traditional Variable Delay Lines is that the delay behaviour is monotonic.

As can be seen in Figure 4, the main element of the structure is a current starved buffer, M0~M5. A current mirror circuit, M2 and M11, is used to control the current through this buffer. The current through M11 is adjusted by the controlling transistors M6~M9, which have been arranged in a binary fashion so that the number of transistors can be minimised. In order to achieve a small delay increment and large delay range, each variable Delay Line includes two cascade stages similar to that of Figure 4, and is designed to achieve an incremental delay of 0.14ps and a delay range of 0~60ps.

The results of the calibration process can be observed in Figure 5, where it can be seen that after a dead time of 1.9ps the input offset of the TDC is linear with maximum resolution of 0.6ps and a dynamic range of ±17ps. Although the proposed flash TDC lacks wide dynamic range, it is ideal for quantising clock Jitter and testing of set-up and hold time violations.

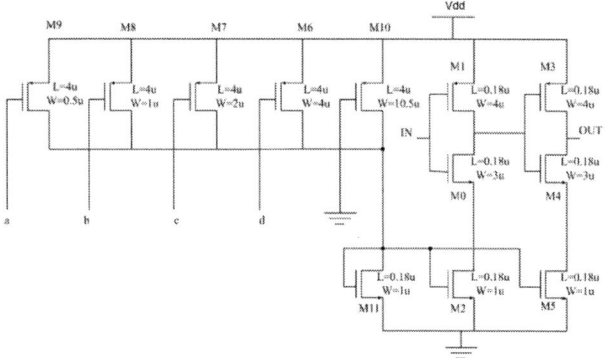

Figure 4. Variable Delay Line [10]

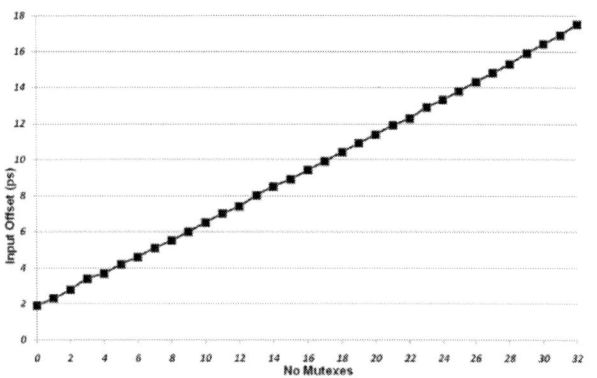

Figure 5. Resolution of each MUTEX

V. CONCLUSIONS

The flash TDC described in this paper offers the possibility to measure time intervals of the order of femto-seconds. The maximum resolution achieved is 0.6ps with a dynamic range of $\pm 17ps$. This method relies on asymmetric MUTEXes with progressive input offset to quantise a step interval rather than independent delay lines. As a result high resolution and low conversion times could be achieved, with relatively small area overheads. Due to the aggressive scaling of VLSI technology, a detailed discussion on the effects of process variation and how it can be overcome it was also presented. The flash TDC circuit has been submitted for manufacturing through IMEC, and the results obtained will be available for discussion during the conference.

REFERENCES

[1] Porat, D.I.: "Review of sub-nanosecond time interval measurements", *IEEE Trans. Nuclear Sci.,* October 1976, **NS-20**, pp. 36-51.

[2] J. Kalisz, R. Szplet, R. Pelka, A. Poniecki, " Single-Chip interpolating time counter with 200-ps resolution and 43-s range,"*IEEE Transaction on Instrumentation and Measurements*, vol. 46, pp. 851-856, 1997.

[3] P.Dudek, S. Szczepanski, j. Hatfield, "A high-resolution CMOS time-to-digital converter utilizing a Vernier delay line," *IEEE Journal of Solid-State Circuits*, vol. 35, pp. 240-247, Feb 2000.

[4] T.E. Rahkonen, J. Kostamovovaara, " The use of stabilized CMOS delay lines for digitalization of short time intervals", *IEEE J. Solid State Circuits*, 1993, vol. 28, pp. 887-894.

[5] M. A. Abas, G. Russell, D. J. Kinniment, "Design of sub-10-picoseconds on-chip time measurement circuit", *Design, Automation and Test in Europe Conference and Exhibition. IEEE Computer Society, Vol.2, 2004*

[6] P. M. Levine, G. W. Roberts, "A High Resolution Flash Time-to-Digital Converter and Calibration for System-on-Chip Testing", *IEE Proceeding- Computers and Digital Techniques*, Vol. 152, No. 3, pp. 415-426, May 2005

[7] D.J. Kinniment, A. Bystrov, A.V. Yakovlev, " Synchronization circuit performance", *IEEE Journal of Solid- State Circuits*, Vol. 37, No. 2, February 2002, pp. 202-209.

[8] Mead, C.A. and L.A. Conway, *Introduction to VLSI Systems*. 1980: Addison Wesley. 396.

[9] Kang, S.-M. and Y. Leblebici, *CMOS Digital Integrated Circuits, Analysis and Design*. 2 ed. 1999: WCB/McGraw-Hill.

[10] M. Maymandi-Nejad, M. Sanchdev, " A digitally programmable delay element: Design and Analysis", *IEEE Transactions on Very Large Scale Intergration (VLSI) Systems*, v.11, issue 5, pp. 871-878, October 2003.

RF Transmitter Architecture Investigation for Power Efficient Mobile WiMAX Applications

Liang Rong, Fredrik Jonsson, Lirong Zheng
Dept. of ECS, School of ICT
KTH Royal Institute of Technology
Stockholm, Sweden
{liangr, fjon, lirong}@kth.se

Mats Carlsson, Charlotta Hedenäs
Catena Wireless Electronics AB
Kronoborgsgränd 19 SE-16446 Kista
Stockholm, Sweden
{mcarlsson, chedenas}@catena.se

Abstract—**Wireless broadband digital communication systems with high spectral efficiency suffer from severe power efficiency problem. Peak-to-Average Power Ratio is reported up to 12dB for WiMAX 802.16e systems implementing OFDM IFFT-1024 and 64-QAM modulation. In this work, outphasing (LINC) and polar transmitter architectures are investigated and compared with direct conversion (DC) architecture. Complete system solution targeting 23dBm output power is evaluated. System level simulation result shows that, with linear power combiner, LINC consumes more power than DC if non-clipping modulation scheme used. And polar system has stringent 3 degree phase matching and 0.5dB gain matching requirements to meet EVM and spectrum mask specifications.**

I. INTRODUCTION

Prevailing wireless digital communication systems are evolving towards highly efficient spectrum usage in both mobile and fixed access. In WiMAX 802.16e-2005 system, OFDM (Orthogonal Frequency Division Multiplexing) and 64-QAM are implemented. However, with Direct Conversion (DC) system (Figure 1), the tradeoff for spectral efficiency is the demanding linearity of the system especially the power amplifier, it is required to linearly output 23dBm average power according to Power Class 2 for mobile WiMAX. With reported PAPR up to 12dB for IFFT-1024/64-QAM system [1], class-A amplifiers will dissipate high percentage power and cause thermal problems, and this situation is even worse for base stations applications.

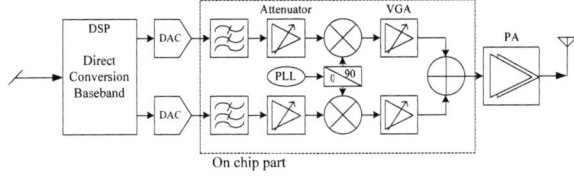

Figure 1. Direct Conversion Architecture Block Diagram

LINC (Linear amplifier with Nonlinear Components) modulation (Figure 2) can avoid power amplifier efficiency obstacle by processing signal into 2 correlated equal envelop, half amplitude signals and combine the amplified signals

---

This investigation is supported by cooperation project between iPack Center, KTH Royal Institute of Technology, Sweden AND Catena Wireless Electronics AB, Sweden.

after PAs, thus nonlinear high efficiency PAs can be used. However, linear power combiner's low efficiency prohibits the direct use of LINC principle, simulation result shows a maximum 7dB loss during the combination of 2 paths. So it is possible that combiner will counteract all the gain achieved by nonlinear amplifier pair.

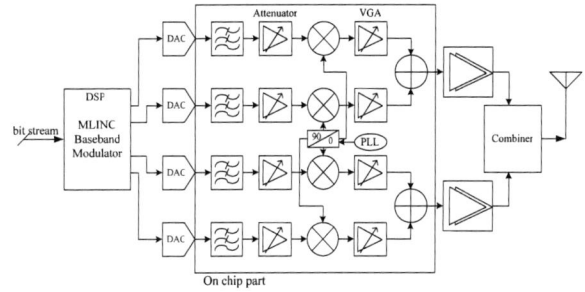

Figure 2. LINC (MLINC) Architecture Block Diagram

Besides these 2 systems, Polar modulation (Figure 3) is gaining more and more attention due to its concise modulation architecture. The separation of phase and envelop signal raise the efficiency of PA without too much increase in system blocks number. Previous work achieving high efficiency of 34% by implementing polar modulation in EDGE system is reported [2]. So it is also a promising candidate to be investigated for broadband system in this work. Envelop tracking on power modulation is often used in polar power amplifier design and simple multiplier model is used here to represent the last stage amplifier.

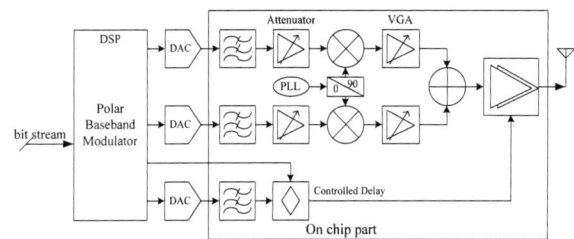

Figure 3. Polar system Architecture Block Diagram

To compare 3 different architectures on the same basis, ADS simulation tool is used because it provides standard test sequence and signal sink modules for WiMAX (WMAN 802.16e) OFDM IFFT-1024/ 64-QAM system, standard EVM and spectrum evaluation is automatically carried out and power value can be calculated by spectrum analysis.

II. WiMAX System Specification

Since WiMAX services are set as a replacement for cable xDSL connection, it gains vast industrial support and a website called WiMAX Forum was founded for standardization purpose. On the other hand, as a broadband wireless air interface, WiMAX is also included as an IEEE 802.16 broadband wireless access standards member. So the specification is distributed in both places. Here a short summary of minimum requirements for WiMAX are cited.

A. Power Class Profile

According to WiMAX Forum power class specification [3], WiMAX mobile system (MS) devices are specified to output up to 23dBm power (class 2 for QAM16) and support a tunable TX dynamic range higher than 45dB. Detailed class category is quoted in Table I.

TABLE I.   Power Class Profile Classification

Class ID	16QAM Tx Pout(dBm)	QPSK Tx Pout(dBm)
Class 1	18≤Tx,Pout<21	20≤Tx,Pout<23
Class 2	21≤Tx,Pout<25	23≤Tx,Pout<27
Class 3	25≤Tx,Pout<30	27≤Tx,Pout<30
Class 4	30≤Tx,Pout	30≤Tx,Pout

As for 64-QAM modulation, 23dBm is also set to be the targeting output power and it is used as a comparison basis in the following context.

B. Frequency Band and Spectrum Mask

After accepted as a formal 3G candidate in ITU-R in Oct. 2007, WiMAX now possess 2 frequency bands including 2.3GHz~2.4GHz (WiBro in Korea) and 2.496GHz~2.69GHz, they clip ISM-2400 band in both sides. To avoid out-of-band emission, WiMAX has very restricted spectrum mask requirements. Here 2 sets of spectrum mask is cited and presented in Table II, one is from ADS WMAN_16e_OFDMA design library and the other is from ITU-R M.1581 [4]. The values are converted to 10 kHz integration bandwidth. From this table, we can find ITU regulations are more restricted.

C. Relative Constellation Error (RCE)

For high order modulation system like 64QAM, EVM is required to be no greater than 3.1%, which converts to RCE of -30dB. Both ADS and IEEE Std 802.16e-2005 have same minimum EVM requirements for all profiles, so here in the system level analysis, highest performance is targeted. The EVM limitation is set to be -30dB to make it suitable for all lower data rate transmission burst types and spectrum mask results are checked to make sure the figure is complying with both requirements. The EVM value is achieved by ADS 802.16e EVM module with package frame of 5.

III. System Simulation and Comparison

A. Direct Conversion System Performance

For direct conversion transmitter, due to the transmission of wideband OFDM signal with large crest factor (Peak-to-Average Ratio), its last stage PA has to be placed at a large back-off position below 1dB compression point. So for DC system, PA's P1dB simulation is compulsory. Besides, imbalance between IQ channels is also a dominant problem. Another signal quality degradation factor is PLL's phase noise. In OFDM system, PLL's noise will be integrated by multiple carriers and it will affect RCE as well. To simplify the simulation, an assumption is made on PA, its OIP3 is 9.6dB higher than P1dB and PA's saturation output power is set to be 6dB lower than OIP3. Simulation result is shown in Figure 4 by sweeping PA's OIP3.

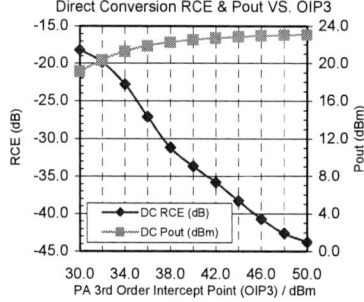

Figure 4.   RCE & Pout vs. PA's OIP3 for Direct Conversion System

From simulation, in order to keep output power of 23dBm and EVM of -30dB, a linear amplifier with OIP3 of 38dBm should be used. While the spectrum mask is violated at 11MHz frequency offset in simulation and this can also be proved by manual calculation. A formal mathematical analysis on multi-tone intermodulation distortion can be

TABLE II.   Spectrum Mask for 10MHz Bandwidth OFDM Signal

Emission Level (dBm)	Frequency Offset (MHz) from Fc								
	5	6	7.144	10	10.572	11	15	20	25
ADS	-8	---	-32	---	-38	---	---	-50	-50
ITU-R M.1581*	-7	-33	NA	-33	---	-45	-48.58	-57	-57

--- No value specified in the offset point. * ITU regulation is for 2496MHz ~ 2690MHz band, value normalized to 10 kHz resolution bandwidth.

found in [5] but the equation is too complex for IM3 value with offset frequency larger than original bandwidth. A simplification of constant subcarrier amplitude is assumed for IM3 calculation here, and since IM3 of this region can only be excited by specific subcarrier pairs as illustrated in figure 5, we just need to accumulate the value in voltage since they are correlated by IFFT algorithm. The simulation and calculation set PA's OIP3 to be higher than 42dBm. This is corresponding to that PA has a P1dB of 32.4dBm, 9.4dB higher than 23dBm output power. This result complies with Crest Factor estimation and provides information for PA selection. And following simulation is carried out after setting PA OIP3 to 42dBm.

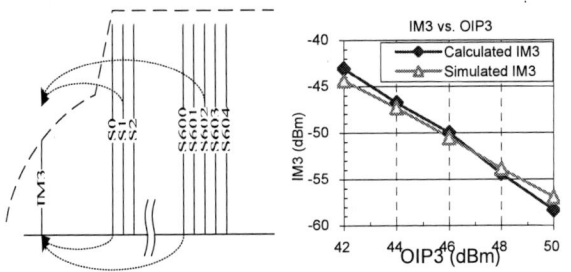

Figure 5.   IM3 Calculation Illustration and Simulation with OIP3

The result for phase imbalance and gain imbalance simulation shows that to keep the spectral margin, they should be kept below 0.8dB and 6 degree for DC system. As stated in [6], phase noise of PLL will be integrated by OFDM signal multiple times. For mobile WiMAX system, since the frequency step for subcarriers is 10.94kHz, phase noise from 10kHz to 100kHz will be of dominant noise source. Here a constant integrated noise value from 10kHz to 100kHz is used for PLL phase noise model. The simulation result is shown in figure 6. To keep EVM of -35dB, phase noise should below -94dBc/Hz.

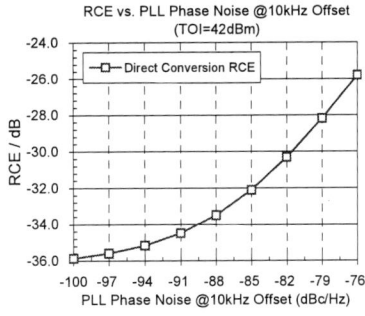

Figure 6.   RCE Degradation by PLL Phase Noise

B.   LINC Modulation System Performance

In order to generate same vector signal as DC system, LINC/MLINC system has two correlated path and the combination of the two equal envelop signal demands highly accurate matching. And the power combiner after PA should be linear combiner like Wilkinson combiner. Since LINC signals may have large angle between two paths, linear combiner's efficiency is low and 7dB loss may happen during the combining process. The distribution of un-clipped LINC signal's vector angle is shown in figure 7. The linear combiner's efficiency can be expressed as equation (1), it is also plotted in figure 7. In this equation, theta is the half value of vector angle.

$$\eta_{comb} = \cos^2 \theta \qquad (1)$$

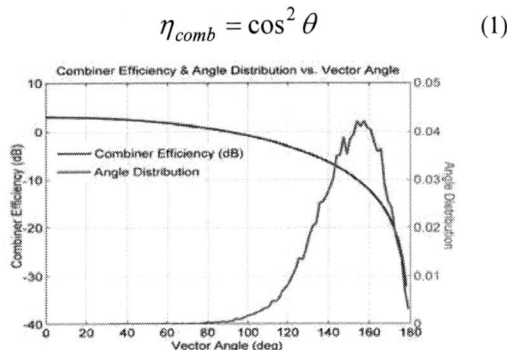

Figure 7.   RCE Degradation by PLL Phase Noise

To solve this problem, MLINC modulation is proposed in [7] and they achieved quite high efficiency. However, due to the un-predictability of signal conversion, MLINC may consume more power in signal processing, thus its efficiency will be a little lower than reported. In this work, a fixed clipping amplitude threshold value is used to test the EVM degradation and power efficiency for LINC system and the result is shown in figure 8.

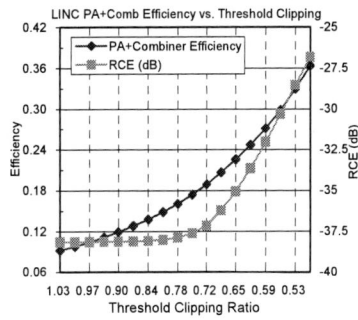

Figure 8.   RCE Degradation by PLL Phase Noise

From simulation, we can achieve 18% efficiency for PA with combiner and maintain same RCE by clipping 75%. And for minimum RCE requirement of -30dB, 30% combination efficiency can be achieved and it is complying with the result of [7], but spectral margin is not promised.

To eliminate out-of-phasing noise, LINC's two correlated vector signals have more restricted matching requirements than un-correlated Cartesian IQ system. With clipping of the signal vector, signals out of interested bandwidth can not cancel each other perfectly and spectrum mask is prone to be violated. In figure 9, the spectral margin result for matching is shown.

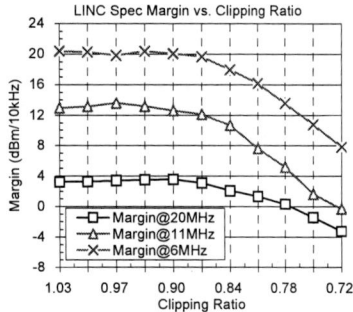

Figure 9.   RCE Degradation by PLL Phase Noise

Even though raising DAC resolution can improve initial spectral margin, in this simulation, clipping ratio is still required to be larger than 78%. Another precaution of LINC system design is that it has not only internal balancing problem between two LINC paths, but also has intra balancing problem inside single LINC path. So the design matching work is much more difficult than DC system.

C. Polar Modulation System Performance

Due to the separation of phase signal and amplitude signal, polar system has little increase in block number and the matching problem between phase and amplitude is of great concern for this architecture [8]. With finite amplifier sensitivity, amplitude signal will also have clipping effect and simulation result is shown in figure 10.

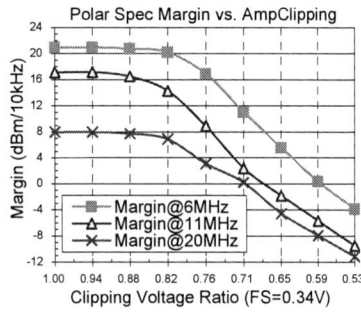

Figure 10. RCE Degradation by PLL Phase Noise

The simulation result shows a clipping ratio of 80% can keep spectral margin of 5dBm with resolution bandwidth of 10kHz and the EVM will still meet requirements. The tolerable phase and gain imbalance error are 3 degree and 0.5dB respectively in simulation, which is about half the value of DC system. Amplitude baseband bandwidth is 3-4 times larger than DC and requires high sampling rate DAC.

D. Power Efficiency Comparison

The total power efficiency for WiMAX system includes baseband and PA blocks, however, the estimation of baseband to RF part can only be compared by block numbers

in system level (# mark in the table). Power estimation and efficiency comparison are summarized in table III.

TABLE III.   POWER EFFICIENCY COMPARISON FOR 3 ARCHITECTURES

	DC	LINC	Polar
DAC #	2	4	3
DAC Resolution	12bit	12bit	12bit
DAC Power (mW)	40	80	>80
Baseband Filter #	2	4	3
Baseband VGA #	2	4	3
Up-Conv. Mixer #	2	4	2
IQ Divider #	1	1	1
RF-VGA #	2	4	2
PA #	1	2*	1**
System Pout (mW)	200	200	200
PA efficiency	12.5%	17%	30%
System efficiency	<12.2%	<15.9%	<26.8%

*   Non-linear PA (class C or above) with combiner.

** Power modulation PA or load modulation PA (class E)

IV.   CONCLUSIONS

With spectral margin limitation and 2 times more blocks, LINC system may not improve efficiency too much comparing with DC system and its matching requirements are more demanding. Polar system shows good efficiency but its matching is also strict. High sampling rate DAC is compulsory and a good polar PA is of first design priority.

ACKNOWLEDGMENT

Special thanks to Mats Carlsson and Charlotta Hedenäs of Catena Wireless Electronics AB Sweden during this work.

REFERENCES

[1] Lloyd S.; "Challenges of Mobile WiMAX RF Transceivers", *IEEE Solid-State and Integrated Circuit Technology*, ICSICT, page: 1821-1824, 2006.

[2] Reynaert P.; Steyaert M.S.J., "A 1.75-GHz Polar Modulated CMOS RF Power Amplifier for GSM-EDGE", *IEEE Journal of Solid-State Circuits*, vol. 40, no. 12, December 2005.

[3] "WiMAX Forum™ Mobile System Profile Release 1.0 Approved Specification (Revision 1.4.0: 2007-05-02)" *http://www.wimaxforum.org/technology/documents/wimax_forum_mobile_system_profile_v1_40.pdf*

[4] "Unwanted Emission Characteristics of IMT-2000 OFDMA TDD WMAN Mobile Stations", 8F/1330-E, 19 June 2007.

[5] Pedro, J.C. and De Carvalho, N.B., "On the use of multitone techniques for assessing RF components' intermodulation distortion", IEEE Transaction on Microwave Theory and Techniques, Vol.47,No.12,Dec.1999.

[6] Masse C., "A direct-conversion transmitter for WiMAX and WiBro applications", "http://www.rfdesign.com", Jan. 2006.

[7] Helaoui M.; Boumaiza S., "A New Mode-Multiplexing LINC Architecture to Boost the Efficiency of WiMAX Up-Link Transmitters", *IEEE Transactions on Microwave Theory and Techniques*, vol. 55, no. 2, February 2007.

[8] Pedro J.C.; Garcia J.A.; Cabral P.M., "Nonlinear Distortion Analysis of Polar Transmitters", IEEE Transactions on Microwave Theory and Techniques, vol. 55, no. 12, December 2007.

Evaluation of Heterogeneous Multiprocessor Architectures by Energy and Performance Optimization

Heikki Orsila, Erno Salminen, Marko Hännikäinen, Timo D. Hämäläinen

Department of Computer Systems
Tampere University of Technology
P.O. Box 553, 33101 Tampere, Finland
Email: {heikki.orsila, erno.salminen, marko.hannikainen, timo.d.hamalainen}@tut.fi

Abstract—Design space exploration aims to find an energy-efficient architecture with high performance. A trade-off is needed between these goals, and the optimization effort should also be minimized. In this paper, we evaluate heterogeneous multiprocessor architectures by optimizing both energy and performance for applications. Ten random task graphs are optimized for each architecture, and evaluated with simulations. The energy versus performance trade-off is analyzed by looking at Pareto optimal solutions. It is assumed that there is a variety of processing elements whose number, frequency and microarchitecture can be modified for exploration purposes. It is found that both energy-efficient and well performing solutions exist, and in general, performance is traded for energy-efficiency. Results indicate that automated exploration tools are needed when the complexity of the mapping problem grows, starting already with our experiment setup: 6 types of PEs to select from, and the system consists of 2 to 5 PEs. Our results indicate that our Simulated Annealing method can be used for energy optimization with heterogeneous architectures, in addition to performance optimization with homogeneous architectures.

I. INTRODUCTION

An efficient multiprocessor SoC (MPSoC) implementation requires automated exploration to find an efficient HW allocation, task mapping and scheduling [1]. Heterogeneous MPSoCs are needed for low power, high performance, and high volume markets [2]. The central idea in multiprocessing SoCs is to increase performance while decreasing energy consumption. This is achieved by efficient communication between cores and keeping clock frequency low.

Mapping means placing each application component to some processing element (PE). Scheduling means determining execution order of the application components on the platform. A large design space must be pruned systematically, since the exploration of the whole design space is not feasible [1]. Fast optimization procedure is desired in order to cover reasonable design space. However, this comes with the expense of accuracy. Iterative optimization algorithms evaluate a number of application mappings for each resource allocation candidate. For each mapping, the application is scheduled and simulated to evaluate the cost of the solution, i.e. the value of the objective function. The objective function may consider multiple parameters, such as execution time, commu-

nication time, memory, energy consumption and silicon area constraints. Figure 1(a) shows the mapping process.

We present an experiment where a set of hardware architectures is generated by random, and applications are mapped on them. Hardware architectures are 2 to 5 PE systems with both singlescalar and superscalar PEs with frequencies from 100MHz to 300MHz. The total area of the system is limited to $8mm^2$. Applications are 300 node acyclic static task graphs (STGs) [3]. Figure 1(b) shows the application, its mapping, and the hardware platform. The application is optimized for each architecture with respect to the energy for that is consumed when the application is run. Resulting energy and execution time values for each architecture are analyzed to find Pareto optimal architectures. Hence, applications are optimized with a single objective (energy), but architectures are analyzed by two objectives.

With the constraints of our experiment, the results show there is a clear trade-off for energy and performance. Low number of PEs means weak performance, but low power. High number of PEs means more performance, but loses energy-efficiency. Also, increasing number of PEs creates demand for automated exploration tools, as the mapping problem becomes more important and increasingly harder.

II. RELATED WORK

Our earlier work [4] evaluated various mapping algorithms to determine optimization convergence rate when application performance was maximized. This paper adds heterogeneous PEs to the problem. Simulated annealing (SA) was found to be an efficient algorithm, and therefore, it is used also in this paper. SA is a probabilistic non-greedy algorithm [5] that explores search space of a problem by annealing from a high to a low temperature state. These methods are also used in our Koski flow [6]. Koski is a high-level design tool for multiprocessor SoCs and applications. Koski utilizes Kahn Process Networks [7] for application modeling.

978-1-4244-2541-9/08 $25.00 © 2008 IEEE

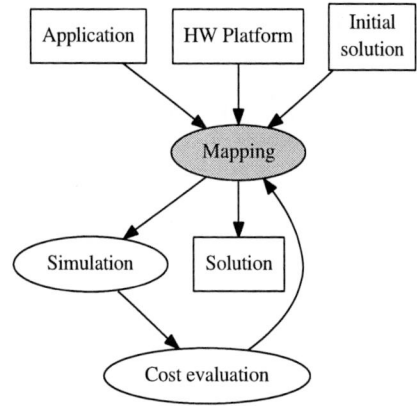

(a) Optimization process. Boxes indicate data. Ellipses indicate operations. This chapter focuses on the mapping part.

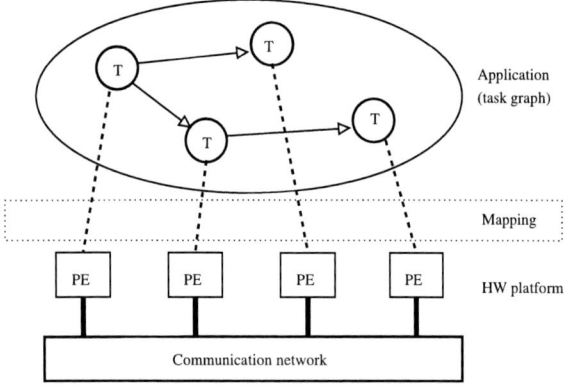

(b) An example MPSoC that is optimized. The system consists of the application and the HW platform. T denotes a task, and PE denotes a processing element.

Fig. 1. Diagram of the optimization process and the system that is optimized

III. EXPERIMENT SETUP

A. Objective And Optimization Algorithm

An experiment was done to investigate trade-offs between energy and application performance. The objective function (1) is to minimize the total energy consumption (static + dynamic). The energy is measured in relative values. It is not a physical energy unit.

$$E = T(P_s + P_d) = T(\sum_{i=1}^{N} A_i f_{max} + k \sum_{i=1}^{N} A_i f_i U_i) \quad (1)$$

where T is the execution time of the application, determined in simulation. N is the number of PEs, A_i is the area of PE i and f_i is its frequency. f_{max} is the maximum frequency of any PE or interconnect, which is at least 200MHz in this experiment because the bus operates at 200MHz. Utilization U_i is the proportion of non-idle cycles of the PE i. The HW architecture defines values A_i and f_i whereas the mapping indirectly defines T and values U_i.

Coefficient k is the factor that changes the relative proportion of static versus dynamic energy. Energy values are comparable when k value is constant. The effect of dynamic power can be eliminated by setting $k = 0$. The static power part P_s of the objective function depends on the number of transistors (relative to A) and their speed (relative to f_{max}). The dynamic power part P_d depends on total capacitance (relative to A), switching frequency f_i, and activity U_i. Supply voltage is assumed fixed.

Simulated annealing algorithm was used to optimize energy (1) by changing application mappings. The algorithm is specified in [8], but modified in two ways. First, the objective function is the application energy on a given platform. Second, the algorithm is run twice for each solution, and the second run always starts from the best solution of the first run. This was done to increase confidence in results, as the SA is stochastic. SA temperature margin value 2 was used to scale initial and final temperatures [4].

B. Simulated HW Platform

The simulated SoC platform for the experiment is a message passing system where each PE has some local memory, but no shared memory. The system is simulated on the behavior level. Each PE and interconnection resource is available for a single action at a time. PE task context switch overhead is 0 cycles, but bus arbitration time is 8 cycles for each transfer.

Figure 1(b) presents simulated HW platform. PEs are interconnected with two shared buses that are independently and dynamically arbitrated. The shared buses limit SoC performance due to contention, latency and throughput. Bus frequency is 200MHz for both buses and they are 32 bits wide. The bus silicon area is $0.1mm^2$ per processor. Each node in the task graph sends a specific number of bytes after computation, thus creating contention on the shared buses. Which ever bus is free at a time is used for communication by using FIFO arbitration, i.e. which ever PE comes first gets the bus.

Table I shows different types of PEs, each presented with a letter. Multiprocessor architectures were varied using these PEs. An architecture consists of 2 to 5 PEs, and the total area has $8mm^2$ upper-bound. Architectures are presented with fingerprint codes from these letters. Parameter p is the average number of instructions per cycle. Frequency f has 3 values: 100MHz, 200MHz and 300MHz. Processor speed is measured in millions of operations per second, which equals $p * f$. A is the area in square millimeters. Each PE can be implemented as a singlescalar or as a superscalar version. The superscalar version can execute $p = 1.8$ instructions per clock. The singlescalar version has area $A = 1mm^2$, superscalar has $A = 2mm^2$. 2 values of p and 3 values of f implies 6 different PEs. A task graph node of n cycles can be computed in $\frac{n}{fp}$ time.

C. Architecture Fingerprinting

Architecture fingerprinting is used to present results. An architecture is characterized by a series of letters from A to F. Letters are labels for different PEs specified in Table I. Letters are assigned in the order of increasing number of operations per second. Letter A is assigned for the slowest PE, and letter F

978-1-4244-2541-9/08 $25.00 © 2008 IEEE

TABLE I

AVAILABLE PROCESSOR TYPES

PE type	f (MHz)	p ($\frac{Ops}{cycle}$)	Speed ($\frac{MOps}{s}$)	A (mm^2)
A	100	1.0	100	1
B	100	1.8	180	2
C	200	1.0	200	1
D	300	1.0	300	1
E	200	1.8	360	2
F	300	1.8	540	2

TABLE II

PROPORTION OF HOW MANY TIMES A GIVEN NUMBER AND TYPE OF PE
WAS IN THE EXPERIMENT'S 141 FINGERPRINTS. TABLE VALUES ARE
EXPLAINED IN SECTION III-D.

PE type	1 PE (%)	2 PEs (%)	3 PEs (%)	4 PEs (%)	5 PEs (%)	PE prop. (%)
A	28.4	10.6	2.8	0.7	0.7	43.3
B	28.4	7.1	1.4	0.0	0.0	36.9
C	30.5	9.2	4.3	1.4	0.7	46.1
D	31.9	8.5	5.7	1.4	0.7	48.2
E	31.9	10.6	2.1	0.0	0.0	44.7
F	29.1	9.2	1.4	0.0	0.0	39.7
Any	0.0	14.2	27.7	33.3	24.8	

TABLE III

ATTRIBUTES AND LIMITS OF THE EXPERIMENT

Attribute	Values
Number of architectures	141
Maximum architecture area	$8mm^2$
Number of PEs in each architecture	2 to 5
Number of 300 node graphs	10
Objective function to optimize energy	$T(\sum A_i f_{max} + k \sum A_i f_i U_i)$, where i is from 1 to N PEs
k values	0, 1, 4
Optimization algorithm	Simulated annealing

E. Applications

The experiment uses ten random task graphs with 300 nodes from the Standard Task Graph set [9]. Random graphs are used to avoid bias in algorithms and results. Nodes of the STG are finite, deterministic computational tasks, and edges denote dependencies between the nodes. Node weights represent the amount of computation associated with a node. Edge weights model the amount of communication needed to transfer results between the nodes. Computational nodes block until their data dependencies are resolved, i.e. when they have all needed data. The edge weights were generated randomly from uniform distribution.

STGs are used because there exists well known efficient and near optimal scheduling algorithms for them [3]. This ensures that the observed differences in optimization results are due to mapping, not scheduling. More complex scheduling properties would diminish accuracy of mapping analysis. However, this experiment is agnostic of the STG structure, and so it could be done with general process networks like Kahn Process Networks (KPN).

F. Experiment Data

Table III shows attributes that were varied in the experiment. Each of the 10 task graphs was optimized and simulated against each architecture. This was done for three values of $k = 0, 1$ or 4 to change static versus dynamic energy balance. Thus, total of $10 * 141 * 3 = 4230$ simulations was run.

G. Software

The optimization software and simulator was written in C language and executed on a 9 machine Red Hat Enterprise Linux WS release 3 cluster, each machine having a single 2.8 GHz Intel Pentium 4 processor and 1 GiB of memory. Jobs were distributed to a cluster with *jobqueue* [10] (version control snapshot *2008-05-30*) by using OpenSSH [11] and rsync [12]. No special clustering software or configurations was used. A total of $8.48 \cdot 10^8$ mappings was evaluated in optimization in 27.4 computation days leading to average of $358 \frac{mappings}{s}$. Rapid mapping evaluation is a benefit of STGs.

IV. RESULTS

Figure 2 plots energy versus execution time for each architecture for $k = 1$ that emphasizes static energy. Figure 3 plots the same data for $k = 4$, i.e. bigger weight on dynamic energy. Energy values are summed and time values are summed for

is assigned for the fastest PE. Each PE in the architecture gets a single letter in the architecture fingerprint. The letters are organized into alphabetical order to facilitate human brain's pattern recognition, i.e. make it easier to see slow, fast and same type of PEs. For example, AAB fingerprint means a three PE architecture with two PEs of type A and one PE of type B. The two PEs of type A are in the beginning of the series to display that there are exactly two instances of A in the architecture and that they are the slowest PEs.

The architecture fingerprint can be extended for heterogeneous interconnections by separating PE selections and interconnection selections with a dash (-). However, the interconnection is the same for all architectures in this paper, and therefore, it is omitted.

D. Random Architectures

141 different architectures were generated. Homogeneous architectures, the architectures with only one type of PE, were inserted manually, and the rest were generated randomly. The total area for each architecture was limited to $8mm^2$. Table II shows the proportion of how many times a given number and type of PE was in all the fingerprints. Rows indicate PE types, and columns indicate proportion of PEs. The last row shows the proportion of architectures with a given number of PEs. The last column shows the proportion of architectures that had at least one PE of that row's type. For example, row A's third column value means that 10.6% of architectures had exactly 2 PEs of type A. Last row's third column value means that 14.2% of the fingerprints had exactly 2 PEs. Row A's last column shows that $0.433 * 141 = 61$ architectures had at least one A type PE. The table has zeroes due to area constraints. For example, an architecture with 4 B type PEs does not exist, because one B type PE takes $2mm^2$, and its associated interconnect area is $0.1mm^2$. Therefore, the total area is $4 * (2 + 0.1)mm^2 > 8mm^2$.

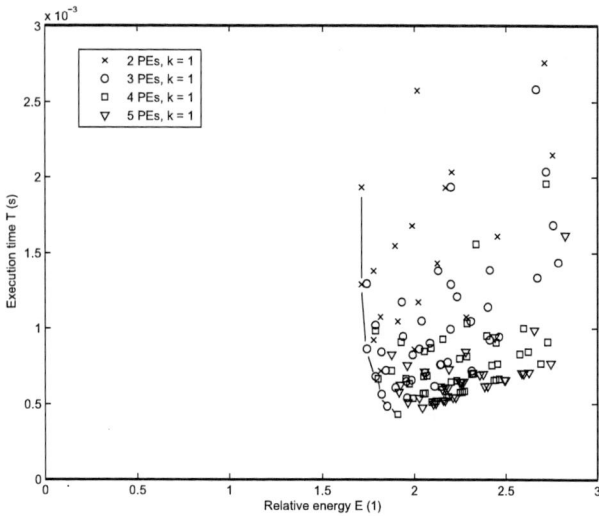

Fig. 2. Energy-time plot for different architectures with $k = 1$

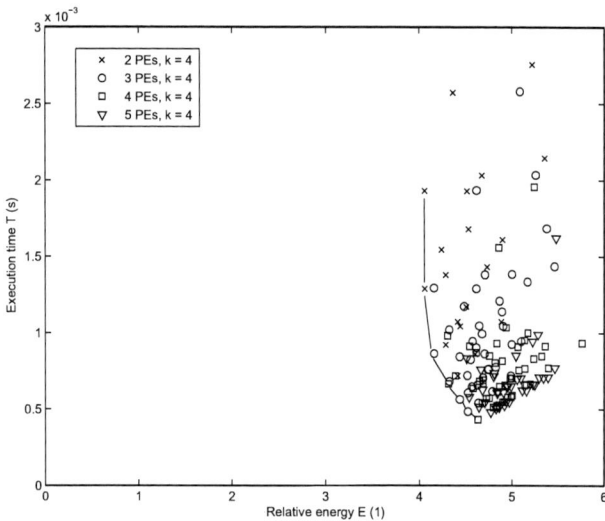

Fig. 3. Energy-time plot for different architectures with $k = 4$

each application. Energy E is the sum of objective function values (1) for a given architecture. Execution time T is the sum of execution times for a given architecture. Time is measured in seconds. Time is comparable even between different k values, but energy is not. A pair (E, T) presents a data point, or architecture, in the figure. E varied in range $[1.7, 3.1]$ for $k = 1$, and $[4.0, 5.8]$ for $k = 4$. Execution time sum T varied in range $[0.43, 3.85]$ms for both cases.

Pareto optimal solution boundary is marked with straight lines. These are not absolute Pareto optimums as not all possible architectures were evaluated. A Pareto optimal architecture is such that improving either energy or execution time leads to worsening the other factor. That is, in the Pareto optimal architectures, there are no two architectures where the other is better in terms of both energy and execution time.

TABLE IV

TOP 10 ARCHITECTURES TOGETHER WITH 3 PARETO OPTIMAL ARCHITECTURES FOR $k = 0$. ARCHITECTURES ARE LABELED WITH RESPECT TO THEIR ARCHITECTURE FINGERPRINTS. PARETO OPTIMAL ARCHITECTURES ARE MARKED WITH *. BEST VALUES IN EACH COLUMN ARE IN BOLD.

Arch. finger-print	E (1)	T (ms)	$\frac{E_{AA}}{E}$ (1)	$\frac{T_{AA}}{T}$ (1)	U_P (%)	U_I (%)	A (mm^2)
CC*	**0.925**	1.926	**1.999**	1.999	**100**	10	**2.4**
DD*	0.925	1.285	**1.999**	2.998	**100**	14	**2.4**
CCC	0.928	1.289	1.993	2.989	**100**	28	3.6
DDD*	0.928	0.860	1.991	4.481	99	41	3.6
CCE	0.935	1.016	1.977	3.789	99	35	4.6
CE	0.935	1.375	1.977	2.800	**100**	13	3.4
DF	0.936	0.917	1.976	4.199	**100**	20	3.4
DDF*	0.936	0.678	1.975	5.677	99	51	4.6
CCCC	0.941	0.980	1.965	3.931	98	52	4.8
CEE	0.941	0.840	1.964	4.583	99	42	5.6
DFF*	0.943	0.561	1.961	6.864	99	62	5.6
FFF*	0.953	0.482	1.939	7.999	98	73	6.6
DFFF*	1.001	**0.428**	1.847	**9.005**	90	**96**	7.8

A. Top Architectures

Table IV, Table V and Table VI show top 10 and all Pareto optimal architectures for cases $k = 0$, $k = 1$ and $k = 4$, respectively. $k = 0$ case is practically pure performance optimization, although measured in energy, but $k = 1$ and $k = 4$ are strictly energy optimization. Energy E and total execution time T are absolute values, and they are comparable to the slowest 2-PE architecture AA. $\frac{E_{AA}}{E}$ is the energy gain over AA architecture, the bigger the better. $\frac{T_{AA}}{T}$ is the speedup over AA architecture, the bigger the better. U_P is the mean PE utilization. U_I is the mean interconnect utilization. A is the area measured in square millimeters.

CC wins energy with all values of k. In $k = 1$ case, it is 1.7% more energy-efficient than the nearest 3 PE solution, CCC. It is 4, 5% more energy-efficient than the nearest 4 PE solution, CCCC. When the role of the dynamic energy increases in $k = 4$, the differences are larger: 2.5% and 6.2% against CCC and CCCC, respectively.

DFFF is the fastest architecture, and also a Pareto optimum. It has the highest performance processors given the area constraints. Note that 5 PE solutions do not have performance advantage over 4 PE solutions due to area constraints. For all values of k, DFFF runs at $4.5\times$ speed compared to the most energy-efficient architecture CC. However, it consumes only 8.2% ($k = 0$) to 14.3% ($k = 4$) more energy.

Most energy-efficient architectures have lower interconnect utilization U_I and higher processor utilization U_P than the fastest architectures. Lower processor utilization in high performance architectures can be explained with high peeks of performance demand that they can satisfy. Low performance architectures have longer task queues during peeks, which balances the load in time, but makes the critical path longer.

Approximately half the architectures are homogeneous in top 10.

Table VII and Table VIII show the proportion of how many times a given number and type of PE was in top 10 least

978-1-4244-2541-9/08 $25.00 © 2008 IEEE

TABLE V

TOP 10 ARCHITECTURES TOGETHER WITH 3 PARETO OPTIMAL
ARCHITECTURES FOR $k = 1$. FIGURE 2 SHOWS ARCHITECTURES
GENERATED FOR $k = 1$ CASE.

Arch. finger-print	E (1)	T (ms)	$\frac{E_{AA}}{E}$ (1)	$\frac{T_{AA}}{T}$ (1)	U_P (%)	U_I (%)	A (mm^2)
CC*	**1.707**	1.926	**1.541**	1.999	**100**	10	**2.4**
DD*	1.708	1.285	**1.541**	2.998	**100**	14	**2.4**
CCC	1.736	1.289	1.516	2.988	99	27	3.6
DDD*	1.737	0.860	1.515	4.479	99	40	3.6
CE	1.773	1.376	1.484	2.800	**100**	13	3.4
DF	1.773	0.917	1.484	4.199	**100**	19	3.4
CCE	1.783	1.016	1.476	3.791	**100**	34	4.6
CCCC	1.784	0.980	1.475	3.929	98	50	4.8
DDF*	1.784	0.678	1.475	5.679	99	51	4.6
DDDD*	1.798	0.663	1.464	5.810	96	73	4.8
DFF*	1.817	0.561	1.448	6.864	99	61	5.6
FFF*	1.847	0.482	1.425	7.998	98	72	6.6
DFFF*	1.907	**0.427**	1.380	**9.028**	90	**95**	7.8

TABLE VI

TOP 10 ARCHITECTURES TOGETHER WITH 3 PARETO OPTIMAL
ARCHITECTURES FOR $k = 4$. FIGURE 3 SHOWS ARCHITECTURES
GENERATED FOR $k = 4$ CASE.

Arch. finger-print	E (1)	T (ms)	$\frac{E_{AA}}{E}$ (1)	$\frac{T_{AA}}{T}$ (1)	U_P (%)	U_I (%)	A (mm^2)
CC*	**4.055**	1.927	**1.228**	1.999	**100**	9	**2.4**
DD*	4.056	1.285	**1.228**	2.998	**100**	14	**2.4**
CCC	4.158	1.290	1.198	2.986	99	26	3.6
DDD*	4.159	0.861	1.197	4.476	99	40	3.6
CD	4.239	1.541	1.175	2.500	**100**	12	**2.4**
CE	4.285	1.376	1.162	2.800	**100**	13	3.4
DF	4.285	0.917	1.162	4.199	**100**	19	3.4
CCCC	4.308	0.980	1.156	3.930	98	49	4.8
DDDD*	4.319	0.663	1.153	5.812	96	71	4.8
CCE	4.325	1.018	1.151	3.785	99	33	4.6
DFF*	4.438	0.561	1.122	6.862	99	60	5.6
FFF*	4.526	0.482	1.100	7.990	98	70	6.6
DFFF*	4.633	**0.430**	1.075	**8.952**	91	**93**	7.8

TABLE VII

PROPORTION OF HOW MANY TIMES A GIVEN NUMBER AND TYPE OF PE
WAS IN TOP 10 ARCHITECTURES WITH $k = 1$. TABLE VALUES ARE
EXPLAINED IN SECTION III-D.

PE type	Once (%)	Twice (%)	3 times (%)	4 times (%)	PE prop.
A	0	0	0	0	0
B	0	0	0	0	0
C	10	20	20	10	60
D	10	20	20	10	60
E	20	0	0	0	20
F	20	0	0	0	20

TABLE VIII

PROPORTION OF HOW MANY TIMES A GIVEN NUMBER AND TYPE OF PE
WAS IN TOP 10 ARCHITECTURES WITH $k = 4$. TABLE VALUES ARE
EXPLAINED IN SECTION III-D.

PE type	Once (%)	Twice (%)	3 times (%)	4 times (%)	PE prop.
A	0	0	0	0	0
B	0	0	0	0	0
C	20	20	10	10	60
D	20	10	10	10	50
E	20	0	0	0	20
F	10	0	0	0	10

$k = 4$ case, it varies between between 21% and 23%. Thus, the energy profile is rather uniform for both cases.

Pareto optimal solutions have 2, 3 and 4 PEs. 2-PE solutions do well due to low energy. 3 and 4 PE systems are do well due to a trade-off between energy and performance.

Figure 2 and Figure 3 show the clustering of solutions in the design space. Pareto optimal solutions constitute mere 5% and 6% of all solutions (7 and 8 out of 141) for $k = 1$ and $k = 4$, respectively. Therefore, automatic exploration is needed even when design space is limited to only 6 types of PEs and 2 to 5 PEs per architecture. It is not feasible to try out these solutions by manual work.

C. Optimization Convergence

Figure 4 and Figure 5 plot ratio $\frac{E_{AA}}{E}$ against mapping iterations for each Pareto optimal solution. The number of iterations it takes to win AA increases as the number of PEs increases. This comes from increased complexity of the mapping problem and the SA mapping algorithm that scales up iterations with respect to architecture complexity, the number of PEs. 4 PE architectures take over 20000 more iterations than 3 PE architectures to reach the level of AA (the gain value 1.0).

Our earlier work [8] presented an automated parameterization method for SA mapping. Originally it was only used for homogeneous architectures and performance optimization. The energy-time trade-offs presented in this paper indicate that the method can also be used for heterogeneous architectures and energy optimization.

In order to reach energy-efficiency of even AA architecture, it takes tens of thousands of mappings for 4 PE systems. Hence, it is a non-trivial problem in most cases. This creates demand for automated mapping (exploration). This may require behavior level simulation due to simulation time,

energy consuming architectures for cases $k = 1$ and $k = 4$.

In $k = 1$ and $k = 4$ cases, 2 PE solutions filled 4 and 5 of the top 10 positions, respectively. 3 PE solutions filled 4 and 3 in those cases. 4 PE solutions filled 2 positions in both cases. 2 and 3 PE solutions seem suitable for low energy applications. However, 3 and 4 PE solutions have high performance.

There are no A and B type PEs in the top 10. This can be attributed to poor performance and energy inefficiency. f_{max} is a determining factor for static energy (1), and it puts processors with frequency less than f_{max} into disadvantage. The minimum value of f_{max} is 200MHz, because the interconnect is clocked at 200MHz. For this reason, A and B types are not favored. However, C and E types have the advantage of not increasing f_{max}. C type was the most common processor among low-energy architectures.

B. Pareto Optimal Solutions

Pareto optimal solutions are labeled with asterisk (*) in Table V and Table VI.

For the $k = 1$ case, static energy proportion for Pareto optimal architectures varies between 52% and 54%. For the

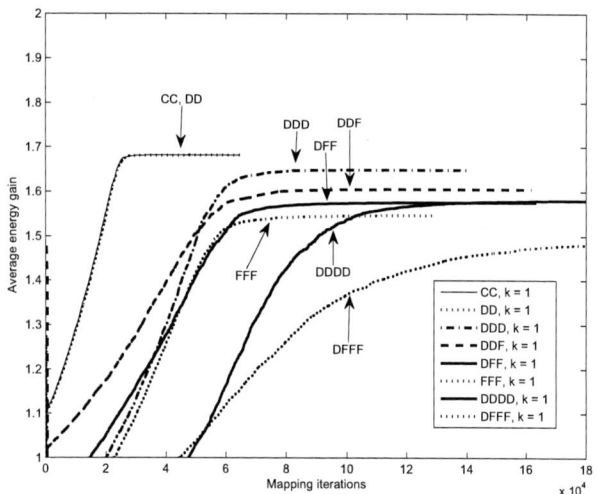

Fig. 4. Average energy gain plotted against mapping optimization iterations for the $k = 1$ case. Gain is computed as reference energy value divided by an energy value. Average gain is normalized to the average best objective value of the AA architecture. DD architecture's value 1.7 means AA consumed 1.7 times the energy of DD.

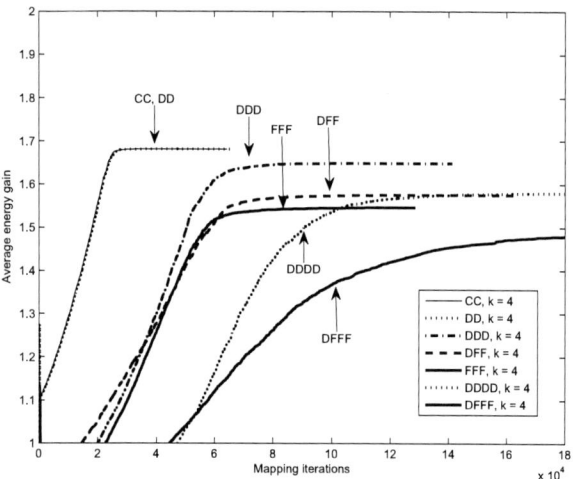

Fig. 5. Average energy gain plotted against mapping optimization iterations for the $k = 4$ case.

as was done in this paper. More accurate simulation may need thousands of CPUs. Fortunately, thousands of CPUs are reachable with current development budgets.

V. CONCLUSION

We evaluated heterogeneous architectures by optimizing both energy and performance for applications. The energy versus performance trade-off was analyzed by looking at Pareto optimal solutions. It was found that both energy-efficient and well performing solutions exist, and in general, performance is traded for energy-efficiency. Results indicated that automated exploration tools are needed when the mapping problem complexity grows, starting already with our experiment setup: 6 types of PEs to select from, and the system consists of 2 to

5 PEs.

Also, the results show that our Simulated Annealing method can be used in energy optimization for heterogeneous architectures, as well as in performance optimization for homogeneous architectures.

In the future, we plan to utilize SA to directly seek an optimal HW allocation and consider the bus or NoC energy more closely.

REFERENCES

[1] M. Gries, *Methods for evaluating and covering the design space during early design development*, Integration, the VLSI Journal, Vol. 38, Issue 2, pp. 131-183, 2004.
[2] W. Wolf, *The future of multiprocessor systems-on-chips*, Design Automation Conference 2004, Proceedings. 41st, pp. 681-685, 2004.
[3] Y.-K. Kwok and I. Ahmad, *Static scheduling algorithms for allocating directed task graphs to multiprocessors*, ACM Comput. Surv., Vol. 31, No. 4, pp. 406-471, 1999.
[4] H. Orsila, E. Salminen, M. Hännikäinen, T.D. Hämäläinen, *Optimal Subset Mapping And Convergence Evaluation of Mapping Algorithms for Distributing Task Graphs on Multiprocessor SoC*, International Symposium on System-on-Chip, 2007.
[5] S. Kirkpatrick, C. D. Gelatt Jr., M. P. Vecchi, *Optimization by simulated annealing*, Science, Vol. 200, No. 4598, pp. 671-680, 1983.
[6] T. Kangas, P. Kukkala, H. Orsila, E. Salminen, M. Hännikäinen, T.D. Hämäläinen, J. Riihimäki, K. Kuusilinna, *UML-based Multi-Processor SoC Design Framework*, Transactions on Embedded Computing Systems, ACM, 2006.
[7] G. Kahn, *The semantics of a simple language for parallel programming*, Information Processing, pp. 471-475, 1974.
[8] H. Orsila, T. Kangas, E. Salminen, T. D. Hämäläinen, *Parameterizing Simulated Annealing for Distributing Task Graphs on multiprocessor SoCs*, International Symposium on System-on-Chip, 2006, pp. 73-76.
[9] *Standard task graph set*, ONLINE: http://www.kasahara.elec.waseda.ac.jp/schedule, 2003.
[10] *jobqueue*. Software. ONLINE: *http://zakalwe.fi/ shd/foss/jobqueue/*
[11] *OpenSSH*, software, ONLINE: *http://openssh.org*
[12] *rsync*, software, ONLINE: *http://rsync.samba.org*

Integrating High Speed Multipliers in Coarse Grain Reconfigurable Arrays

Stavros Georgiopoulos[1], Grigoris Dimitroulakos[2], Costas E. Goutis[1]

[1]ECE Department, VLSI Laboratory, University of Patras, Patras, Greece
[2]Department of Computer Science and Technology, University of Peloponnese, Tripolis, Greece
sgeorgiop@upatras.gr, dhmhgre@uop.gr, goutis@ece.upatras.gr

Abstract— The efficiency of a Coarse Grained Reconfigurable Array architecture in terms of performance and hardware cost is hard to be determined. Until now, few case studies have been published to determine the impact of the architecture parameters on the Instructions per Cycle and the architecture area. However, none of those have considered the impact of multipliers embedded in the Processing Elements of Coarse Grain Reconfigurable Array architectures. This paper focuses on multipliers both from the compiler and the architecture perspective. An already existing exploration framework has been used for our study. It consists of two parts: a) an existing retargetable compiler from which the mapping efficiency is estimated and b) from the parametric realization of the coarse grained reconfigurable array in hardware description language (VHDL). The latter is used as input in the Synopsys Design Compiler for the estimation of the area and clock frequency of each architecture instance. The system has been realized using the 0.13μm process of ASIC technology. The experiments report the system area, clock frequency and performance for different embedded multipliers.

I. INTRODUCTION

Coarse Grain Reconfigurable Array (CGRA) architectures [1]-[5] have been proposed for accelerating computation intensive parts of algorithms residing in several scientific domains. These kinds of applications have high amounts of inherent operation and data parallelism. The large number of PEs available in CGRAs which are organized in a 2-Dimensional (2D) array and connected with a configurable interconnect network, can be used to exploit this parallelism and thus accelerate the applications' loops.

Computation intensive applications have high amounts of parallel multiplication operations. CGRAs instruction repertoire must support multiplication, in order to achieve the required acceleration of applications. However, multiplication units have a large critical path and area requirements. An existing work [5] explored the effect on performance when a limited number of multipliers are integrated in the CGRA. Though, the impact of multipliers on the system's clock frequency and area has not been considered. Moreover, different multiplier implementations have not been studied in their evaluation.

In this paper, the performance and area for CGRAs integrating different multipliers, are studied with the use of an exploration framework consisting of: 1) an existing retargetable mapping methodology [4] based on a modulo scheduling technique and 2) by a parametric CGRA architecture template that has been described in hardware description language [4] (VHDL). The latter is used for estimating the clock frequency

and the area of each considered CGRA architecture instance using the Synopsis Design Compiler in the 0.13μm process of ASIC technology. Our experiments refer to four different CGRA architecture scenarios with respect to the multiplier implementations inside each PE. In specific, we made use of the following multipliers: an array, a Booth-Wallace, a Booth-Wallace two-level pipelined and a special implementation of the same Booth-Wallace where each multiplication is split into three distinct operations. This study examined the Instructions Per Cycle (IPC), clock frequency, area and performance in the basis of a design space exploration scheme.

The paper is organized as follows: Section II reviews the existing work, Section III describes the CGRA architecture template, Section IV reviews the mapping methodology, Section V describes the multipliers' architecture, Section VI presents the results while section VI concludes the paper.

II. RELATED WORK

Quite a few works [4]-[7] have been published exploring the design space of mapping applications to CGRAs. The approach [5] studies the value of IPC for different architecture alternatives. Only three approaches, [4], [6], [7] considered evaluation metrics (such as area or power) relative to the hardware implementation of a CGRA. The KressArray [7] is a CGRA architecture template where each PE has two input and two output registers while its operation is data-driven. KressArray explorer is a design exploration system aiding the identification of architectures with optimized performance/ power trade off. However, area or clock frequency issues have not been studied in this work. In [4], area, clock frequency, IPC and performance exploration has been conducted on a parametric architecture template. Nonetheless, the exploration has not focused on the CGRA multipliers inside each PE.

The work in [6] investigates the register file architecture effect on the area and IPC of a CGRA system (ADRES). The architecture has a global register file where it has dedicated connections to the PEs, as well as distributed register files. In this exploration, multiplication has been modeled as one cycle operation hence, the clock frequency of the system, for all experiments, has been upper-bounded by the frequency of the multiplier's circuits. So, the system's clock frequency behavior was not explored. By modeling the multiplication as a multicycle operation, higher frequency operation is achieved not constrained by the multiplier circuits thus, permitting clock frequency exploration. In [5], the DRESC flow was applied in nine kernels for finding the most efficient ADRES instance for

978-1-4244-2541-9/08 $25.00 © 2008 IEEE

executing them in terms of IPC. For this purpose, a series of experiments concerning different interconnection topologies, heterogeneous functional units, different memory interface and variations of the register file structure were conducted. However, the area and clock frequency were not studied in that work.

III. RECONFIGURABLE ARCHITECTURE TEMPLATE

The considered CGRA architecture is shown in Fig.1 and consists of the following parts: a) a 2D array of PEs connected via an interconnection network, b) a data memory interface including a set of buses and the scratch-pad memory and c) a configuration memory where the overall configuration code of the program's critical loops is loaded. Moreover, the overall system includes a shared data memory which can be a typical SDRAM chip and holds the overall amount of program data. A typical Instruction Set Processor (ISP) (such as ARM) is present responsible for executing applications' non-critical parts and controlling the CGRA operation as described in [8]. The design specification of the 2D array and the data memory interface is parametric enabling the exploration on different architectures.

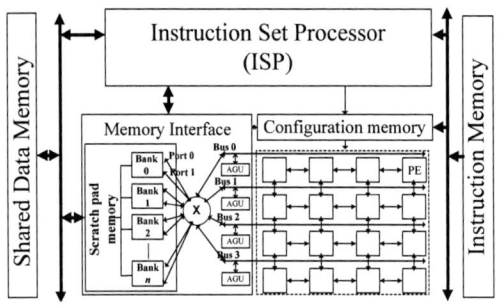

Figure 1. CGRA Architecture Template

The CGRA's data memory interface consists of: a) the memory buses where in the architecture considered, the PEs residing in a row share a common bus connection to the scratch-pad memory (Fig.1), b) an Address Generation Unit (AGU) is associated to each bus for the addressing, c) the scratch pad memory [9] which serves as a fast global foreground memory

for loading data in the PEs and d) the switch which acts as an interface between the scratch-pad's banks and the buses.

Furthermore, the PEs' interconnection network is defined as a graph enabling the instantiation of different interconnection network configurations. Additionally, each PE consists of one Reconfigurable Functional Unit (RFU) which can be configured to perform a specific word-level operation in every cycle. The operations supported by the RFU are the operators of the C Language (except division). Other operations such as divisions are analyzed to operations supported by the PE instruction repertoire. The CGRA's configuration memory (Fig.1) stores the whole configuration for setting up the execution of the application's loops. Context Caches distributed in the PEs enable their reconfiguration on a cycle by cycle basis. The configuration contexts can also be loaded from the configuration memory at the cost of extra delay.

IV. MAPPING METHODOLOGY

This section briefly describes the mapping method (Fig.2a) used by the proposed exploration framework. The reader is referred to [4] for a detailed description. The input to the mapping stage is the applications' critical loops identified by the methodology script described in [8]. The developed CGRA mapper advances other existing ones since: a) it concurrently encounters scheduling, register allocation and register spilling phases as these are highly related and b) reduces data bandwidth bottleneck by exploiting data reuse opportunities.

Before the mapping stage, a series of pre-mapping steps are applied for the purpose of analysis and to give an optimized startup to the mapper. These include: a) memory hierarchy optimizations [10] to increase the impact of the incorporated bandwidth optimization technique, b) loop normalization, c) loop unrolling in order to saturate the CGRA PEs and d) static and dynamic profiling which is used for the mapping of variables in the scratch pad. The front-end of the SUIF2 Compiler [11] has been used to produce the Intermediate Representation (IR) for the mapping phase. Using the IR and the CGRA architecture description, the input algorithm is mapped to the CGRA.

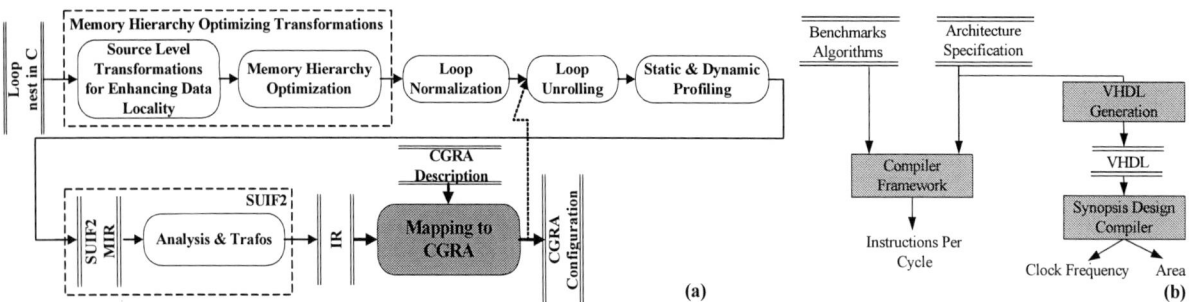

Figure 2. a) Mapping methodology flow for CGRAs and b) CGRA Exploration Framework

V. MULTIPLIERS ARCHITECTURE

This section describes the architecture and motivation behind the integration of high performance multipliers in the CGRA architecture. The starting point is the use of a Wallace tree to reduce the number of carries being transmitted. More specifically, the full adder residing in the multiplier is used to

apply a 3:2 compression, taking 3-bit long partial products as inputs and outputs 2-bit long results, thus achieving compression of the intermediate values. In Fig.3a, the process of compressing the partial products by a 3:2 factor [14] is presented. A straight-forward 16-bit multiplier would require 16 levels of addition, without the use of a Wallace tree. In the latter case, only 6 levels of addition are needed. However, better results can be further

978-1-4244-2541-9/08 $25.00 © 2008 IEEE

achieved by applying the Booth technique which accomplishes further compression.

Even by using the Booth-Wallace multiplier, the circuit delay may become a burden when more performance and speed are needed. To overcome this obstacle, pipelining is applied by inserting successive rows of latches in selected places inside the multiplier, so as the delay between them to be equal or less than the total circuit delay, like it is shown in Fig.3a. By the use of a single row of latches in the Booth-Wallace multiplier, we achieved doubling of the operation frequency for the pipelined version of the multiplier, bearing in mind that an additional clock cycle is needed for the first operation. The pipelined Booth-Wallace multiplier performs better in the case of successive multiplications. On the other hand, when the frequency of multiplications is lower, pipelining delays the result extraction by an additional clock cycle.

The fourth approach consists of splitting the multiplication operation into two equal parts. In the first part, multiplication is

conducted between the multiplicand and the higher rank bits of the multiplier, whereas in the other one, multiplication between the lower rank bits of the multiplier and the multiplicand is executed. As a result, a 16x16bit multiplication can be separated into two distinct 16x8bit multiplications which are faster and less area consuming. Finally, the result of the first multiplication should be leftward shifted by 8 bits and then added to the result of the second multiplication. The procedure is presented in Fig.3b. It should be further noticed, that by splitting multiplication into two parts, every multiplication node in the dependence graph should be replaced by an equivalent group of three nodes that is two multiplications and one addition (Fig.3c). Subsequently, the two 16x8 multiplications can be executed in the same or different PEs. However, for that case, it remains to be investigated whether the increase on the number of instructions to be mapped and scheduled is counterbalanced by the increase in system operation frequency achieved by applying this technique.

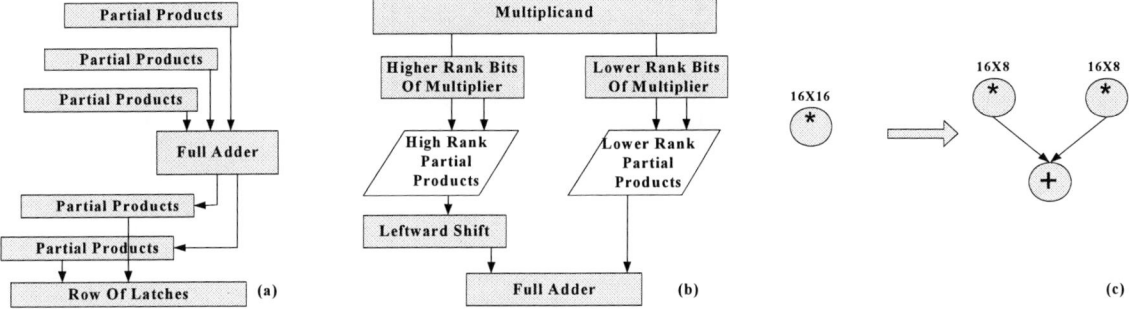

Figure 3. a) Wallace tree with the use of pipeline, b) Booth-Wallace using multiplication splitting technique and c) Multiplication transformation in the Dependence Graph

VI. EXPERIMENTS

The experiments are conducted on a 4x4 CGRA architecture with 16 PEs where each PE is connected only to its nearest neighbors. Additionally, each PE has a register file of 16 words size with two input and four output ports. There is one RFU in each PE that can execute any of the operations described in Section III in one clock cycle, except for multiplication that is multicycle. The granularity of the FU is 16 bits which is also the word size. We assume that the Context Caches have 32 context words size. In addition, the direct connection delay among the PEs is zero cycles. Finally, the configuration for the memory interface in the experiments assumes that: a) two buses per row are dedicated for transferring data from the scratch-pad memory as in [3], b) the scratch-pad memory consists of 4 banks each having size of 1Kbytes, c) each bank has 2 ports with read/write capability, d) the buses are properly multiplexed so that each PE accesses every bank and e) each bus transfers one word per scratch pad's memory cycle which equals to one CGRA cycle.

The exploration framework (Fig.2b) integrates an existing compiler methodology [4] and the Synopsys Design Compiler [12]. The input to the compiler is a set of 16 DSP applications written in C code and derived from the Texas Instruments DSP Benchmark suite [13] and the CGRA architecture specification. Beyond that, the CGRA architecture description in VHDL is fed into the Synopsys Compiler for deriving the area and clock frequency using the 0.13μm process of ASIC technology.

The exploration considers four different PE multipliers architectures, three of which are based on the Booth-Wallace technique [15]. This type of implementation is well-known for its high performance and low area requirements. The fourth architecture refers to the array multiplier used as a reference point. In specific, a) the array multiplier [15] is modeled as a 4 cycle operation, b) the Booth-Wallace multiplication is considered as a two clock cycle operation, so it is obvious that the RFU of each PE is binded for the same duration, c) the Booth-Wallace multiplication with two-level pipelining can be issued in every clock cycle while reserving the RFU for other operations (except for multiplication) for two cycles and d) the last architecture which is implemented on the basis of splitting the multiplication into three distinct operations (two multiplications and one addition) each one requiring one clock cycle to execute. This is the Half Booth-Wallace multiplier. It must be noted that from this design approach, the number of operations in the IR has been increased by an average ratio of 80% for the set of algorithms used in the exploration. The latencies of the aforementioned multipliers were deducted by applying appropriate time analysis on the CGRA's critical path.

In Fig.4a, the area requirements for every architecture scenario are compared for both the system and the multipliers. It is deduced that the area for the whole system is nearly the same for the first three scenarios (Array Multiplier, Booth-Wallace, Booth-Wallace Pipelined), whereas there is a 25% decrease for

978-1-4244-2541-9/08 $25.00 © 2008 IEEE

the Half Booth-Wallace architecture. Moreover, the maximum multiplier area needed for their hardware implementation is almost 9% of the total CGRA area, which constitutes a reasonable area penalty. On the other hand, the Half Booth-Wallace multiplier demands 60% less space to be implemented on than the Booth-Wallace. Fig.4b depicts an illustration of both the system's and the multiplier's operation frequency. Starting commenting on the system behavior, it should be noted that the first three scenarios operate on almost the same frequency, whereas the fourth one (Half-Booth Wallace) shows a 20% decrease since in that case multiplication is a one cycle operation. From the multipliers' point of view, like it should have been expected, the Booth-Wallace Pipelined scenario exhibits the highest operating frequency, followed by the Half Booth-Wallace showing a 20% decrease and the Booth-Wallace, which only achieves almost half of the frequency of the Half Booth-Wallace. On the other hand the two pipeline stages (each

one of them lasting for one clock cycle) of Booth-Wallace Pipelined explain the doubling of frequency. Finally, the Array Multiplier shows the worst operating frequency and runs twice slower than the Booth-Wallace.

Fig.4c shows a diagram of the IPC, which is obtained by the compiler framework for the considered design scenarios while in Fig.4d the normalized performance bar graph has been constructed. The IPC peaks at roughly 12 instructions per cycle for the Half Booth-Wallace architecture and the Booth-Wallace Pipelined. As expected, the Array Multiplier shows the worst IPC figure, since it is a 4 cycle operation. Although the Booth-Wallace Pipelined and the Half Booth-Wallace exhibit the same IPC, the latter delivers half the performance of the Booth-Wallace Pipelined due to lower operating frequency and the higher number of instructions to schedule. However, it is three times faster than the array multiplier implementation.

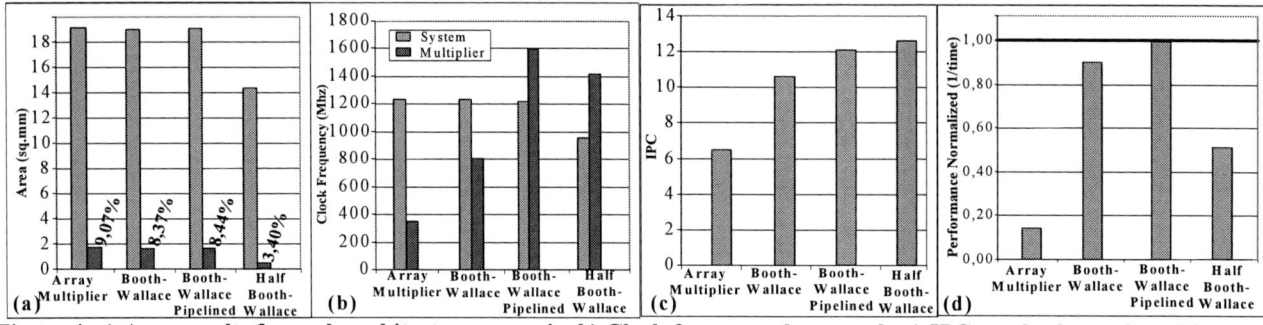

Figure 4. a) Area results for each architecture scenario, b) Clock frequency bar graph, c) IPC results for each architecture scenario, d) Normalized performance bar graph

VII. CONCLUSIONS

This paper presented an exploration that focused on the impact of multipliers on the area and performance of CGRA architectures. Till now multiplier implementations inside PEs were not studied in relation with CGRA architectures. Moreover, existing studies left clock frequency issues unexplored. We have shown that even by using high performance multipliers in each PE the area requirements range from 3% to 10% of the overall circuit area. Also, sophisticated multipliers can even double the achieved performance ratio.

ACKNOWLEDGEMENT

This work was supported by the project PENED 2003 which is funded in 80% by the European Social fund and in 20% by the Greek state-Greek Secretariat for Research and Technology.

REFERENCES

[1] R. Hartenstein, "A decade of reconfigurable computing: A visionary retrospective", *Proc. of ACM/IEEE DATE '01*, pp. 642-649.

[2] Pact Corporation, *"The XPP white Paper"*, Technical report, www.pactcorp.com, 2005.

[3] H. Singh, L. Ming-Hau, *et.al.*, "MorphoSys: An Integrated Reconfigurable System for Data-Parallel and Communication-Intensive Applications", in IEEE Trans. on Computers, vol. 49, no. 5, pp. 465-481, May 2000.

[4] G.Dimitroulakos, M.D. Galanis, Nikos Kostaras and C.E Goutis "A Unified Evaluation Framework for Coarse Grained Reconfigurable Array Architectures", in Proc. ACM Proceedings of the 4th Conference on Computing Frontiers, pp.161-172, 2007, Ischia, Italy

[5] B. Mei, A. Lambechts, D. Verkest, *et.al*, "Architecture Exploration for a Reconfigurble Architecture Template", in IEEE Design & Test of Computers, vol.22 no.2 pp. 90-101, March-April 2005

[6] Zion Kwok, S.Wilton, "Register File Architecture Optimization in coarse grained reconfigurable architecture", In Proc. 13th Annual IEEE Symp. on Field Programmable Custom Computing Machines 2005, pp 1-10.

[7] Reiner W. Hartenstein, Th. Hoffman, and U. Nageldinger, "Design-Space Exploration of Low Power Coarse Grained Reconfigurable Datapath Array Architectures", in Proc. PATMOS 2000 LNCS 1918, pp 118-128.

[8] M.D. Galanis, G. Dimitroulakos, and C.E. Goutis, "Partitioning Methodology for Heterogeneous Refonfigurable Functional Unit", in Journal Of SuperComputing,Vol.38, Num1,pp.17-34, Oct.2006

[9] P. R. Panda, N. Dutt, and A. Nicolau, *"Memory Issues in Embedded Systems-on-Chip: Optimizations and Exploration"*, Kluwer Academic Publishers, 1999.

[10] Sven Wuytack, J. P. Diguet, Francky Catthoor and Hugo De Man, "Formalized methodology for data reuse exploration for Low-Power Hierarchical Memory Mappings", In IEEE Transactions on VLSI Systems, Vol.6, No. 4 Dec 1998

[11] M. W. Hall *et al.*, "Maximizing multiprocessor performance with the SUIF compiler", Computer, vol. 29, pp. 84-89, 1996.

[12] http://www.synopsys.com, Synopsys 2008

[13] Texas Instruments Inc., www.ti.com, 2005

[14] Lakshmanan, Masuri Othman *et.al*, "High Performance Parallel Multiplier using Wallace-Booth Algorithm",IEEE International Conference on Semiconductor Electronics, pp. 433- 436, 2002

[15] Behrooz Parhami, "Computer Arithmetic Algorithms and Hardware Designs", Oxford University Press, USA,2000

Analyzing Models of Computation for Software Defined Radio Applications

Heikki Berg, Claudio Brunelli
Nokia Research Center
FIN-33720 Tampere, Finland
Email: heikki.berg@nokia.com, claudio.brunelli@nokia.com

Ulf Lücking
Nokia Research Center
FIN-00180 Helsinki, Finland
Email: ulf.lucking@nokia.com

Abstract—**Applying design principles and methodologies constituted in the software domain and being adapted to the complete execution environment provides new perspectives for future multi-radio computers. In order to share the underlying hardware resources efficiently, the overall system architecture and related programming model has to support dynamic behavior and extensive changes in the configuration during run-time. The requirements for such a multi-radio computer are demanding, as there will be various radio access stacks with inhomogeneous characteristics executing in parallel. This implies a configuration and control framework, besides the different protocol stacks, that is aware of the managed system in every state and is capable of dynamically scheduling different dataflow graphs corresponding to the applications running on the underlying system. This paper presents the main concepts behind such a reactive system, focusing in particular on the proposed model of computation, giving an overview on the software architecture and related problems to be solved.**

I. INTRODUCTION

A reactive system like a Software Defined Radio (SDR) computer can be seen as a heterogeneous system consisting of software components mainly combining hierarchical finite state machines (FSM) and data flows (DF). Efficient sharing of resources of this particular execution environment is guaranteed by the implementation of control & configuration framework, which provides services and signaling for monitoring the local conditions of the device and enables the system to reason about its own structural and behavioral capabilities. In particular, the framework defines a set of services allowing unified access to any radio system: user data services, reconfiguration services, multi-radio scheduling services and resource management services [1].

User data services include both control services for managing radio connections, instantiating the relevant protocol stacks, as well as the actual data transfer (uni- and bidirectional). These services are visible at the multi-radio access API where incoming and outgoing data to the networking stacks are defined as separate flows of data, which are then multiplexed sharing the same radio resources in the physical layer. Reconfiguration services are used to set up and reconfigure the set of active radios by managing the involved components. Services for multi-radio scheduling ensure that simultaneously active radio systems follow prior defined rule sets to avoid radio interference to each other. Resource management services map tasks to the relevant hardware resources.

Contradictory to this particular control framework and its incorporated radio protocols, the physical layer (PHY) signal processing of Radio Access Technologies (RAT) is going to be mapped to computing nodes in the form of directed graphs. Those graphs naturally reside in the DF domain and are composed of various computational intensive algorithms. Due to the required flexibility and dynamic behavior of the system at runtime, the combination of different computational models causes problems in terms of predictability and determinism. In this paper we review some key computational models, such as Statically Schedulable Dataflow (SSDF), Cyclo Stationary Dataflow (CSDF), Heterochronous Dataflow (HDF) proposed for digital signal processing and analyze their applicability for PHY signal processing in software defined radio. Two most promising candidates, due to their properties, are the Cyclo Dynamic Dataflow (CDDF) and Parametrized Cyclo Stationary Data Flow (PCSDF).

II. MODELS OF COMPUTATION

Dataflow is a widely used computational model for specifying applications in the domain of digital signal processing. Important properties of dataflow graphs are their capability of describing parallelism and data dependencies in a natural way, without introducing side effects by additional control of the application. In dataflow, a program is represented as a directed graph, in which vertices, called actors, represent computations and arcs represent the channels to transport the processed data. These channels queue data values, in the form of tokens, which are passed from the output of one actor to the input of another. When an actor is executed, it consumes a certain number of tokens from its inputs, and produces a certain number of tokens at its outputs. The transported data has to be buffered by intermediate FIFOs, in order to decouple the execution of producer and consumer tasks. It is important to be able to assess the consistency of the graph, that is, whether the number of tokens produced on each graph edge corresponds to the number of consumed tokens from the same edge in one cycle of actor firings which return connecting FIFOs to their original state. More specifically consistency means that a sequence can be executed in bounded memory and time and that the graph does not deadlock.

A. Statically Schedulable Dataflow

Statically Schedulable Dataflow (SSDF) is a restricted version of dataflow in which the number of produced and consumed tokens is fixed and known at compile-time, thus the schedule of actor firings can be computed as well as consistency and schedule can be analyzed at compile time [2].

B. Boolean-controlled Dataflow

Boolean-controlled dataflow (BDF) is an extension to the SDF, where the number of tokens produced or consumed on an edge either is fixed, or is a two-valued function of a control token present on a control terminal of the same actor [3]. This allows conditional execution of the graphs with different numbers of input and output tokens for the dataflow. In some cases a quasi-static schedule can be computed, where each firing is annotated with the run-time conditions under which the firing should occur. Because only topology information of the graph is available, it is sometimes impossible to prove automatically the consistency of the BDF graph. The BDF model has subsequently been extended by Buck to Integer Controlled Dataflow [4].

C. Cyclo Static Dataflow

In Cyclo Static Dataflow (CSDF), production and consumption of tokens are sequences of constant integers within a minimum period, where it is still possible to construct a valid schedule with bounded buffer lengths at compile-time [5]. A phase of an actor can be fired, if all previous phases of the actor are already executed and sufficient data is present at the input buffers. This procedure is repeated recursively until all phases are scheduled for each actor of the graph. The state then is equal to the initial state and the schedule can be repeated infinitely.

D. Cyclo Dynamic Dataflow

The Cyclo Dynamic Data Flow (CDDF) model allows the use of symbolic variables that reflect relevant properties of actor behavior [6]. A symbolical variable makes some relevant property of the internal behavior of the actor explicit. A CDDF control token can determine the token transfer at an actor port, and the next actor phase to be executed, however the control tokens must be present at the moment that the actual actor phase is determined and the token consumption can not depend on symbolic variable. CDDF is constructed so that all firing rules are derivable from the consumption. In SDF and BDF the actor schedule depends only on availability of tokens. For CDDF the run-time scheduler must be able to read the values

of the control tokens to determine the actor phase and related firing rules. Consistency can be evaluated probabilistically using bounded set of actor phases, with application specific triggering conditions and consumption rates.

E. Heterochronous Data Flow

Heterochronous Data Flow (HDF) model proposes hierarchical finite state machines with multiple concurrency models, where the idea is to decouple the concurrency model from the hierarchical FSM semantics [7]. This allows FSMs to be embedded in various concurrency models. In HDF an actor can have a finite number of actor phases with associated firing rules. This is somewhat similar to CSDF model but the phase (state) sequence between firings is not cyclic. Unlike CDDF the actor phase cannot be dependent on the token present in the control port. When HDF system starts execution the initial phase is in effect, thus actors can solve their balance equations to find iteration. Each phase must remain constant for the duration of the iteration of complete graph.

F. Parameterized Data Flow

Parameterized dataflow modeling differs from dataflow modeling techniques in that it is a meta-modeling technique: parameterized dataflow can be applied to any underlying "base" dataflow model that has a well-defined notion of graph iteration (invocation) [8]. Consistency issues in PSDF are based on disciplined dynamic scheduling principles that allow every PSDF graph to assume the configuration of an SSDF graph on each graph invocation. Such scheduling leads to a set of local synchrony constraints for PSDF graphs and PSDF subsystems that need to be satisfied for consistent specifications. This is a clear restriction since parameters which affect to the dataflow ports can be changed only once per activation of parent graph, rather than once per activation of the actor.

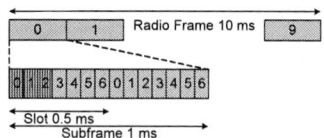

Figure 1 LTE Radio Frame Structure

III. LTE EXAMPLE

A. LTE Downlink Physical Layer

A 10 ms LTE radio frame depicted in figure 1 and consists of 10 subframes of time duration 1 ms [9]. One subframe consists of two slots of 7 OFDM symbols, when system parameters have been set for 15 kHz subcarrier spacing and normal cyclic prefix. Single frequency domain OFDM symbol for 20 MHz bandwidth consists of 2048 subcarriers of which only 1200 centermost (except dc-subcarrier) subcarriers are used for transmission. Cyclic prefix of 160 samples for the 1st, and 144 samples for the remaining six OFDM symbols of a slot are added to the time domain OFDM symbols in the

978-1-4244-2541-9/08 $25.00 © 2008 IEEE

transmitter. 12 consecutive subcarriers of a subframe form a resource block. Thus one subframe contains 100 resource blocks for user allocation. Pilot subcarriers, which are used for channel estimation, are transmitted known subcarriers of OFDM symbols 0, 4, 7 and 11.

Physical control format indicator channel (PCFICH) is completely multiplexed into OFDM symbol 0. PCFICH carries information about the number of OFDM symbols used for transmission of physical downlink control channel (PDCCH) in a subframe. Physical downlink shared channel (PDSCH) is transmitted in the remaining subcarriers and symbols of a subframe, which are not occupied by control channels. In order to demodulate and decode payload channels transmitted for user equipment (UE), UE first needs to demodulate and decode PDFICH. PDCCH can be decoded based on the information in PDFICH. Based on PDCCH information UE can decode payload channels transmitted for it.

PDFICH demodulation can start immediately when OFDM symbol 0 is received and buffered, and channel estimation for the symbol is completed. If more than one OFDM symbol is used to carry PDCCH, all OFDM symbols up to symbol number 4 need to be received and buffered, because of channel estimate interpolation.

Coded PDSCH transport block is mapped to the resource elements of allocated resource blocks in the order of filling all allocated subcarriers of an OFDM symbol first before proceeding to the next OFDM symbol. If coded transport block exceeds the maximum code block size, the transport block is segmented into separately decodable segments. Demodulation and decoding of code block segments require that channel estimation up to the last OFDM symbol used by the codeblock is available.

B. LTE Receiver

A possible configuration of a LTE receiver used in our analysis is depicted in figures 2 and 3. Figure 2 shows the graph configuration up to buffering of frequency domain OFDM symbols. Frequency offset correction derotates complex samples of both antennas in order to correct any residual frequency error. Frequency offset correction value is updated periodically. This can be done for instance once per subframe, thus the number of complex samples processed for each antenna and one correction value is $2*((160+2048) + 6*(144+2048))$. Cyclic prefix remove operation selects 2048 samples of each OFDM symbol for FFT operation. Reordering shuffles subcarriers from their bit reversal order after FFT to natural subcarrier ordering and removes guard bands. Remaining 1200 subcarriers per one OFDM symbol are stored into an intermediate sample buffer, from where channel estimator and demodulator can read correct subcarriers for each physical channel (PCFICH, PDCCH and PDSCH) to be decoded. Channel estimator triggers on each OFDM symbol which has pilot carriers and produces interpolated channel estimates up to that OFDM symbol. Intermediate buffering of

payload subcarriers as well as channel estimates is required, because resource blocks transmitted to UE are not known until PCFICH and PDCCH have been decoded and interpreted. Figure 3 shows the graph for demodulation and decoding of PDSCH channels. The demodulation and decoding configurations can be different for each code block.

IV. APPLYING MODELS OF COMPUTATION TO LTE RECEIVER

A. Frontend of the Receiver

Let us consider using SSDF programming model for the LTE receiver frontend of the LTE System. Buffering the complete subframe before the first actor in the graph of figure 2 before firing the actor would create already 1 ms latency, thus it is impractical. Alternatively one could consider, for instance, queuing multiple identical frequency-offset compensation values for each OFDM symbol of a subframe. As the symbol size in samples is not the same for each OFDM symbol, SSDF could not be applied but CSDF could, since OFDM symbol size forms a periodic pattern. Cyclic prefix remove block selects 2048 samples of each OFDM symbol for the FFT operation. For this block also the number of input tokens forms a periodic pattern, and the block would fit to the programming model of CSDF.

Strictly speaking SSDF could be applied for the first actor, for instance, by zero padding cyclic prefixes of OFDM symbols 1, 2, 3, 4, 5 and 6 to 160 samples by "A/D Interface", thus creating uniform OFDM symbol sizes, and making SSDF applicable for the complete graph. Since the signal processing of the graph is data independent, we have easily shown that any dataflow graph which belongs to the family of statically schedulable graphs can be utilized in the receiver frontend.

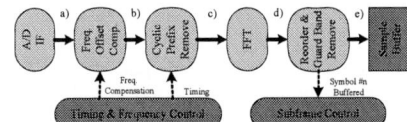

Figure 2 Frontend of the LTE receiver.

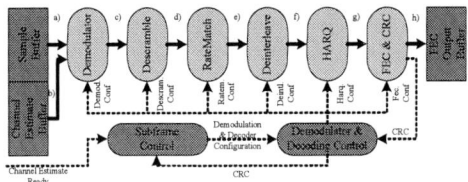

Figure 3 Demodulation and decoding of PDSCH code blocks

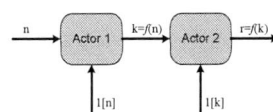

Figure 4 Configurable graph with CDDF principles

978-1-4244-2541-9/08 $25.00 © 2008 IEEE

B. Demodulation and Decoding

Graph illustrating the demodulation and decoding of PDSCH is depicted in figure 3. The connections between actors (or algorithms) are static, however the demodulation and decoding parameters including code block segment lengths, can be different between code blocks [9]. Therefore any of more or less static data flow computational models SSDF, CSDF, HDF or even PSDF are difficult, or even impossible, to apply. It would be possible to create a new graph in run-time for each code block segment length; this would be impractical and cause waste of memory. Alternatively, the number or size of the tokens passed between the actors could always be configured to be of maximum size required by the use case, which would create expensive overhead when actual amount of passed data is low. A third alternative would be to store the actual data to shared memory accessible by connected actors and the actual token would only contain a pointer to the data as well as the length of the data vector. This would require management of separate shared memory. Strictly speaking these examples do not fit to any of the static programming models mentioned above, because execution time of the actor is still data dependent and can not be analyzed statically. However, SSDF principles can be applied to the scheduling analysis of the worst use case and to decide to which computational node each task (actor) is deployed.

BDF is difficult to utilize as the number of tokens produced or consumed on an edge either is fixed, or is a two-valued function of a control token present on a control terminal of the same actor. As such does not offer capabilities required to implement data dependency of LTE demodulation and decoding, at least intuitively.

CDDF seems to offer the minimal set of functionality needed to implement configurable queue of actors, required for implementing demodulation and decoding of LTE as depicted in figure 4. In the figure the first actor has two inputs: one input for data and one for control. A single control token with value n, makes explicit information on internal behavior of the actor available to the scheduler: actor 1 is expecting to read n tokens from data port. The number of tokens produced to the output port is a function of consumed tokens $k=f(n)$. Comparing this to figure 3 we can immediately see the similarity.

Parametrized CSDF is applicable as well. As it is possible to change input-output rates of actors once per firing of the parent graph, one could consider running the init graph for figure 3 once the multiplicity and configuration of code block segments is known, and configure one actor phase for each code block. The parent graph would execute the demodulation and decoding sub graph only once, which would execute all phases. The semantics of PCSDF can be used for the modeling, however we are not able to get the benefit from the local synchrony, as the demodulation and decoding graph body is executed only once for each init configuration.

V. CONCLUSIONS & FUTURE WORK

Statically schedulable dataflow principles are best fitted to digital signal processing simulation tools. In these tools usually a static use case, or configuration, is simulated at a time and therefore dataflow can be computed in compile or at least iteration time of the complete graph.

Practical implementation of data dependent computational model, for instance in mobile terminals, can not fully be based on statically scheduled data flows. However, SSDF principles can be applied to the scheduling analysis of the worst case runtime scenario in order to decide deployment of each task (actor) to computational nodes. CDDF seems to offer the minimal set of functionality needed to implement configurable queue of actors, required for implementing signal processing functionality in transceivers. PCSDF is applicable as well; but all the benefits of local synchrony are not available due to the nature of the application. Although we used LTE receiver signal processing as our example, the same conclusions can be drawn for other modern radio systems.

In our future work we want to study suitable models of computation for software defined radio, in case graphs of multiple radio systems with hard real time requirements are deployed into shared computing nodes and study required load balancing and scheduling algorithms, when the concurrently running radio systems change, start and stop, at runtime.

REFERENCES

[1] Berg, H., A. Ahtianen, A. Parssinen, J. Westmeijer, U. Lücking, "Architecting Software Radio", 2007 Software Defined Radio Technical Conference and Product Exposition, November 5-9, 2007 - Denver, Colorado

[2] E. A. Lee and D. G. Messerschmitt, "Static Scheduling of Synchronous Data Flow Programs for Digital Signal Processing," IEEE Trans. on Computers, January, 1987.

[3] J. T. Buck, "Scheduling Dynamic Dataflow Graphs with Bounded Memory Using the Token Flow Model," Technical Memorandum UCB/ERL 93/69, Ph.D. Thesis, Dept. of EECS, University of California, Berkeley, CA 94720, 1993.

[4] J. T. Buck, "Static scheduling and code generation from dynamic dataflow graphswith integer-valued control streams", Conference Record of the Twenty-Eighth Asilomar Conference on Signals, Systems and Computers, 1994.

[5] M. Engels, G. Bilsen, R. Lauwereins, and J. Peperstraete. Cyclo-static dataflow: Model and implementation. In Asilomar Conf. Sig. Sys. and Comp., Pacific Grove, California, Oct. 1994.

[6] P. Wauters, M. Engels, R. Lauwereins, J.A. Peperstraete "Cyclo-dynamic dataflow", EUROMICRO Workshop on Parallel and Distributed Processing, Braga, Portugal, January 1996

[7] A. Girault, B. Lee, E. A. Lee, "Hierarchical Finite State Machines with Multiple Concurrency Models", IEEE Transactions on Computer-aided Design of Integrated Circuits and Systems, Vol. 18, No. 6, June 1999

[8] B. Bhattacharya, "Parameterized modeling and scheduling for dataflow graphs", Master's thesis, Department of Electrical and Computer Engineering, University of Maryland, College Park, November 1999

[9] 3GPP, "TS 36.2xx Physical Layer Specifications, Release 8", http://www.3gpp.org/ftp/Specs/html-info/36-series.htm

Multi-Objective Genetic Optimized Multiprocessor SoC Design

Mohammad Arjomand
Computer Engineering Department
Sharif University of Technology
Tehran, Iran
arjomand@ce.sharif.edu

Hamid Sarbazi-Azad
Sharif University of Technology &
Institute for Research in Fundamental
Sciences (IPM), Tehran, Iran
azad@sharif.edu

S. Hamid Amiri
Computer Engineering Department
Sharif University of Technology
Tehran, Iran
s_amiri@ce.sharif.edu

Abstract— In this paper, we introduce a new Multi-Objective Genetic Algorithm (MOGA) for mapping a given set of intellectual property onto a Network-on-Chip architecture such that for a specific application total communication cost and energy consumption become optimized while bandwidth constraints are satisfied. As the main theoretical contribution, we first introduce a generic queuing model to estimate performance and an experimental energy consumption model during the design phase, with acceptable accuracy. Then, an efficient genetic algorithm employs these models to propose a Pareto optimal front for an application and an arbitrary topology. Experimental results show that the proposed algorithm is very fast which results in a new approach for mapping MPSoC cores on chip.

I. INTRODUCTION

As current IC features have trend toward nanometer sizes, most embedded systems evolve much more IP cores which compute complex functions or store huge amount of data. This will lead to systems with cores realizing different functions running at different clock frequencies which result in embedded Multiprocessors System-on-Chips (MPSoCs).

The most challenges associated with MPSoC design is its inter-node interconnects. Transmission delay through traditional synchronous bus-based architectures does not scale well as computation delay decreases. Networks on-chip is an alternative design as they provide scalable, modular, reliable and asynchronous communication medium for large size MPSoC architectures [2]. Many MPSoCs are designed for one or limited specific applications. Then SoC traffic is usually heterogeneous in term of bandwidth or traffic patterns and will be available at design phase. The design of NoC has multiple trades-off between several important architectural choices, such as topology, routing and switching strategy and efficient mapping of target application tasks onto network tiles [8]. It is shown that topology customization has a great effect on infrastructure delay as well as considerable power saving. Routing paradigm can be deterministic or adaptive. Adaptive routers have more complicated structures, but can manage high traffic load better. The final design issue relates to an optimal power/performance aware mapping of cores such that some design constraints are satisfied.

Since exhaustive design space exploration takes a long time, there exists interesting works which customize NoC structures for an application. Authors in [7, 8] investigate topological mapping of IPs on the network cores with optimal energy consumption and bandwidth requirement satisfaction. In [1], a genetic algorithm (GA) is used to obtain an optimal trade-off between performance and power.

Through these methods, evaluating objective function is done through time consuming simulations. Guz et al. [5] use an analytical model for network delay estimation to efficiently explore design space for mappings with least link capacity considering network delay requirements.

In this research we propose a new Multi-Objective Genetic Algorithm (MOGA) to explore design space for more proper solutions from different perspectives. This algorithm is applicable for different network topologies with various routing schemes. To evaluate a typical mapping, a new analytical model for a wormhole router configuration is desired. Analytical model saves exploration time in comparison with time consuming simulation experiments [1] while providing acceptable accuracies. This queuing model traces traffic pattern on each router's link. The outline of this work is as follows. Section 2 expresses mapping as a multi-objective problem. In section 3, a generic analytical model for latency and energy consumption per traffic flow under arbitrary traffic pattern is proposed. In Section 4, optimization process is formulated by a multi-objective genetic algorithm. Experiments for real video applications are presented in section 5. Section 6 concludes the paper.

II. MULTI-OBJECTIVE DESIGN PROBLEM

Most recent NoC topologies emerge from macro networks. On-chip design has more limits on resources which impose constraints on design strategies. Power dissipation issues have grown to such an importance that now constraints total chip performance. On the other hand, each router limits maximum accessible bandwidth on its ports. Thus, NoC should be customized such that overall power or performance becomes optimal and also bandwidth requirements are satisfied. Topology customization and application task mapping are addressed in this study. Although, results are presented for mesh and hypercube networks, this methodology can be extended to other topologies. Mesh size is the same as the number of tasks while hypercube's dimensionality is chosen with least unassigned nodes.

The mapping problem can be expressed as following. A target application is described by a set of concurrent tasks assigned to IP cores. A graph with edges labeled with maximum required bandwidth (in MByte/Second) are used for this purpose. To get the best power/performance trade-off, a designer determines an optimal topology customization and IP core placement onto tiles. For a typical mapping M and traffic scenario S, an evaluation consists of two objective functions $Delay(S,M)$ and $Energy(S,M)$.

978-1-4244-2541-9/08 $25.00 © 2008 IEEE

III. PERFORMACE AND ENERGY MODELLING

Wormhole routing has increasingly become the common routing scheme for NoCs. This is due to minimal latency and buffer space, and relatively simple implementation. Up to now, most studies on wormhole routing in NoCs have relied on time-consuming simulation. An alternative for simulation for design space exploration is analytical modeling of performance metrics [6]. In this study, an accurate analytical model is proposed to estimate peer-to-peer delay in NoCs. This model is a modification of the proposed method in [9] to consider different configurations for routers with multiple virtual channels under heterogeneous traffic pattern. VCs are used to implement a deadlock free routing algorithm as well as higher link bandwidth utilization. To estimate energy consumption, a gate level experimental model at is used.

Suppose the with P ports and V virtual channels per port. Message service time through a router for zero load, involves header and body flits transmission time. Header flit encounters one hop routing, switching and wire delays, while body flits move through the route reserved by the header in a pipeline fashion. So, the zero load service time for L flit message through a router is

$$T = D_{Header} + D_{Body} = T_R + L(T_S + T_W) \quad (1)$$

For each source/destination flow (S,D), traffic transmission rate is denoted by $X(S,D)$. If deterministic routing algorithm (like dimension order routing) is used, traffic rate on each link is predictable. This rate on virtual channel v of port j in router i is denoted by $\lambda_{i,j,v}$ and is given by

$$\lambda_{i,j,v} = \sum_S \sum_D X(S,D) R(S,D,i,j,v) \quad (2)$$

where R has value 1 if associated flow traverses this virtual channel. Otherwise, it will be 0 (At source node, a message is driven to any virtual channel with the same probability). For each message entering virtual channel v of port j, average queuing waiting time, $\tau_{i,j,v}$ is composed of service time for previous messages in the same buffer, service time of other buffers that attempt to obtain the same output port and residual service time of the current serviced message, R [9]. Therefore,

$$\tau_{i,j,v} = T \times N_{i,j,v} + \sum_{\substack{k=1,p \ m=1,V \\ (k,m) \neq (j,v)}} \sum C_{j,v,k,m} \times T \times N_{i,k,v} + R \quad (3)$$

In this equation, $C_{j,v,k,m}$ represents the contention probability of a message at header of the current queue with virtual channel m of port k. $C_{j,v}$ is contention probability matrix of virtual channel v of port j with other buffers and is given by

$$C_{j,v} = \left[c_{j,v,m,n} \right]_{P \times 1}. \quad (4)$$

We can formulate $\lambda_{i,j,v}$ with the following matrix notation [9]

$$\lambda_{i,j,v} = \frac{N_{i,j,v}}{TC_{j,v} + R} \quad (5)$$

where $N_{i,j,v}$ is average queue size. Then for entire routers, we can write

$$T\Lambda CN + \Lambda \overline{R} = N \quad (6)$$

3-D matrix Λ contains arrival rates for all buffers through a network. Also, C and R matrixes with size $P \times V$ are multi-dimensional contention and residual functions for each buffer. An M/G/1 queuing system is used to model buffer specifications. One specification is average residual service time of a message at the arrival time of another message. This can be modeled as

$$R_{i,j,v} = \frac{1}{2} \lambda_{i,j,v} \overline{T}^2 \quad (7)$$

where $\overline{T}^2$ is the second moment of service time and is equal to T^2 as T is fixed.

$C_{i,v1,j,v2}$ is calculated as following [9]. $f_{i,v1,j,v2}$ is the probability that a message goes from port i virtual channel v_1 to output port j virtual channel v_2 (since routing is deterministic, this value is predictable) and can be written as

$$f_{i,v1,j,v2} = \frac{f_{i,v1,j,v2}}{\sum_k \sum_m f_{i,v1,k,m}} \quad (8)$$

Then, we have

$$c_{i,v1,j,v2} = \sum_k \sum_m f_{i,v1,k,m} \times f_{j,v2,k,m} \quad (9)$$

Using Little theorem, the average queuing waiting time of virtual channel v of port j in router i, $W_{i,j,v}$ is given by

$$W_{i,j,v} = \frac{N_{i,j,v}}{\lambda_{i,j,v}} \quad (10)$$

Now, for each flow (S,D) with H hops to traverse, the average source to destination delay, $L_{S,D}$ considering waiting times in intermediate queues and injection channel is given by equation 11. To model bandwidth multiplexing effects by virtual channel on latency, peer-to-peer delay is scaled by factor $\overline{V}$, representing the average degree of virtual channel multiplexing that takes place at a given physical channel.

$$L_{S,D} = \left(Hop(T_R + (L-1)(T_S + T_W)) + \sum_{(i,j,v) \in \Pi_{S,D}} W_{i,j,v} \right) \overline{V} \quad (11)$$

$\Pi_{S,D}$ is deterministic route from source S to destination D. To evaluate $\overline{V}$ a Markov model similar to [10] is used. Average message latency for overall application is calculated by averaging over all flows with message generation rate per flow as weight

$$L = \frac{\sum L_{S,D} X(S,D)}{\sum X(S,D)} \quad (12)$$

To validate our performance model an event driven wormhole NoC simulator is used to run a typical application. Packets are generated by a Poisson process according to task generation rates. As temporal traffic is generated by a pseudo random process, we use steady state simulation to get acceptable confidence level. So, a batch mean method with total simulation length divided into ten bunches is used. The metrics for the first batch is discarded to avoid its distortion effects. For an application, simulation will be terminated if confidence level 95% with interval 5% is achieved. Results for a typical mapping of four video decoders, driven from simulation and proposed analytical model are shown in Figure 1. Results are given for two mesh and hypercube

978-1-4244-2541-9/08 $25.00 © 2008 IEEE

networks. Messages have 32-flit length and routers have 2 virtual channels per physical channel. It is obvious that analytical model results have an acceptable variation (8% in average) from simulations'.

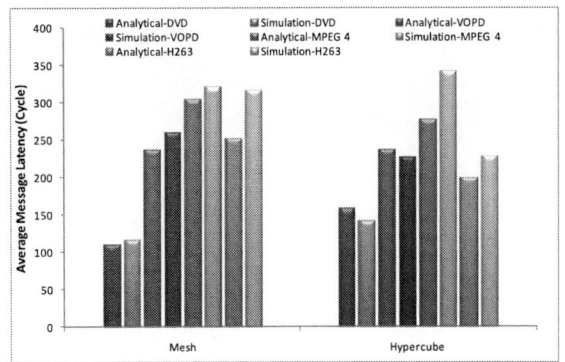

Figure 1. Model comparison with simulation for Mesh and Hypercube networks for four video decoders under typical mappings

The average flit energy, per flow with H hops to traverse, is given by

$$E_{Flit} = w \times (H \times E_{Sw} + (H+1) \times E_{Link}) \qquad (13)$$

where w is flit width in bits and E_{SW} and E_{Link} are the average energy consumed in a typical switch and a link when 1-bit data is transferred. E_{Link} can be estimated as $\alpha V^2\, dC_{Wire}/2$ where d is the link length in millimeter and C_{Wire} is 1 millimeter wire capacitance. In this equation, α is activity factor during data transmissions. We assumed that all links have 1 millimeter length and 0.18um CMOS technology is used. Also, activity factor is chosen to be 0.15 and operating frequency is 100 MHz, while supply voltage is 1.8V. Gate-level energy analysis based on switching activity, makes sense that E_{sw} consists of dynamic and static energy with 1.27uJ and 0.17uJ for 1-bit data transfer under light and medium traffic loads, respectively. Also, link energy dissipation for transferring each flit is estimated to be 0.67 pJ. Then, the average energy dissipation for all flows is

$$E = \frac{\sum E_{S,D}\, X(S,D)}{\sum X(S,D)} \qquad (14)$$

where $E_{S,D}$ is aggregate energy consumed per flow.

IV. MULTI-OBJECTIVE GENETIG MAPPING ALGORITHM

Genetic algorithm is well suited to solve multi-objective optimization problems [5]. It can simultaneously explore and exploit solution space. Up to now, several MOGAs have been introduced in the literature. Between Pareto ranking solutions, NSGA-II is an effective one to approximate true Pareto front for underlying problem [4].

Weighted average energy and delay are cost functions for genetic algorithm with maximum bandwidth requirement as design constraint. For each mapping, cost functions are evaluated from proposed analytical models. For a traffic scenario, a mapping is feasible, if links' bandwidth constraints are satisfied. A mapping is said dominating another, if at least for one objective function it has a better

solution. For the remaining objectives, it should have the cost equal or less than others. If a mapping exactly dominates i IPs, it has a rank $i+1$ for Pareto-optimal front in NSGA-II. So, solutions from Pareto rank 1 are usually optimal with low diversity [3]. If matrix notation B defines bandwidth requirements, feasibility is satisfied if each entry in B is less or equal than entry in constraint matrix, T. Constraints are handled by designing an extra objective function g as

$$\qquad (15)$$
$$g = \sum_i \sum_j \max\!\left(b_{i,j} - t_{i,j}, 0\right)$$

g is 0 if all links bandwidth constraints are satisfied. To shorten execution time, variation operators (crossover and mutation) in MOGA will not investigate constraint satisfaction for generated offspring (except the last population).

Each IP mapping is related to one chromosome in genotype domain and each gene is related to only one core. So, we use permutation representation [5]. Variation operators must maintain this permutation for offspring. hence we use Edge Crossover [5]. However neighboring in regular Edge Crossover is not enough for regular networks. So, a modification on neighborhood definition and gene allocation is done to fill this gap. Mutation operator for a typical network is Swapping or Inversion. The reason is that they do not change parents neighboring relation much and also maintain permutation. Details of the used GA are shown in Table 1.

TABLE 1. DETAILS OF PROPOSED GENETIC ALGORITHM

MOGA	NSGA-II
Representation	Permutations
Recombination	Modified Edge crossover
Recombination rate	0.9
Mutation	Swap/Inversion
Mutation rate	0.1
Parent selection	Binary tournament
Population size	200
Generation	200

V. EXPERIMENTAL RESULTS

In this section, we compare results from application mapping using proposed MOGA method with those by random mapping (as exhaustive design space exploration requires long time). Random solution includes 5000 feasible mappings.

Typical applications with real time or minimum delay constraints in congestion with low-energy design of on-chip networks are desirable. Four video applications including VOPD, MPEG-4, H263 and DVD decoders are evaluated. Each application maps on mesh and hypercube architectures to decide on most suitable infrastructures. In all GA results, only Pareto optimal front with rank 1 is shown. Because all solutions from other rankings are dominated by rank 1 solution. Results for MPEG-4 application are reported wider while for other applications only the best ones are shown.

The MPEG-4 decoder application task graph with 12 concurrent tasks is shown in Figure 2-a [6]. Our algorithm found the optimal solutions for mapping over mesh with size 3x4 and 4-dimentional hypercube. Pareto optimal front for each architecture in conjunction with random mappings are shown in Figures 2-d and 2-e. As a typical solution, we show the performance optimal solutions for the mesh and energy optimal solution for hypercube in Figures 2-b and 2-c, respectively. It is obvious that tasks with maximum flow bandwidth are seated close to each other. For all applications, performance and power optimal solutions for different topologies with details of execution time and number of feasible mappings in the last generation are reported in Table 2. Low execution times are due to inherent speed of NSGA-II and analytical modeling. Choosing mesh as network infrastructure has a great effect on design exploration time while hypercube has the mapping with better average delay.

VI. CONCLUSION AND FUTURE WORKS

In this paper, analytical performance and experimental power models for transferring flits over chained routers were developed. We then proposed an efficient NSGA-II algorithm employing these models to realize a high-speed exploration in design phase for optimal mapping and network topology selection, considering multiple objectives for a given MPSoC applications. Further work can consider more accurate router models and more detailed traces from real applications.

REFERENCES

[1] Asica G. et al., "Multi-objective Mapping for Mesh-based NoC Architectures," Proceedings of *International Conference on Hardware/Software Codesign and System Synthesis (CODES + ISSS)* pp. 182-187, September 2004.

[2] Benini L. et al., "Networks on Chips: A New SoC Paradigm," *IEEE Transactions on Computer* vol. 35, no. 1, pp. 70-78, January 2002.

[3] Bertsekas D., Gallager R., "Data Networks," *Prentice Hall Inc.* 1987.

[4] Deb K, et al. "A Fast Elitist Nondominated Sorting Genetic Algorithm for Multi-objective Optimization: NSGA-II," *Proceedings International Conference on Parallel Problem Solving from Nature* pp. 849-858, September 2000.

[5] Eiben A. E. et al., "Introduction to Evoloutionary Computing," *Springer* 2003.

[6] Guz Z., et al., "Network Delay and Link Capacities in Application-Specific Wormhole NoCs," *VLSI Design* 2007.

[7] Hu J. et al., "Energy- and Performance-Aware Mapping for Regular NoC Architectures," *IEEE Transactions on Computer Aided Design of Integrated Circuits and Systems* Vol. 24, No. 4, pp. 551-562, April 2005.

[8] Muraly S. et al. "Bandwidth-Constrained Mapping of Cores onto NoC Architectures," *Proceedings of Design, Automation and Test in Europe Conference and Exhibition (DATE)* Vol. 2, pp. 896-901, February 2004.

[9] Ogras U. et al., "Analytical Router Modelling for Network-on-Chip Analysis," *Proceedings of Design, Automation and Test in Europe Conference and Exhibition (DATE 07)* pp. 1096-1101, 2007.

[10] Ould-Khaoua M. and Sarbazi-Azad H., "An Analytical Model of Adaptive Wormhole Routing in Hypercubes in The Presence of Hot Spot Traffic," *IEEE Transactions on Parallel and Distributed Systems* Vol. 12, No. 3, pp. 283-292, March 2001.

TABLE 2. MAPPING FOUR VIDEO APPLICATIONS ON DIFFERENT NETWORK INFRASTRUCTURES

Application		Mesh				Hypercube			
		Delay (cycle)	Energy (pJ)	Num. Mapping	Exe Time (sec)	Delay (cycle)	Energy (pJ)	Num. Mapping	Exe Time (sec)
VOP Decoder	Eng.	235.35	99.66	147	13.27	205.55	99.65	150	21.56
	Per.	231.52	100.19			201.77	99.76		
MPEG4 Decoder	Eng.	303.8	97.68	136	12.31	251.46	100.74	144	26.06
	Per.	257.79	118.35			230.01	103.66		
H263 Decoder	Eng.	281.74	94.09	183	14.29	211.57	93.78	167	28.83
	Per.	249.88	114.97			199.63	100.46		
DVD Decoder	Eng.	106.41	62.05	168	8.38	99.38	56.69	137	18.6
	Per.	92.89	65.46			96.63	59.71		

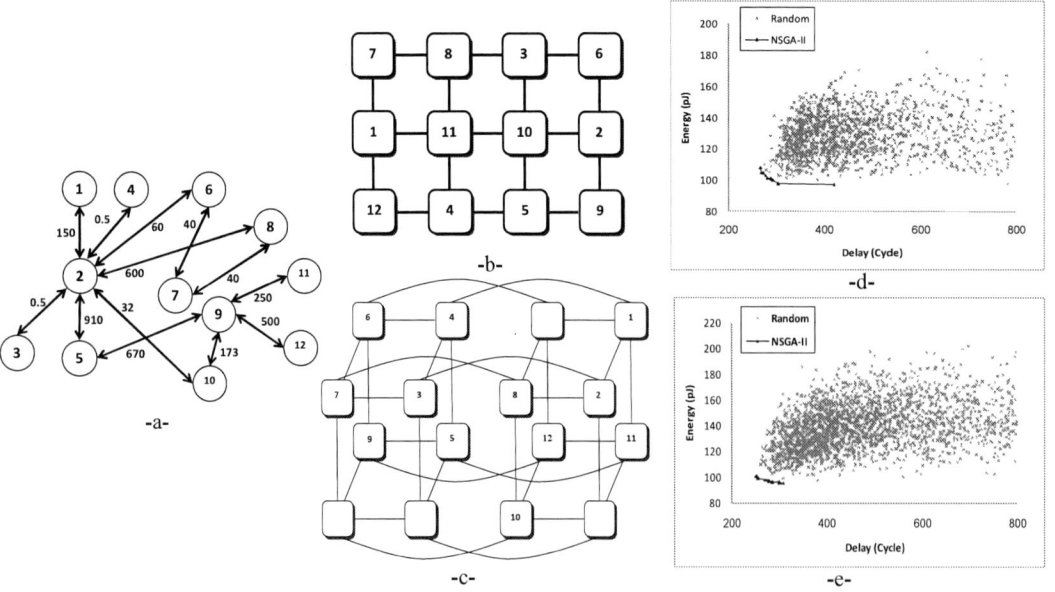

Figure 2. MPEG-4 application mapping using proposed MOGA –a- MPEG-4 application graph [5] –b- Performance aware mapping over mesh –c- Power aware mapping over Hypercube –d- Pareto optimal front vs. random mapping over Mesh –e- Pareto optimal front vs. random mapping over Hypercube

A 110 dB, 3-mW Fourth-order Σ-Δ Modulator for Atmospheric Pressure Sensor

Taeyoon Kim[1, 2*], Wonki Park[2], Heesun Ahn[3], Kyongwon Min[2],
Sangyong Lee[2], Jongchan Choi[2], Chulwoo Kim[1], Kynnyun Kim[2], Sungchul Lee[2]
[1]Department of Electronics and Computer Engineering, Korea University, Seoul, Korea
[2]SoC Research Center, Korea Electronics Technology Institute, Seongnam, Korea
[3]Department of Electronics and Information Engineering, Chonbuk National University, Jeonju, Korea
[*]daloveme@korea.ac.kr

Abstract— In this paper, a 110 dB, 1.024 MHz fourth-order single-loop sigma-delta modulator has been presented with an over-sampling ratio of 128 and an overload factor of -6 dB for a bandwidth of 4 kHz. In particular, this Σ-Δ modulator is well suited for atmospheric pressure sensor. The whole modulator consumes only 3-mW from a single 3.3V supply in a 0.35-μm CMOS technology and chip size is 1.68 mm².

I. INTRODUCTION

Sigma-Delta (Σ-Δ) modulation based analog-to-digital (A/D) conversion technology is a cost effective alternative for high-resolution converters which can be ultimately integrated on digital signal processor ICs. High-order Σ-Δ modulators are the most suitable A/D converters for low-frequency, high-resolution applications, in view of their inherent linearity, reduced anti-aliasing filtering requirements, and robust analog implementation. Σ-Δ modulators are, therefore, the best candidates for implementing the A/D converters in sensor applications, where the signal bandwidth is typically small, the required resolution is high and the environment where the circuit has to operate can be quite harsh. In particular, in sensor applications, a high resolution is required. So far, complex mash, multi-loop or multi-bit architectures with dynamic element matching have been used to achieve resolutions higher than 16 bits [1]-[3]. These solutions, however, consume large power and occupy large area. For distributed sensor applications, therefore, a single loop, single bit Σ-Δ modulator seems to be the best suited, since the area and the power consumption are intrinsically lower and the single-bit output can be easily transmitted without any digital post-processing. The signal-to-noise ratio (SNR) due to quantization noise, assumed white and additive, for such a Σ-Δ, a modulator is approximately given by [4], [5]

$$SNR = \frac{6(2L+1)M^{2L+1}}{\pi^{2L}}$$

where L is the order of the modulator and M is the oversampling ratio ($M=f_s/(2B)$, with f_s denoting the sampling frequency and B the signal bandwidth).

Considering the resolution required in distributed sensor (SNR > 100 dB) with f_s = 1.024MHz and B = 4kHz, (1) leads us to the choice of the fourth-order Σ-Δ modulator (L = 4) presented in this paper, whose specifications are summarized in Table I.

II. Σ-Δ ARICHITECTURE

Single-loop topology is preferable for low-voltage low-power design since it is less sensitive to circuit nonidealities, e.g., OTA dc gain and switch on-resistance. The Σ-Δ ADC is known for its high tolerance for circuit nonidealities compared to other ADC architecture. A fourth-order single loop topology was chosen in this design as shown in Fig. 1. The loop coefficients are set to [0.2 0.2 0.5 0.5] [6]. Unfortunately, the quantization noise is not only parameter that determines the SNR of Σ-Δ modulator, since the thermal noise and the building block nonidealities are, in many cases, quite significant. Considering a switched capacitor implementation, therefore, nonideal simulation was done using the SIMULINK toolbox, which takes into account the most important nonideal effects (kT/C noise, operational amplifier finite gain, bandwidth, noise, and slew rate, as well as clock jitter). A typical noise power spectral density of the Σ-Δ modulator with nonideal building blocks is shown in Fig. 2. The behavioral simulation was done by setting all of the OTA gains to 80dB and the oversampling ratio to 128.

TABLE I. MOST IMPORTANT FEATURES OF THE FOURTH ORDER, SINGLE LOOP, SINGLE BIT Σ−Δ MODULATOR.

Parameter	Value
Order of the modulator (L)	4
Sampling frequncy (f_S)	1.024 MHz
Signal bandwidth (B)	4 kHz
Oversampling ratio (M)	128
Signal-to-noise ratio (SNR)	> 100 dB

978-1-4244-2541-9/08 $25.00 © 2008 IEEE

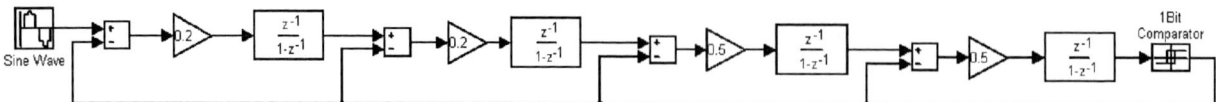

Fig. 1 Fourth-order, single-loop, single-bit, Σ-Δ modulator architecture.

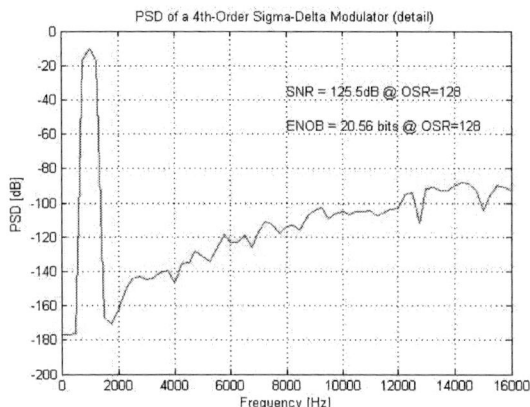

Fig. 2 Simulated typical noise power spectral density of the Σ-Δ modulator with nonideal building block.

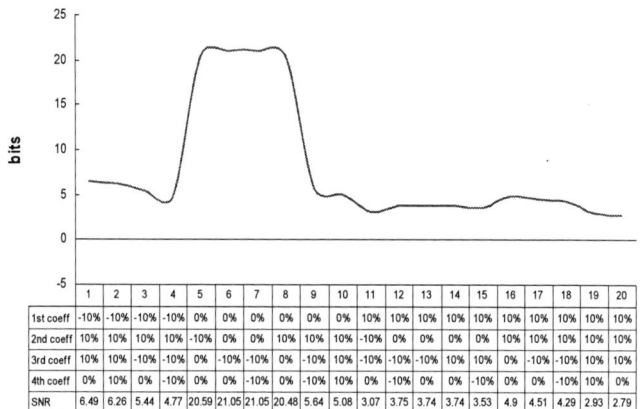

	1	2	3	4	5	6	7	8	9	10	11	12	13	14	15	16	17	18	19	20
1st coeff	-10%	-10%	-10%	-10%	0%	0%	0%	0%	0%	0%	10%	10%	10%	10%	10%	10%	10%	10%	10%	10%
2nd coeff	10%	10%	10%	10%	-10%	0%	0%	0%	10%	10%	10%	10%	-10%	0%	0%	0%	10%	10%	10%	10%
3rd coeff	10%	10%	-10%	-10%	0%	-10%	-10%	0%	-10%	10%	-10%	-10%	-10%	10%	10%	0%	-10%	-10%	10%	10%
4th coeff	0%	10%	0%	-10%	0%	0%	-10%	0%	-10%	10%	0%	-10%	0%	-10%	0%	0%	-10%	10%	0%	
SNR	6.49	6.26	5.44	4.77	20.59	21.05	21.05	20.48	5.64	5.08	3.07	3.75	3.74	3.74	3.53	4.9	4.51	4.29	2.93	2.79

Fig. 3 Simulated SNR versus capacitance mismatch of each integrators

More importantly, this single loop topology is quite tolerant to the inaccurate coefficients caused by capacitance mismatches. Among them, the first and second coefficients are very important because these decide NTF (Noise Transfer Function) in Fig. 3.

The OTA composes the main building block of the Σ-Δ modulator. It determines the main power consumption of the modulator. The requirements for the OTA are mainly output swing because the output swing is of great importance in low-voltage designs. For a kT/C noise-dominated modulator, the dynamic range can be written as

$$DR = \frac{P_{in_{max}}}{P_{kT/c}} = \frac{V_{in_{max}}^2 \cdot OSR \cdot C_s}{2kT} \qquad (1)$$

where $V_{in_{max}}$ is the maximum input amplitude of the modulator and Cs is the sampling capacitance of the first integrator. The maximum input amplitude of the modulator is defined by the output swing of the OTA. In order words, the maximum input amplitude has to do with reference voltage, as can be seen in Fig. 4. For certain dynamic range, an increase in the output swing can result in a large reduction of sampling capacitance and hence power consumption.

Fig. 5 shows the obtained SNR versus the OTA gains in the topology. The minimum gain requirement for OTAs is 50dB to ensure 120-dB SNR of modulator. This gain requirement is drawn only from the noise shaping consideration. However, taking the distortion into consideration, the higher the OTA gain, the better the distortion performance.

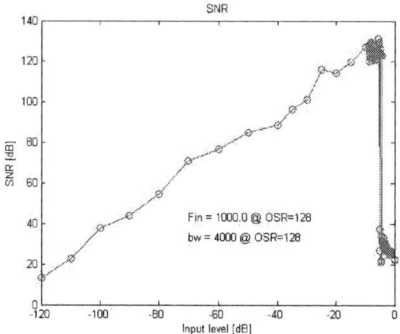

Fig. 4 Simulated SNR versus ratio of input amplitude and reference voltage.

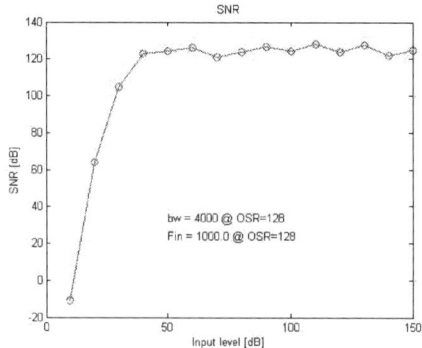

Fig. 5 Simulated SNR versus OTA dc gain of the fourth-order single loop Σ-Δ modulator.

978-1-4244-2541-9/08 $25.00 © 2008 IEEE

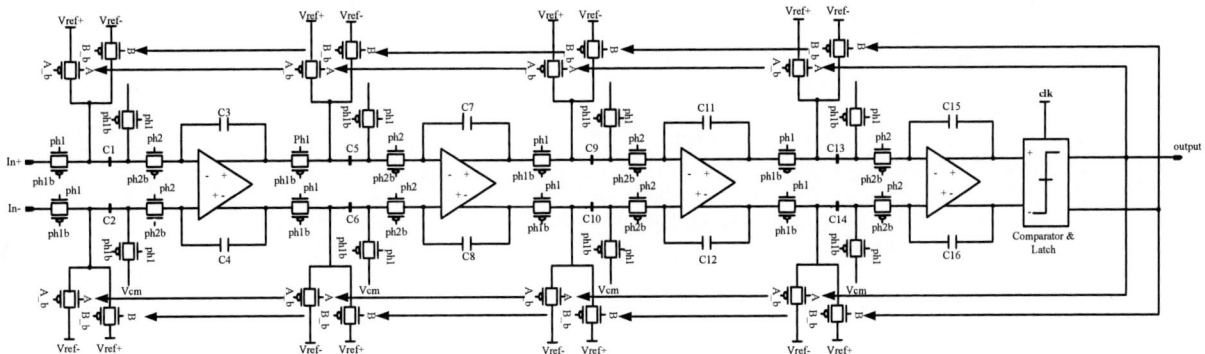

Fig. 6 Switched-capacitor implementation of the fourth-order, single loop, single-bit Σ-Δ modulator

III. IMPLEMENTATION

The switched capacitor implementation of the fourth-order, single-loop, single-bit Σ-Δ modulator is illustrated in Fig. 6. The circuit, based on a rail to rail fully-differential architecture, reflects the block diagram shown in Fig. 1.

A. OTA design

The main requirement for the operational amplifiers used in a switched capacitor Σ-Δ modulator is a speed. Indeed, in a switched capacitor circuit, if the operational amplifier is fast enough to allow the circuit to settle completely within half of the clock period, the nature of settling will not impact the overall circuit performance. Therefore, in the Σ-Δ modulator, rail to rail fully differential folded cascode operational amplifier is used as shown in Fig. 7. A large output voltage swing is achieved with the use of high-swing cascade current source and active loads. The switched-capacitor CMFB circuit is presented in Fig. 8. The precharged capacitor C_b senses the output common-mode voltage and shifts the voltage to proper level in code CMFB. Capacitor C_a periodically recharges the sensing capacitor. The main features of this CMFB circuit are its power efficiency and high performance.

The frequency response of the OTA with load capacitance is simulated. The gain reaches 85dB and the GBW is 100MHz while the phase margin if kept 85 degrees.

B. Switch driving and other circuits design

All switches are implemented with transmission gates. As the circuit is working on its rated supply voltage, there is no need to use any clock bootstrapping circuit to boost the driving voltage. Simple inverters are employed to drive switching transistors. The maximum driving voltage is same as the supply voltage. Hence, no internal node of the whole circuits is exposed to a voltage higher than V_{dd} or lower than V_{ss}, which is essential for high reliability operation of the circuits.

The on-chip clock generator makes non-overlapping clock. The external clock input signal is buffered and then two non-overlapping clock signals are generated. To avoid the signal dependent charge injection, two delayed clocks., C1d and C2d, are also generated [8].

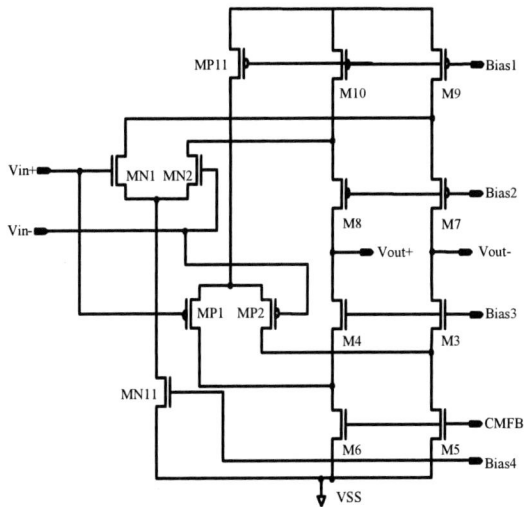

Fig. 7   Schematic of the operational amplifier used in the Σ-Δ modulator.

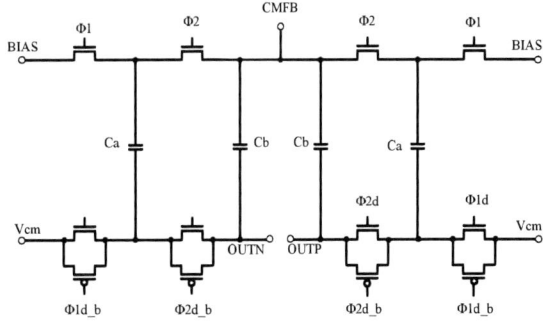

Fig. 8   Schematic of switched-capacitor CMFB circuit.

IV. RESULT

The high-resolution Σ-Δ modulator has been designed in a 0.35-μm CMOS process. The simulated noise power spectral density of the Σ-Δ modulator, obtained by fast FFT on 4096 samples with Hamming window and no averages, is shown in Fig. 9. The achieved noise power in the signal band is as low as -120 dB, which results in a maximum SNR of 110.9 dB calculated using the procedure described in [9], corresponding to 18.14bits of resolution which is sufficient for the considered application. The Σ-Δ modulator consumes 3-mW from a single 3.3-V power supply. The most important features of the circuit are summarized in Table II.

V. CONCLUSION

In this paper, a fourth-order single-bit single-loop switched-capacitor Σ-Δ modulator design for atmospheric pressure sensor is presented. The circuit operates with a sampling frequency of 4 kHz and an oversampling ratio of 128. A prototype of the modulator has been fabricated in a double poly, three metal 0.35-μm CMOS technology. The 1.4 x 1.2 mm^2 chip consumes 3-mW from 3.3-V supply and achieves a peak signal-to-noise ratio of 110.9 dB from simulation corresponding to a resolution of 18.14bits. Table III shows the performance comparison of published Σ-Δ modulators. The achieved performance is suitable for the considered application.

ACKNOWLEDGMENT

This work was supported by IT R&D program of the MIC/IITA [2006-S-054-02], Development of CMOS based MEMS processed multi-functional sensor for ubiquitous environment project

TABLE II.   FEATURES OF THE 4TH ORDER SINGLE LOOP SINGLE BIT Σ−Δ MODULATOR

Parameter	Value
Technology	0.35 μm CMOS
Power supply voltage	3.3 V
Power consumption	3 mW
Input voltage range (peak-to-peak, differentail)	1.04 V
Signal bandwidth (B)	4 kHz
Sampling frequency (f_s)	1.024 MHz
Signal-to-noise @ Full scale signal	110.9 dB
Resolution	18.14 bits

TABLE III.   PERFORMANCE COMPARISON

Architecture	BW (kHz)	VCC [V]	SNR [dB]	Power [mW]	Area [mm²]
6th (1-b) [10]	48	5	113	190	10.5
4th (1-b) [11]	20	3	99	5.6	1.16
4th (1-b) [This work]	4	3.3	110.9	3	1.68

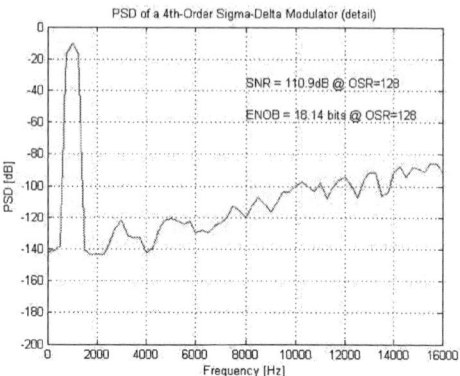

Fig. 9   Simulated output spectrum of an 1 kHz sinusoidal input

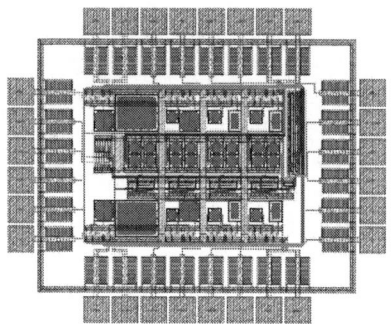

Fig. 10   Chip layout

REFERENCES

[1]   D. A. Kerth, D. B. Kasha, T. G. Mellissinos, D. S. Piasecki, and E. J. Swanson, "A 120dB linear switched-capacitor delta-sigma modualtor," in *ISSCC Dig. Tech. Papers*, San Francisco, CA, 1994, pp. 196-197.

[2]   K. Y. Leung, E. J. Swanson, K. Leung, and S. S. Zhu, "A 5 V, 118 dB delta-sigma analog to digital converter for wideband digital audio," in *ISSCC Dig. Tech. Papers*, San Francisco, CA, Feb. 1997, pp. 218-219.

[3]   O. Nys and R. Henderson, "A monolithic 19-bit 800 Hz low power multi-bit sigma-delta CMOS ADC using data weighted average," in *Proc. ESSCIRC*, Neuchâtel, Switzerland, Sept. 1996, pp. 252-255.

[4]   M. W. Hauser, "Principles of oversampling A/D conversion," *J. Audio Eng. Soc.*, vol. 39, pp. 3-26, Jan./Feb. 1991

[5]   S. Norsworthy, R. Schreier, and G. Temes, Eds., *Delta-Sigma Data converters. Theory, Design and Simulation*. New York: IEEE, 1997.

[6]   A. Marques, V. Peluso, M. Steyaert, and W. Sansen, "Optimal parameters for Delta-Sigma modulator topologies," *IEEE Trans. Circuits Syst.*, vol. 45, pp. 1232-1241, Sept. 1998.

[7]   T. B. Cho and P. R. Gray, "A 10 b, 20 Msample/s, 35 mW pipeline A/D converter," *IEEE J. Solid-State Circuits*, vol. 30, pp. 166-172, Mar. 1995.

[8]   D. Haigh and B. Singh, "A switching scheme for switched capacitor filters which reduces the effects of parasitic capacitances associated with switch control terminals," in *Proc. IEEE Int. Symp. Circuits and Systems*, May 1983, pp 586-589.

[9]   P. Malcovati, S. Brigati, F. Francesconi, F. Maloberti, P. Cusinato, and A. Baschirotto, "Behavioral modeling of switched-capacitor sigma-delta modulators," *IEEE Trans. Circuits Syst. 1*, vol. 50, pp. 352-364, Mar. 2003

[10]   C. B. Wang, S. Ishizuka, and B. Y. Liu, "A 113-dB DSD Audio ADC Using a Density-Modulated Dithering Scheme," *IEEE J. Solid-State Circuits*, vol. 38, no. 1, pp. 114-119, Jan. 2003.

[11]   Youngkil Choi, Hyungdong Roh, Hyunseok Nam, and Jeongjin Roh, "Design of a 99-dB DR single-bit fourth-order high-performance delta-sigma modulator", *ITC-CSCC 2007*, Jul. 2007.

A STATE BASED FRAMEWORK FOR EFFICIENT SYSTEM-LEVEL POWER ESTIMATION OF OF CUSTUM RECONFIGURABLE CORES

Ali Ahmadinia, Balal Ahmad, Tughrul Arslan

School of Engineering and Electronics, University of Edinburgh

{a.ahmadinia, b.ahmad, t.arslan}@ed.ac.uk

ABSTRACT

This paper presents a new system level power estimation methodology based on transaction level modeling for costum reconfigurable cores. The methodology can lead to significant improvement in trade-off between accuracy and efficiency of power estimation at system level. A SystemC based simulation environment is presented that allows rapid introduction of a power model into the executable specification of a sophisticated reconfigurable hardware design. The proposed environment allows efficient power estimation of custom reconfigurable cores through state based power modeling, leading to a viable solution for early power aware design. The simulator has been applied to SystemC module of a custom reconfigurable core for Viterbi decoding. Power figures have been compared with the results obtained by state of the art industrial tools.

1. INTRODUCTION

Low power is one of the major design challenges for custom reconfigurable SoC design. For example in mobile and wireless devices, it is necessary for maximizing standby time, active time, and battery life. Early power-aware design validation is a crucial step to achieve a power-optimized design. Multi-core reconfigurable SOC designs require accurate power analysis at the system level to allow performance and architectural feature trade-offs to be made within a constrained power envelope.

Although SystemC is considered the most promising language for high level functional modeling, it does not support power modeling capabilities. We present a novel power estimation framework which instruments SystemC for power characterization, modeling and estimation. This approach targets reconfigurable systems that their architectures can be reconfigured dynamically. The proposed power modeling is based on a state machine which fits well in a reconfigurable hardware concept. Reconfigurable design enables us to develop multi-standard systems by reconfiguring custom reconfigurable embedded cores in the system. These characteristics of our power estimation methodology make it a suitable solution for power-aware digital communication designs. For this reason, our proposed methodology is demonstrated by a custom reconfigurable Viterbi decoder core, which is typically used in wireless applications. Since it is entirely based on SystemC, our methodology allows consistent power modeling for different abstraction levels.

2. RELATED WORK

Numerous methods have been proposed for estimating power at system level. Some of the techniques are dedicated to particular components on a system. For processor cores, the work in [4] is based on monitoring the activity of various components within the processor's micro-architecture and uses this information to estimate power consumptions. [3] describes a mode based approach and different modes (e.g. active mode, idle mode) are distinguished. Power models for on chip buses are described in [6]. [4] proposes a system level approach for power modeling of Network on chip. For component with regular implementations, such as memories, analytical models have been proposed to estimate power consumption under given access patterns [9]. The modeling technique presented in [7] uses a hybrid approach to estimate the power consumption. It uses the gate level power estimates as the input data to the executable system level model to calculate system level power consumption. Similar modeling approach was used to estimate the system level power consumption of the peripherals components [8] and IP cores used for multimedia applications [10]. [1] proposes a framework for power estimation of heterogeneous IP cores in a whole system.

In contrast with all these approaches, we present a novel state based methodology is applied which gives an efficient power estimation for reconfigurable cores embedded within system on chip devices.

3. POWER ESTIMATION FLOW

Power estimation consists of static and dynamic power consumption. Dynamic power dissipation [11] is mainly caused due to the switching activities of the circuits ($P_{dynamic}$). Dynamic power dissipation depends on operating frequency, higher the operating frequency leads to more frequent switching of the circuits, thus increases the power consumption. Static power dissipation is related to the logical states of the circuits rather than switching activities. In CMOS logic, only possible source of static power dissipation [11] is the leakage current ($P_{leakage}$). Then we can define the total power consumption with the following equation:

$$(1) \quad P_{all} = P_{dynamic} + P_{leakage}$$

Furthermore, $P_{dynamic}$ can be rewritten as:

$$(2) \quad P_{dynamic} = \frac{1}{2}\alpha C V^2 f$$

Where α is the capacitance switching (charging &discharging), C is the total capacitance, f is the average frequency and V is the voltage swing.

Commercial power estimation tools mostly use gate level simulation to estimate the power consumption of hardware components. The power consumption of each hardware block can be defined [12]:

Power Consumption = Gate Count x µW/MHz x Activity x Frequency (3)

978-1-4244-2541-9/08 $25.00 © 2008 IEEE

The gate counts for a block can be estimated by using synthesis tools with technology library. The important aspect of the power estimation is assignment of switching activity levels for each gate, as each gates of a design have different switching activity levels. RTL simulator tools such as Cadence Verilog XL, Mentor Graphic Modelsim, can automate this process by generating a SAIF (Switching Activity Interchange Format) file. The SAIF, was created by Synopsys to standardize a power format, is mainly composed of the toggle count (the number of logic-0 to logic-1 and logic-1 to logic-0 transitions of each net in the net-list, per unit of time) and the state probability.

Power compilers estimate the power using the following three inputs: gate level netlist, SAIF file, and the technology library. Power compilers back annotate ports of the design to extract exact switching activity, and the default switching activity is used for non-annotated ports. In the next section, we provide a switching activity file similar to SAIF file by SystemC simulation, which is significantly faster than gate level simulation.

4. SYSTEMC POWER ESTIMATION

The major problem for power estimation at system level is identifying a relationship between functional behavior of the system with that of real circuit events. Depending on different behavioral aspects of various components within the system, different characterization techniques are developed for the modules. These will have an impact on the accuracy of the whole system.

SystemC Transaction level models (TLM) are increasingly being used for SoC architectural analysis and early SoC development. TLM reduces simulation time while exploring and validating implementation alternatives higher levels of abstraction. Growingly IP providers are providing such models for users. Incorporating power estimation techniques into a SystemC functional model designed to run embedded software would be a fast way to get power related information while performing architectural and performance analysis. The proposed SystemC based system level power estimation captures power related activity, where real power figures are assigned to each power related activity.

4.1. State based Power Estimation in SystemC

A power model utilizes an algorithm that allows energy consumption estimation for a system, on the basis of suitable applied data. In this work, a state machine based power model is used while each state represents average power consumption of the component in a specific functional mode configuration. The characterization through power states might be achieved from energy analysis on a low level module description. Typically these low level analyses could be carried out once and for all. The steps to realize power states are as follows:

- Determination of power states which cover the SystemC module functionality.
- Dedication of the appropriate power model to each power state.
- Identification of operating conditions in which each power state is valid.

All these steps can be realized through customized instructions related to the environment. For implementing the third step, the user has to define an appropriate state machine which establishes the current power state. This task is carried out by monitoring the run-time evolution of the monitored sc_module. For this purpose, the state machine can have access

to the sc_module input signals and can be activated by triggering events set by the user. Each monitored sc_module is associated to its own state machine.

A monitoring system is used for making power models of costum reconfigurable cores. The functional model has source code available and it could be modified into a power model in order to trigger its corresponding monitor, which will generate power related transactions. For different hardware blocks, model developers have to figure out which information can be used to trigger its FSM power model. The designer provides information needed to monitor the power behaviors and the monitor writes out the transactions into the transaction database. The activity database stores all monitored signal switching in the sc_module during functional simulation. The power state models initially are based on the technology dependent power estimation such as leakage power. To assign an appropriate power value relative to speed (in μW/MHz) for the switching of each signal type, library vendor's manuals can be used.

This approach computes the signal activities of a SystemC module by monitoring all writing operations to store signal value changes in an activity list (See Fig. 1). More precisely, during the simulation a monitored sc_module is always in a particular operative condition, hence in a particular current power state. If the current power state changes, due to change of the operative conditions, the estimator calculates a partial energy estimation related to the period during which the past power state has been valid. This estimation is achieved by applying the power model associated to the past power state. For this purpose, the power estimator considers and uses all the data required by the power model, both static and dynamic data. In the last step, the current power state is updated to the new power state, which will be valid till the sc_module remains in the operative conditions associated to it. At the end of the simulation, the total energy estimation will be given by the sum of the partial estimations related to the triggered power states.

4.2. Reconfigurability

The large amount of SoC fabrication cost has demanded increasingly a single SoC device with upgradeability feature to prolong SoC product life cycle. Embedded custom reconfigurable cores are one way of solving the above problem, where cores with targeted and tailored reconfigurability replace

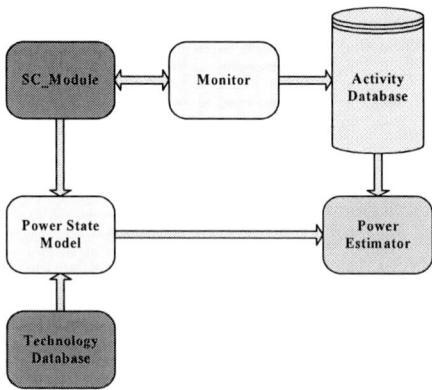

Fig.1. SystemC State-based Power Estimation Flow.

conventional IP cores. Due to the computational complexity of future SoCs, instruction set processors are not sufficient and dedicated data path blocks may only handle the processing requirements. The use of dedicated data path structures might be restrictive if standards are not finalized at the time of the implementation and if a fast time-to-market is required. In such cases reconfigurable cores becomes a strong candidate of choice. Such cores allow conforming to multiple or migrating international standards, and tasks of the target system can be upgraded by adopting a more sophisticated/efficient algorithms. In addition, their functionality can be enhanced some time after the development of the first version of the system.

Our power modeling approach allows using more than one power model for each monitored module within the system (sc-module) during a simulation session. Hence, we can model reconfigurable cores to have a dedicated state for each configuration. This is possible due to the developed power state-based approach, which allows using the most suitable power model according to the operative conditions of a monitored sc_module. In the simplest case, an sc_module can be characterized by only a single power state (and then by a single power model) good for all the possible operational modes. This represents a significant improvement in comparison with instruction-based strategies proposed in [6], [8], where only a single power model is used for all the possible instructions.

Our methodology provides flexibility and accuracy for power modeling of reconfigurable modules, since it allows applying the most suitable power model for each working phase. As shown in Fig. 2, we can define a power state for each configuration mode, and these states are all part of the whole power states of the module including power states of the static section.

4.3. Theoretical FSM-based Power Modeling

Finite State Machine (FSM) based power model will be used to represent power related states. Power consumption in each component of a system can be defined as follows:

Definition 1 (Power State Characteristics)
Given a set of power states $S=\{s_1, s_2,..., s_n\}$.
For each state s_i in S, the following terms are defined:
$P_i(tr_j)$ is the power consumption of state s_i in transaction tr_j
$P_i(s_j)$ gives the amount of power dissipation to transfer from state s_i to state s_j
lk_i is the dissipated leakage power of state s_i
All transactions are given as a set $Tr=\{ Tr_1, Tr_2,..., Tr_m \}$.

Now, for the given sets of power states and transactions, the total power consumption P_{all} can be computed as follows:

$$P_{all} = \sum_{i=1}^{n}\sum_{j=1}^{m} P_i(tr_j) + \sum_{i=1}^{n}\sum_{j=1}^{m} P_i(s_j) + \sum_{i=1}^{n} lk_i \quad (4)$$

The above power parameters are the key elements, which will contribute to power consumption of a power model. These will be stored in a format of lookup tables or formulated into equations. All these parameters are technology dependent.

5. CASE STUDY

For evaluation of efficiency and accuracy of the proposed power estimation framework, a custom reconfigurable Viterbi decoder is chosen as a real world application in telecommunication. The implemented reconfigurable Viterbi module can decode for different constraint lengths *(K=3, K=5, and K=7)*. It would be synchronized with the input sequence at the beginning of the burst sequence decoder (i.e. reset to the beginning known state).

Viterbi decoder uses forward and reverse state metrics

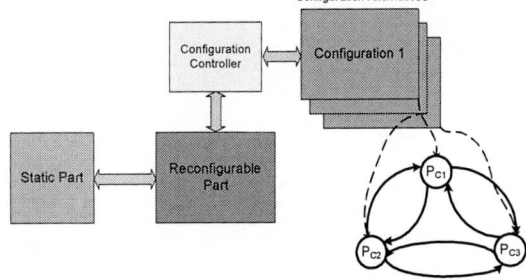

Fig.2. Power Modeling of a Reconfigurable Module.

processing. To improve the latency, sliding windows can be used [1]. Here, we use two reverse processors B1 and B2 in parallel with one forward processor FP. Four blocks of memory are used for path history unit which is read by trace back processors as they calculate the survivor path in the reverse traversing of the trellis. The write and read control is provided by the state machine. Reverse processor can start cold in any state (initializing each state as equivalent), but after few iterations (equal to window length: WL) the state metrics are as reliable as if the process had been started at the final node of trellis. Let B2 be the dummy reverse processor that starts from state S_1 and after reverse traversing the trellis for a WL, provides the start state for the actual reverse processor B1.

Fig. 3 shows the four basic states of Viterbi decoder with start state as S_1. It also shows the working of FP, B1 and B2 as they write and read path history memories. After 4 WLs, the cycle repeats. First decoded bits are output continuously after latency of 3 WLs from state S_4.

For power modeling of our custom reconfigurable Viterbi core, we used a hierarchical power state machine. We have a power state machine corresponding to each constraint lengths configuration, and in the lower level, we model power consumption of each configuration based on the reconfigurable Viterbi decoder state machine as mentioned before.

The Viterbi decoder uses an N state trellis diagram. For each stage L of trellis, branch metrics (BM) are computed using the soft input symbols. State metrics for stage $L+1$ are updated using the SMs for stage L of trellis. The number of state and stages are dependent on Constraint Length (K) of Viterbi decoder which is the configurable parameter of this reconfigurable Viterbi: $N=2^K$ and *Number of Stages=K*. Therefore, we use a power model for each configuration according to the number of states and stages of trellis diagram, while for lower constraint length configurations, more states

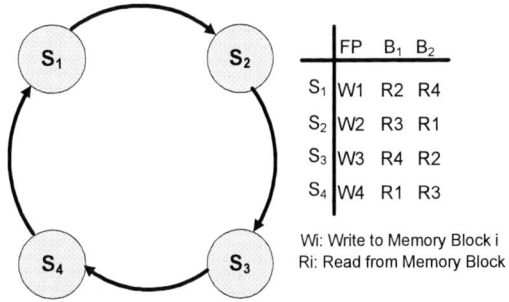

Fig.3. Power State Machine of a Reconfigurable Viterbi Decoder.

and stages of Viterbi architecture will be in idle switching

978-1-4244-2541-9/08 $25.00 © 2008 IEEE

activity mode. Fig. 4 shows the state machine of power

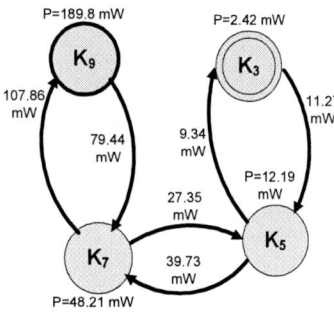

Fig.4. SystemC State-based Power Estimation Flow.

consumption in a reconfigurable Viterbi core, while each state and transition is associated with average power consumption.

6. EVALUATION RESULTS

A new design power estimation framework was constructed. This involved power estimation of a costum reconfigurable Viterbi decoder core. As mentioned earlier, system level power estimation has a lower accuracy compared to other abstraction levels. Power estimation at lower abstraction design levels, such as, RTL, gate-level, or the real silicon could be used as the reference for system level result. Therefore, we have compared the estimated results with gate-level simulated results of the platform. For gate level based power estimation, Synopsys Power Compiler is used. We have applied 2000 random samples to calculate the power consumption with Synopsys

Table 1. Accuracy and Efficiency of reconfigurable Viterbi core with different constraint lengths.

Constraint Length	Estimated Power (mW)	Simulated Power (mW)	Error (%)	Speed Up (X)
3	2.12	2.42	14.2	1251
5	10.83	12.19	12.6	1674
7	43.76	48.21	10.2	1866
9	174.53	189.80	8.8	2039

Power Compiler in 130nm technology.

Table 1 illustrates the efficiency of different configuration modes of Viterbi decoding core and the simulated versus estimated power consumption and shows the error percentage of estimated power consumption, which is within a 8.8-14.2% worst case range. The power consumption increases moderately with larger constraint lengths, and the accuracy of estimated power also increases by increasing constraint length. Although, the estimated power consumption is more accurate in longer Viterbi constraint lengths, which indicates the need for further refinement of the estimation in order to have a more precise power model of idle parts of the Viterbi decoder with shorter Viterbi constraint lengths.

7. CONCLUSION

This paper has presented a methodology capable of evaluating crucial tradeoffs (energy, speed and power) at system-level based on SystemC TLM for emerging custom reconfigurable embedded cores targeting multi-standard devices for wireless applications. Considerable speed-up has been achieved in

evaluating the energy efficiency when integrating processors with custom reconfigurable core modules on the proposed framework. A SystemC simulator has been presented as the only means of collecting system transactions, providing for a practical and fast evaluation of energy and power when combining events with power models. We have demonstrated that the robustness and reuse achieved through using the proposed approach considerably simplify both the creation of power models and the top-down data gathering required for power analysis. No changes in the SystemC kernel were required and the power-enabled components fit directly into the existing framework.

Providing configurable power model with distinct accuracy and efficiency characteristics reduces computational effort towards power estimation. The proposed power estimation framework has proved valuable for early power aware design of custom reconfigurable systems.

8. REFERENCES

[1] N. Bansal, K. Lahiri, A. Raghunathan, and S. T. Chakradhar. Power monitors: A framework for system-level power estimation using heterogeneous power models. In Proc. Of the 18th Int. Conf. on VLSI Design, pages 579–585, 2005.

[2] S. Benedetto and G. Montorsi. Soft-output decoding algorithms for continuous decoding of parallel concatenated convolutional codes. In Proc. of ICC'96, pages112–117, Dallas, June 1996.

[3] L. Benini, R.Hodgson, and P. Siegel. System-level power estimation and optimization. In Proc. of the Int. Symposium on Low power Electronics and Design, pages173–178, 1998.

[4] A. Bona, V. Zaccaria, and R. Zafalon. System level power modeling and simulation of high-end industrial network-on-chip. In Proc. of the Conference on Design, Automation and Test in Europe, page 30318, 2004.

[5] D. Brooks, V. Tiwari, and M. Martonosi. Wattch: a Framework for architectural-level power analysis and optimizations. In Proc. of the 27th annual international symposium on Computer architecture, pages 83–94, 2000.

[6] N. Dhanwada, I.-C. Lin, and V. Narayanan. A power estimation methodology for systemc transaction level models. In : Proceedings of the CODES+ISSS conference 2005, pages 142–147, 2005.

[7] T. Givargis, F.Vahid, and J.Henkel. A hybrid approach for core-based system-level power modeling. In Proceedings of ASP-DAC conference, pages141–146, 2000.

[8] T.Givargis, F.Vahid, and J.Henkel. Instruction based system level power evaluation of system-on-a-chip peripheral cores. IEEE Transaction on VLSI systems, 10(6):856–863, 2002.

[9] M. B. Kamble and K. Ghose. Analytical energy dissipation models for low-power caches. In Proc. of the Int. Symposium on Low power Electronics and Design, pages143–148, 1997.

[10] M. Onouchi, T.Yamada, K. Morikawa, I. Mochizuki, and H.Sekine. A system-level power-estimation methodology based on ip-level modeling, power-level adjustment, and power accumulation. In Proc. of the 2006 conference on Asia South Pacific design automation, pages547–550, 2006.

[11] Gray K.Yeap, "Practical Low Power Digital VLSI Design", Kluwer Academic Publishers, 1998, ISBN 0-7923-8009-6.

[12] Jim Flynn, Brandon Waldo, " White paper: Power Management in Complex SoC Design",   http://www.synopsys.com/sps.

Low Noise Amplifier Architecture Analysis for UWB System

Peng Wang[1,2], Fredrik Jonsson[1], Dian Zhou[2], and Li-Rong Zheng[1]

[1]Royal Institute of Technology (KTH), ECS/ICT, ELECTRUM 229, SE-164 40 Kista-Stockholm, Sweden
Email: {pengw, fjon, and lirong}@kth.se

[2]School of Microelectronics of Fudan University, 825 Zhangheng Road, Shanghai 201203, China
Email: zhoud@fudan.edu.cn

Abstract— This paper analyzes the architecture of wideband low noise amplifier (LNA) for multi-band orthogonal frequency division multiplexing modulation (MB-OFDM) ultra-wideband (UWB) system. Noise matching and input impedance matching are compared among different LNA architectures. Power consumption and area for different kinds of LNA architectures are also compared through the figure of merit (FOM).

Index Terms—LNA, MB-OFDM UWB, Noise Figure, Input Matching, FOM.

I. Introduction

Federal Communication Commission (FCC) has opened up to 7500MHz frequency spectrum ranging from 3.1-10.6GHz since 2002. The UWB WPAN physical (PHY) layer standard divides the whole available ultra wideband spectrum between 3.1-10.6GHz into 14 sub-bands belonged to 6 band groups respectively [1].

Low noise amplifier is usually the first on-chip active stage of a radio frequency (RF) receiver. According to Friis functions, the specifications of both low noise factor and high gain are required. In [1], it recommends LNA have 15dB gain with noise figure (NF) less than 4dB through the whole 7.5-GHz bandwidth. The defined power consumption which must be less than 250mW [2] for the UWB system requires the RF front-end modules being compact and operating in a low power mode. Hence, wideband LNA becomes one of the biggest challenges in UWB system. In this work, NF, input impedance matching and gain are analyzed for different kinds of wideband LNA architectures, and a comprehensive comparison is made by evaluating FOM through different LNA architectures.

The organization of this paper arranges as follows. First, different kinds of UWB LNA architectures are analyzed for noise and input impedance matching. Next, FOM among different architectures of LNA are compared, in which the power consumption, and IIP3 (linearity) are included. Conclusions are given in the last chapter.

II. Analysis of Different Wideband Low Noise Amplifiers

Several circuit topologies have been proposed for developing broad-band LNA. The distributed amplifier (DA) is well known for its capability of providing very wide bandwidth, but it suffers from high power consumption, medium gain and significant chip area [3]. Shunt feedback is very popular for wideband application [4]. It provides good impedance matching and gain flatness. However, low NF, high gain and low power consumption can be hardly achieved simultaneously across a large frequency range. Inductor degenerated common-source LNA with LC-ladder as the input matching network is also a good choice for UWB application. It has low NF, low DC power, and high gain across a broad band, but it consumes a large chip area due to passive components for input matching [5]. Common-gate amplifier can realize wideband input impedance matching by modifying the input transistor size and bias current [6]. In spite of its advantages, common-gate amplifier suffers from poor power gain and NF. Gm-boosting common-gate amplifier is developed to overcome the drawbacks mentioned above [7]. Using the transformer as the input matching and feedback is also proposed to realize wideband matching [8-9]. Due to the high DC power and poor NF, the resistive terminated LNA as the input matching network is unavailable for UWB system.

A. Inductor degenerated CS-LNA with LC-ladder

This kind architecture is typical for narrow band. However, when LC-ladder matching network is added working as a high-pass-filter/band-pass-filter in the front of LNA, it can realize a wideband input impedance matching for LNA. Due to the feedback from the gate-drain capacitor C_{gd}, the isolation is not very good for this architecture. Hence a cascade stage is introduced in this architecture to reduce the miller effect and support good isolation. The scheme and equivalent small-signal circuit of LNA is shown in Fig.1. The noise analysis is as follows according to [10]:

$$\overline{i^2_{ng}} = 4kT\delta g_g \Delta f, \quad \overline{i^2_{nd}} = 4kT\gamma g_{d0} \Delta f \qquad (1)$$

Where $g_g = \omega^2 C^2_{gs} / 5g_{d0}$, g_{d0} is the zero-bias (V_{DS}) drain conductance of the device, γ is a bias-dependent factor 0.67~1.33, for short channel devices, the value can be more than 2, depending on the bias conditions, and δ is the gate noise coefficient, the value is 1.33~4. But the ratio δ/γ stays nearly constant at 2. The two sources of noise are partially correlated by their correlation coefficient c expressed in (2), which is theoretically computed to be 0.395j, with technology scaling it is slightly higher than 0.4. The noise figure under power constraint is given as follows:

$$c = \overline{i_{ng} i_{nd}^*} / \sqrt{\overline{i_{ng}^2 i_{nd}^2}} \qquad (2)$$

978-1-4244-2541-9/08 $25.00 © 2008 IEEE

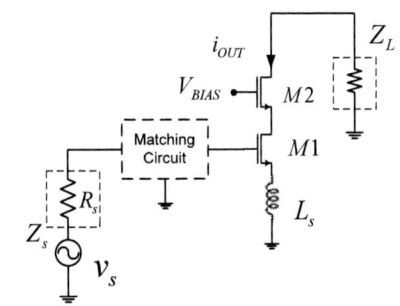

(a) Scheme of an inductor degenerated cascade LNA

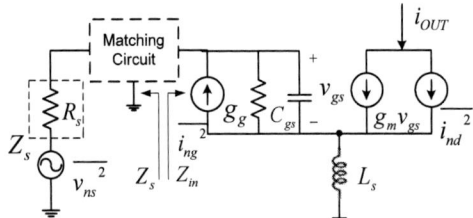

(b) Small-signal equivalent circuit

Fig. 1. (a) Scheme of an inductor degenerated cascade LNA. (b) Its small-signal equivalent circuit.

$$F = F_{\min} + \frac{R_n}{G_s}\left|Y_s - Y_{opt}\right| = F_{\min} + \frac{\gamma}{\alpha g_m R_s}(1 - \frac{R_s}{R_{opt}})^2 \quad (3)$$

Where, $F_{\min} \approx 1 + 2.4\gamma\omega/(\alpha\omega_T)$, R_n is the equivalent noise resistance, Y_{opt} is optimum source admittance corresponding to $F_{\min}$, G_s and Y_s are source conductance and admittance seen from gate of the transistor, $\alpha = g_m/g_{d0}$ becomes lower than 1 in short-channel device. After tedious calculation, the real and imaginary part of Z_{opt} can be given:

$$\text{Re}\{Z_{opt}\} = \frac{\alpha\sqrt{\delta/(5\gamma(1-|c|^2))}}{\omega C_{gs}\{\alpha^2\delta/(5\gamma(1-|c|^2) + (1+\alpha|c|\sqrt{\delta/5\gamma})^2\}} \quad (4)$$

$$\text{Im}\{Z_{opt}\} = -\omega L_s - K/(\omega C_{gs}) \quad (5)$$

Where K is a technology dependant parameter, as device sizes scale down, its value approaches to 1 (K≈0.72 for 0.18um CMOS technology).

In order to obtain a wideband noise input matching, the input source impedance Z_s seen from the gate of the input transistor should be the complex conjugate of Z_{in} while be equal to Z_{opt}. The expression for $\text{Re}\{Z_{in}\}$ and $\text{Im}\{Z_{in}\}$ are as follows: $\text{Re}\{Z_{in}\} = g_m L_s / C_{gs} = \omega_t L_s \quad (6)$

$$\text{Im}\{Z_{in}\} = \omega L_s + 1/(\omega C_{gs}) \quad (7)$$

Where, ω_t is the cut-off frequency of the transistor. On one side, $\text{Im}\{Z_{in}\}$ is approaching to $\text{Im}\{Z_{opt}^*\}$ when the technology is scaling down. Particularly at high frequency, the effect of the second term in (5) and (7) fades away, further

improving the matching. On the other side, $\text{Re}\{Z_{in}\}$ matching with $\text{Re}\{Z_{opt}\}$ in a wideband is more challenging, because the former is constant and bias-dependant while the latter is frequency dependant. Proper choice of Ls is important.

Gain analysis is given below. The impedance looking into the amplifier is equal to R_s in-band, and it is very high out-of-band. The current flowing into M1 is v_{in}/R_s. At high frequency, the MOS transistor acts as a current amplifier, the current gain being $\beta(\omega) = g_m/\omega C_{gs}$. As a consequence, the output current is $v_{in}g_m/\omega C_{gs}R_s$, the overall gain is therefore: $A_v = v_{out}/v_{in} = -g_m Z_L/(\omega C_{gs}R_s) \quad (8)$
Always the shunt-peaking resistor is used as the load to extend the bandwidth and compensate the current gain roll-off at high frequency. When there is the mismatch among $\text{Re}\{Z_{opt}\}$, $\text{Re}\{Z_s\}$, and $\text{Re}\{Z_{in}\}$, the gain and noise figure are expressed: $\quad G = G_{\max} \cdot \left[4R_s R_{in}/(R_s + R_{in})^2\right] \quad (9)$

$$NF = NF_{\min} + R_n(R_s - R_{opt})^2/(R_s R_{opt}^2) \quad (10)$$

Where R_s is the real part of input source impedance, its value should be a compromise between R_{in} and R_{opt}. The inequality among these three resistive values should be $R_{in} \leq R_{opt} \leq R_s$. Also the non-50Ω signal-source impedance can realize a very good noise figure and gain over a wideband [15].

B. Shunt-Feedback Wideband LNA

The shunt-feedback is usually used for extending the bandwidth of the amplifier. Fig. 2 (a) shows the typical shunt-feedback amplifier topology. Without adding the shunt-feedback, the input impedance is: $Z_{in} = 1/(j\omega C_{gs})$.

Referring the small-signal equivalent circuit shown in Fig.2, the input impedance with the feedback is:

$$Z_{inF} = 1/\{[1 + g_m(r_o \parallel Z_L)]/[Z_F + (r_o \parallel Z_L)] + j\omega Cgs\}(11)$$

Where, Z_F is the feedback impedance, and r_o is the output resistor of the transistor. The input conjugate matching impedance can be shifted toward Z_{opt}. The gain is expressed as follows: $A_v = [g_m - (1/Z_F)]/[Z_L \parallel r_o - (1/Z_F)] \quad (12)$
Because of the Z_F, the correlation admittance Y_c is found as:

$$Y_c = 1/\left[R_f + j\omega C_{gs}(1 + \alpha|c|\sqrt{\delta/5\gamma})\right] \quad (13)$$

Where, $R_f = \text{Re}\{Z_F\}$. The noise analysis is based on two-port noise model, derived as:

$$F = F_{\min} + (R_n/G_s) \times \left[(G_s - G_{opt})^2 + (B_s - B_{opt})^2\right] \quad (14)$$

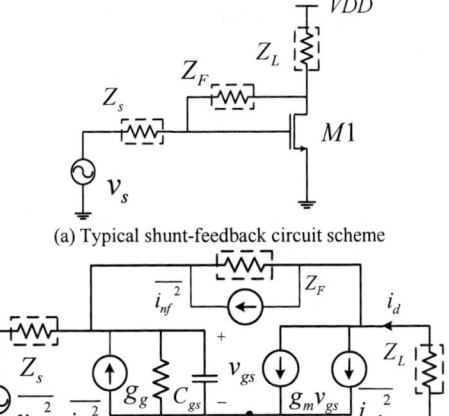

(a) Typical shunt-feedback circuit scheme

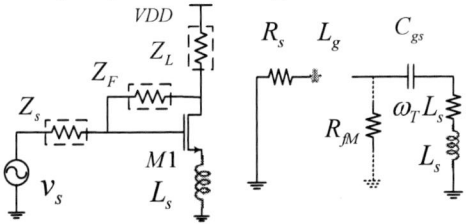

(b) Small-signal equivalent circuit for typical shunt-feedback circuit

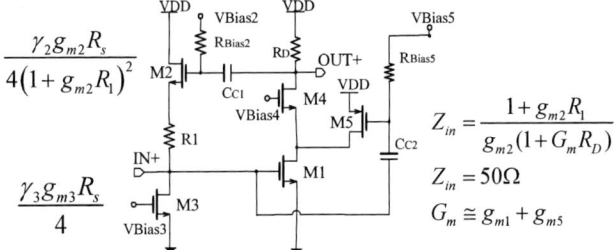

(c) Improved shunt-feedback schematic based on the typical scheme

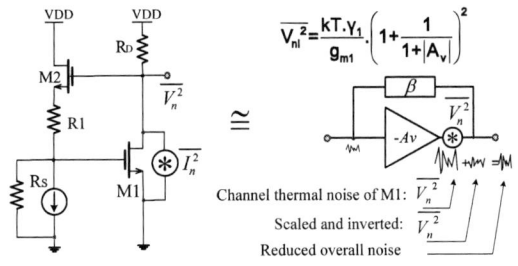

(d) Common-drain feedback: Zin-Gain tradeoff is isolated from each other

(e) Common-drain feedback: Noise cancellation

Fig. 2. (a) Scheme of a typical shunt-feedback CMOS amplifier. (b) Its small-signal equivalent circuit. (c) Its improved circuit. (d) Common-drain as the feedback for shunt-feedback architecture. (e) Noise cancellation for (d)

$$F_{min} \cong 1 + 2\left[\sqrt{\frac{\delta(1-|c|^2)\gamma}{5}\left(\frac{\omega}{\omega_T}\right)^2 + \frac{\gamma g_{d0}}{g_m^2 R_f}} + \frac{\gamma g_{d0}}{g_m^2 R_f}\right] \quad (15)$$

From (11)-(15), it is obvious the input impedance, gain, and noise figure are coupled by Z_F. The size of R_f is limited by the input impedance. However, the higher R_f, the lower

impact on NF from R_f. To decouple the Z_F impact on NF and input impedance, the degenerated inductor is introduced into the shunt feedback architecture as shown in Fig. 2 (c). The noise keeps the same expression but the input impedance is equal to $\omega_T L_s$. It results in the much larger R_f than the conventional shunt-feedback. The bandwidth in this kind architecture is: $BW_{-3dB} = \omega / Q_{WB}$, where,

$$Q_{WB} = 1/\left\{\left[R_s + \omega_T L_s + (\omega L_g)^2 / R_{fM}\right] \cdot \omega C_{gs}\right\} \quad (16)$$

Where, $R_{fM} = R_f /(1-A_{vo})$, A_{vo} is the open loop gain. Compared with the narrowband quality factor, in (16) the term $(\omega L_g)^2 / R_{fM}$ is added due to the feedback resistor, which increases the noise figure. Since the NF is coupled with bandwidth by $R_f (R_{fM})$, to realize the high gain, good NF, and very wide bandwidth is difficult. Fig.2 (d-e) shows the inductor-less common-drain feedback architecture to decouple the input impedance, gain and NF [9]. NF is expressed by:

$$F = 1 + (1+\frac{1}{1+|A_v|})^2 (\frac{\gamma_1}{4g_{m1}R_s} + \frac{\gamma_5}{4g_{m5}R_s}) + \frac{\gamma_3 g_{m3}R_s}{4}$$
$$+ \frac{\gamma_2 g_{m2}R_s}{4(1+g_{m2}R_1)^2} + \frac{R_1}{4R_s(1+|A_v|)^2} + \frac{R_D}{4R_s(1+|A_v|)^2}$$

C. Common-gate Wideband Amplifier

The common-gate amplifier has the highest potential to achieve the wide-band input matching. The scheme of common-gate LNA (CG-LNA) is shown in Fig. 3 (a). The input impedance of CGLNA is derived from Fig. 3 (a),

$$Z_{in} = \frac{1}{g_m + 1/Z_{s-in}(\omega) + [1 - g_m Z_o(\omega)]/[r_o + Z_o(\omega)]} \quad (17)$$

Where,

$$Z_{s-in}(\omega) = j\omega L_s \| (1/j\omega C_{gs})$$
$$Z_o(\omega) = (1/j\omega C_{gd}) \| Z_L$$

The NF is expressed as follows by assuming $g_m R_s = 1$,

$$F = 1 + \frac{\gamma}{\alpha} + \frac{\delta\alpha}{5}(\frac{\omega}{\omega_T})^2 \approx 1 + \frac{\gamma}{\alpha} \quad (18)$$

For the short channel devices, according to the ratio value mentioned above, the minimum NF is 4.8dB. It is too high to be fit for UWB application. The g_m-boosting method shown in Fig. 3 (b) can overcome this drawback to improve the NF.

$$F_{gm-boosting} = 1 + (\gamma/\alpha)/(1+A) \quad (19)$$

Where, $A = C_c /(C_c + C_{gs}) \approx 1$ due to $C_c \gg C_{gs}$, the input impedance matching is changed into $1/R_s = (1+A)g_m$. Because the trans-conductance is decreased by the same factor, the power consumption is also decreased by $1+A\approx 2$.

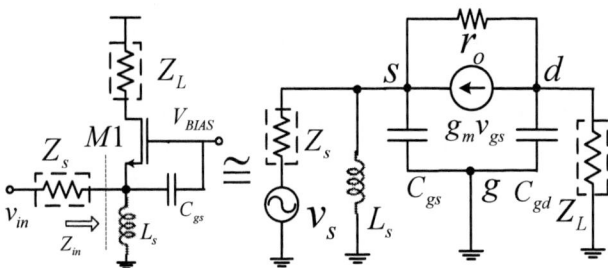

(a) Common-gate scheme and small-signal equivalent circuit

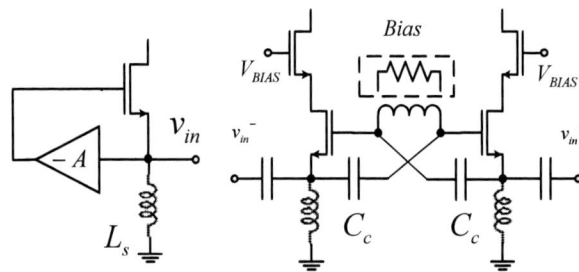

(b) Boost the equivalent trans-conductance to improve NF for CGLNA

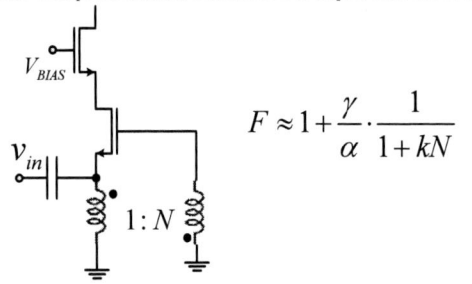

$$F \approx 1 + \frac{\gamma}{\alpha} \cdot \frac{1}{1 + kN}$$

(c) Using transformer to boost gate drive

Fig. 3. (a) Scheme of common-gate amplifier and its small-signal equivalent circuit. (b) gm-boosting using feedback. (c) Gm-boosting using transformer.

Fig. 3 (c) shows another way to boost gate drive. The NF is shown in Fig. 3 (c), where, k is the magnetic coupling coefficient. There is no current gain for CGLNA. The gain of common-gate LNA depends on the ratio of load to source impedances: Z_L / Z_s, in which Z_s is usually set by input matching (50Ω).

III. COMPARISON OF DIFFERENT ARCHITECTURES

These three kinds wideband LNA are basic architectures. Other new architectures are based on these three architectures to improve the performances (gain/NF/bandwidth) by using transformer feedback, load impedance shunt peaking, thermal noise cancelling, and inter-stage gain roll-off compensation. A FOM is proposed to characterize the overall performance of the broad-band LNA as follows:

$$FOM = \frac{|S21| \cdot BW[GHz]}{|F_{ave} - 1| \cdot Power[mW]}$$

(20)

The comparison among different LNAs in different CMOS technology by FOM is displayed in Table I. According to the analysis above, LNA architectures can be selected for different wideband system application by trading off the gain, bandwidth, NF.

TABLE I
FOM COMPARISON FOR DIFFERENT WIDEBAND LNA IN CMOS TECHNOLOGY

FOM	Shunt-feedback	LC-ladder	CGLNA	DA	Trans-former
0.18um	4.21[13]	8.6 [5]	2.75 [22]	0.98[24]	4.33 [8]
FOM*IIP3	1.33 [13]	1.72 [5]	16.6 [22]	N/A	0.34 [8]
0.13um	12 [9]	2.18 [12]	N/A	N/A	34.7 [11]
FOM*IIP3	4.67 [9]	1.81 [12]	N/A	N/A	5.5 [11]
90nm	3.3 [14]	N/A	N/A	N/A	N/A
FOM*IIP3	1.32 [14]	N/A	N/A	N/A	N/A

IV. CONCLUSION

The architectures of wideband LNA are analyzed through the gain, noise figure, and input impedance matching in this paper. Performance comparison among different LNA architectures is given by FOM. This paper gives the LNA design trend according to different UWB architecture application.

REFERENCES

[1] "High Rate Ultra Wideband PHY and MAC Standard," Standard ECMA 368, 2nd Edition, Dec. 2007.

[2] Ruey-Lue Wang et al, "A 0.18-um CMOS UWB Low Noise Amplifier for Full-Band (3.1-10.6GHz) Application", IEEE Asia Pacific Conf. on Circuits and Systems (APCCAS), pp. 363 – 366, Dec. 2006.

[3] R.-C. Liu, K.-L. Deng, and H. Wang, "A 0.6-22GHz broad-band CMOS distributed amplifier," in Dig. IEEE Radio Freq. Integrated Circuits Symposium, Jun. 2003, pp. 103-106.

[4] Yi-Jan Emery Chen, Yao-I. Huang, "Development of Integrated Broad-Band CMOS Low-Noise Amplifiers", IEEE Transactions on Circuit and Systems, Vol. 54, NO.10, pp. 2120-2127, Oct. 2007

[5] Yi-Jing Lin et al., "A 3.1-10.6 GHz Ultra-Wideband CMOS Low Noise Amplifier With Current-Reused Technique", IEEE Microwave and Wireless Components Letters, Vol. 17, NO.3, pp. 232-234, Mar. 2007.

[6] Yang Lu et al., "A Novel CMOS Low-Noise Amplifier Design for 3.1 to 10.6-GHz Ultra-Wide-Band Wireless Receivers", IEEE Transactions on Circuits and Systems, Vol. 53, NO.8, pp. 1683-1692, Aug. 2006.

[7] W. Zhuo, X. Li, S. Shekhar, S. H. K. Embabi, J. Pineda de Gyvez, D. J. Allstot, and E. Sanchez-Sinencio, "A Capacitor Cross-Coupled Common-Gate Low-Noise Amplifier", IEEE Transactions on Circuits and Systems-II: Express Briefs, Vol. 52, NO.12, Dec. 2005.

[8] Chang-Tsung Fu, Chien-Nan Kuo, "3~11-GHz CMOS UWB LNA Using Dual Feedback for Broadband Matching", IEEE Radio Frequency Integrated Circuits (RFIC) Symposium, 2006 Publication Date: 11-13 June 2006.

[9] R. Ramzan, et al., "A 1.4V 25mW Inductorless Wideband LNA in 0.13um CMOS," IEEE International Conference on Solid State Circuit, ISSCC 2007, pp. 424-613, Feb. 2007.

[10] T. H. Lee, The Design of CMOS Radio-Frequency Integrated Circuits, 2nd ed., Cambridge Press, 2004.

[11] M. T. reiha, J. R. Long, "A 1.2V Reactive-Feedback 3.1-10.6GHz Low-Noise Amplifier in 0.13um CMOS," IEEE J. Solid-State Circuits, Vol. 42, NO.5, pp. 1023-1033, May 2007.

[12] A. Bevilacqua, et al, "A fully integrated differential CMOS LNA for 3-5GHz UWB wireless receivers," IEEE Microw, and Comp. Lett., Vol. 16 NO.3, pp.134-136, Mar. 2006.

[13] J.-Y. Hu, Y.-L. Zhu, H. Wu, "An Ultrai-Wideband Resistive-Feedback Low-Noise Amplifier with Noise Cancellation in 0.18um Digital CMOS," IEEE Topical Meeting on Silicon Monolithic Integrated Circuits in RF Systems (SiRF), pp. 218-221, Jan. 2008.

[14] C.-S. Wang, C.-K. Wang, "A 90nm CMOS Low Noise Amplifier Using Noise Neutralizing for 3.1-10.6GHz UWB System," European Solid-State Circuits Conference (ESSCIRC), pp. 251-254, Sept.

[15] L. Belostotski, J. W. Haslett, "Sub-0.2 dB Noise Figure Wideband Room-Temperature CMOS LNA with Non-50Ω Signal-Source Impedance," IEEE J. Solid-State Circuit, Vol. 42, NO.11, Nov. 2007.

Impact of Power-Management Granularity on The Energy-Quality Trade-off for Soft And Hard Real-Time Applications

Aleksandar Milutinović
University of Twente,
The Netherlands
Email: a.milutinovic@utwente.nl

Kees Goossens
NXP Semiconductors & Delft University of Technology
The Netherlands,
Email: kees.goossens@nxp.com

Gerard J.M. Smit
University of Twente,
The Netherlands,
Email: g.j.m.smit@utwente.nl

Abstract—In this paper we introduce the concepts of *work* of tokens (e.g. video frames) in an application, and *slack* arising from variations in work. Slack is used for dynamic voltage and frequency scaling in combination with a conservative power-management policy that never misses deadlines, for hard real-time applications, and with a non-conservative policy for soft real-time applications. We evaluate both policies for a number of different *granularities* (frequency of activation of the power manager) on an MPEG4 application, on *energy and quality* (deadline misses).

We conclude that for soft real-time applications, there is a clear optimum in the energy, which depends on the work histogram of the application. The conservative policy has no deadline misses, and is only negligibly more expensive in terms of energy than the non-conservative policy. Finally, the granularity of both policies can be very coarse (128 frames) to reduce the power manager activation frequency, which has an insignificant energy cost.

I. INTRODUCTION AND SCOPE

Power management (here, energy minimisation) is imperative to increase the battery life time of nomadic devices such as mobile phones, but also for tethered devices such as set-top boxes to increase their life time e.g. through reduced thermal stress.

In this paper we perform an analytical study of slack (spare capacity) in a SOC, and how it can be used by several dynamic-voltage-and-frequency-based power-management policies. In addition, we vary the granularity (frequency) of power management. We consider the energy and quality (number of deadline misses) impact of the policies on soft and hard real-time applications, through an evaluation using an MPEG4 decoder mapped on an ARM processor.

In Section II, we introduce the applications of interest, work and slack, our energy model, and power management (policies). Section III introduces conservativeness of a policy (when it is safe to use for hard real-time applications) and its granularity. Section IV describes our experiments and their results, in particular the impact of the policies and their granularities on the energy-quality trade-off. After reviewing related work in Section V, we conclude in Section VI.

II. MODEL

A. Application model

In this paper we focus on power management of a single tile, consisting of a programmable *processor* with local memories and peripherals. Although our power management policies are compatible with multiple such tiles in a multi-processor SOC, we will not further consider inter-tile power management in the remainder. Each tile has its own frequency and voltage domain that can be set independently to a voltage-frequency *operating point* at run time. The benefits and costs of scaling are discussed below.

We consider soft and hard *real-time streaming applications*. In general, such applications operate on sequences of tokens that each

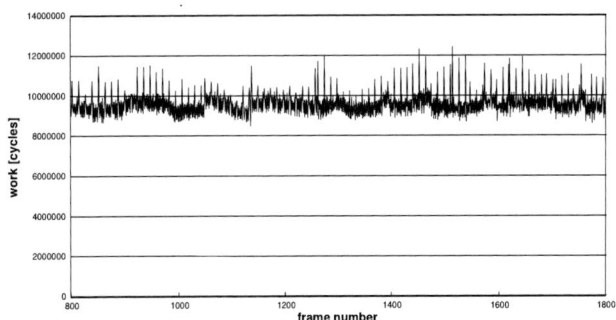

Fig. 1. Work per frame for part of an MPEG4 sequence.

have a deadline by which they should be produced. Hard real-time applications do not allow deadline misses (late production of tokens), whereas soft real-time applications allow a limited number of deadline misses, but at the cost of a quality degradation. In our case, tokens are compressed video frames, and the deadlines define when they should be displayed. The frame rate f_{FR} determines the regular spacing $T = 1/f_{FR}$ of deadlines in time. We assume that the input data and output space of the application are always available. In Section IV we comment on the buffer utilisation within the application.

B. Work and slack

The *work* w_i of a frame i is the number of processor cycles required to fetch, process, and store it. The *total work* of a sequence of frames is the sum of work of the individual frames. We assume that work depends only on the input token(s), and is independent of the operating point of the processor. This holds when the input and output tokens of task, as well as its instructions, are stored in the local memories of the tile [1].

Work for different input tokens may vary, e.g. the work for a frame depends on the complexity of its decoding, which strongly depends on whether it is an MPEG I or P frame. I frames require considerably more work than B and P frames, as we shall see later. The *worst-case work* of a sequence of frames is $wcw = Max_{j=0}^{\infty} w_j$. The time to finish the work of frame i at a frequency f_i is the *actual-case execution time* $acet_i = w_i/f_i$. Figure 1 shows work per frame of part of an MPEG4 sequency, further discussed in Section IV.

In order not to miss any deadline, f_i should be high enough. In fact, there are two different kinds of deadlines. A *relative deadline* requires that the actual execution time of a frame i is less than required by frame rate. With a regular execution, it must be less than the frame rate: $acet_i \leq T = 1/f_{FR}$. *Relative slack r* is the difference

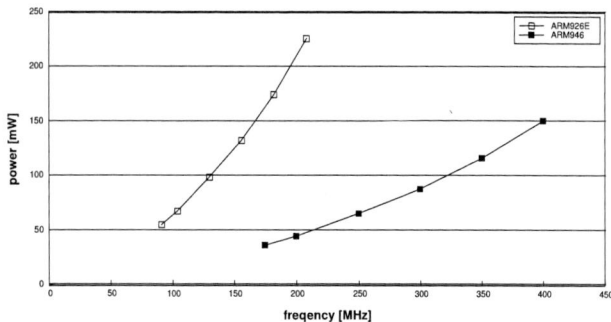

Fig. 2.   Power Model: a) ARM 926E, b) ARM 946.

of the worst-case and the actual-case execution time of frame i at the maximum operating frequency f_{Max}: $r_i = T - acet_i$. The *absolute deadline* of a frame f_i is the absolute time at which it must be produced (displayed). The *absolute slack* is defined correspondingly: $s_i = (i+1)T - \sum_{j=0}^{i} acet_j$.

When a deadline is not met, it is a *miss*. There are fewer relative than absolute misses because a single hard frame can cause several successive frames to miss their absolute deadlines, even though they do not miss their relative deadlines. We focus on absolute deadlines in the sequel, because they are important for the user (e.g. frame rate), and are harder to ensure.

C. Energy model

In common with many other power management strategies, we use slack to reduce the operating point (frequency and voltage) of the processor, and thus save energy. In this paper we assume that the process technology used is optimised to minimise leakage, and we can only affect the *dynamic energy*, which is dominant in SOCs.

Dynamic *power* is given by $P_{dyn} = \alpha C V^2 f = \alpha C V^2 w/t$, where α is the switching activity, C is the switched capacitance, and V and f define the voltage-frequency operating point. Alternatively, w is the work performed in time t. The *energy* spent is then $E_{dyn} = P_{dyn}t = \alpha C V^2 w$. To minimise energy, the voltage must be scaled to the lowest value supporting the frequency required to meet a deadline.

A processor can run at a minimum (maximum) frequency f_{Min} (f_{Max}), requiring a minimum (maximum) voltage $v(f_{Min})$ ($v(f_{Max})$). The power model $P(f)$ used in this paper, related to the function $v(f)$ is computed as follows. Our starting point is frequency-power measurements of an ARM926E board, which are shown in Figure 2. The ARM926E is not powerful enough to execute our application in real time. The ARM946 is, but the public data on its power characteristics are insufficient. For this reason, we correlated the maximum operating point (peak power) of the ARM926E with the maximum operating point of an ARM946 as given in [2].

D. Dynamic power management and policies

We assume that the SOC under consideration has been dimensioned at design time to minimise the energy consumption, and focus on dynamic power management. Dynamic voltage and frequency scaling (DVFS) power management defines voltage-frequency operating points at run time according to a *policy* to trade processor performance for energy. A *transition* occurs whenever the operating point is changed, to increase the performance and energy, or decrease them, as required.

Dynamic power management, and DVFS in particular, has several costs, in terms of area of the DVFS infrastructure, and reduced processor performance (assuming the policy is implemented in software).

Both result in a power cost. In some implementations, processors must be idle during transitions, again lowering the processor performance. Additionally, it takes time to change the voltage of a processor due to its capacitance, which means that transitions are not instantaneous, resulting a minor loss of performance or energy. Our experiments take these costs into account, as described in Section IV.

III. POLICY CONSERVATIVENESS AND GRANULARITY

A policy is *conservative* if it does not introduce any deadline misses (i.e. lowers the quality of the result) compared to operating at f_{Max}, and non-conservative otherwise. Conservative policies are required for hard real-time applications, whereas soft real-time applications can tolerate occasional deadline misses and could use non-conservative policies. We use DVFS with policies based on run-time observation of already available slack or prediction of future slack to reduce energy. The *proven slack* of a frame i is the cumulative slack of the frames before it, i.e. the absolute slack of frame $i-1$. Proven slack can be detected at run time. There may be future slack that is unproven at the start of frame i's work. We will use a (hypothetical) *perfect predictor* to compute future unproven slack for frame i before it is executed. This is a useful baseline for later comparisons because no real predictors can do any better.

We define the *granularity* of a policy as the shortest time between successive transitions. The aim of this paper is to investigate the impact of the granularity of the policy on the energy reduction, taking into account the transition overhead. We will scale the granularity from 1 frame to the length of the entire sequence of frames. Given a certain granularity N we use two policies: *perfect predictor* and *proven slack*. The former accurately predicts the cumulative amount of work of the next N frames and scales the performance of the processor to the average frequency for those frames ($f_{avg_i} = (\Sigma_{j=0}^{N-1} w_{i*N+j})/(NT)$ for group i). In other words, the last of the N frames will never miss its deadline, but preceding frames might. For $N > 1$ this policy is not conservative, therefore allowed for soft real-time applications, but not for hard real-time applications. The proven-slack policy assumes that the next N frames all require the worst-case work, but uses all the proven slack of previous group to reduce the frequency of the processor (but never scaling below f_{Min}): $f_{max_i} = (N Max_{j=0}^{\infty} w_j)/(NT + s_{i-1})$ for group i. The first N frames have no proven slack, and hence run at the frequency required for the worst-case of the entire sequence. Hence, all frames will meet their deadline, and this policy is always conservative, and suitable for both soft and hard real-time applications.

For $N = \infty$ (i.e. the length of the input sequence), the processor operates at f_{avg} with the perfect-predictor policy, and at f_{max} using the proven-slack policy. The former is the traditional minimum energy operating point, when deadlines are ignored (running a best-effort application at its average requirement). The latter is our baseline (no misses, no power management, no overhead) for later comparisons. The final point of interest is the lowest frequency f_{min}, the maximal frequency at which all deadlines are missed. When varying the granularity N for both policies, the total energy will vary, as will the number of misses. This is an instance of an *energy-quality* or *cost-performance* plot. In the results section we shall present this plot, and draw some conclusions on the relative performance of the policies.

When $N > 1$, frames may be produced early, in which case they are stored in memory. We do not increase the energy for the policy because only the time at which data (frames) are produced changes, but not their number (or size). Hence the energy consumed by the interconnect between processor and memories is unchanged.

978-1-4244-2541-9/08 $25.00 © 2008 IEEE

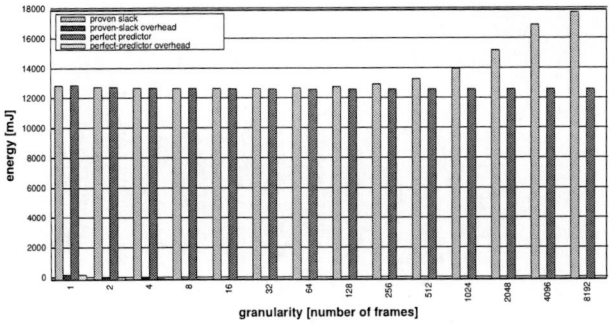

Fig. 3. Application & power-management energies, for different granularities.

Regarding the impact on memories, we assume that data is stored in sufficiently large buffers in either SRAM or DRAM. For DRAM we assume that unused banks are not switched off. As a result, the buffer filling (due to early or late data production) does not change the energy consumed by the memory.

IV. EXPERIMENTAL RESULTS

Our application is an MPEG4 decoder running on an ARM946 running at 86 MHz. It decodes an input stream of 207 seconds, with I and P frames, at 25 frames per second, and a resolution of 176x144 pixels. The measured number of cycles per frame were measured, and define the work per frame. On the basis of this, we analytically evaluate our two policies for a number of different granularities. A transition results in a 20 μsec inactive period, and a 1 msec execution of the power manager, with the associated energy cost.

Figure 3 shows the energy consumed by the application and the power manager for different granularities, for both policies. The rightmost bar corresponds to the entire sequence, i.e. f_{avg} for the perfect-predictor policy and f_{max} for the proven-slack policy. The energy savings w.r.t. operating at f_{max} are around 30% for 1-128 frames, at a cost of 2% for the power manager. Above 128 frames the proven-slack policy uses linearly more energy. The energy used by the perfect-predictor policy decreases lightly, but at the cost of increasingly missing deadlines, as we shall see below.

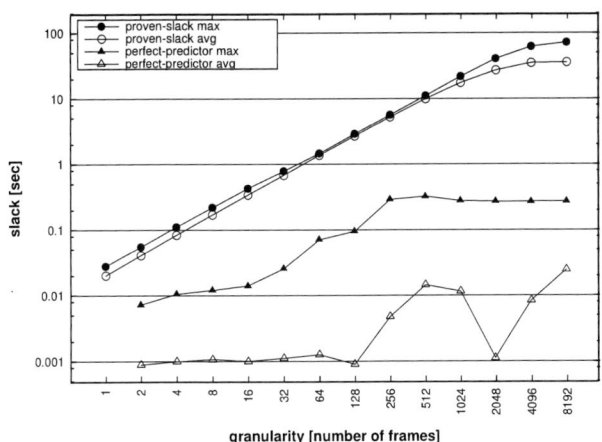

Fig. 4. Remained slack versus granularity.

Figure 4 shows the average slack (($\sum_{i=0}^{S-1} s_i)/S$, for a sequence of S frames) and worst-case slack ($Max_{i=0}^{S-1} s_i$) for different granularities, for both policies. For the proven-slack policy the average slack saturates at the difference between worst-case work and average-case

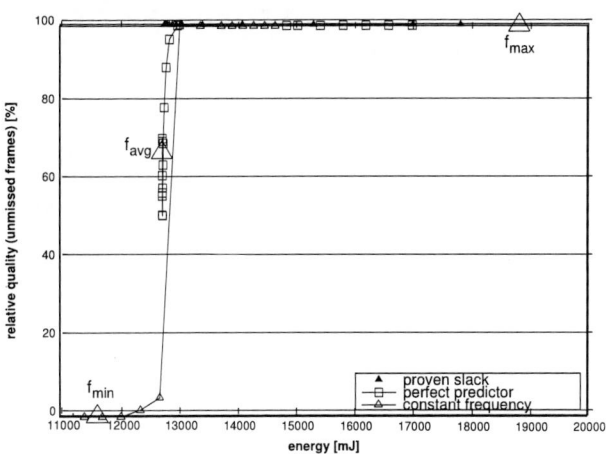

Fig. 5. Energy-quality trade-off for different policies and granularities.

work, while maximum slack keeps increasing. This indicates that the policy cannot always exploit the accumulated slack (e.g. because the processor cannot be scaled below f_{Min}), which therefore reaches large values. The perfect-predictor policy, on the other hand, uses slack effectively because it never accumulates to large values.

However, the perfect-predictor policy obtains its lower energy at the cost of deadline misses. Figure 5 shows the energy versus the relative quality (number of frames that are produced on time). Both policies are plotted for various granularities. The three reference points are also indicated: the baseline f_{max} (always running at the frequency of the worst-case frame, i.e. no, misses, no power management, no overhead), f_{min} (the maximal frequency at which all deadlines are missed), f_{avg} (the frequency for minimum energy, when only the global deadline is met). The *constant-frequency* policy, which has no power management (overhead), connects these three points.

Note that the proven-slack policy is indeed conservative, because it provides 100% quality (no misses) for all energies (operating points). The perfect predictor, however, starts missing deadlines around 13000 mJ, and drops to 0% quality (f_{min}) at 11600 mJ. The transition from good to bad takes place in a very narrow band: a 95% quality improvement costs only 3% additional energy. This is positive, because it is clear where the optimum is (13000 mJ). In fact, this results from the distribution of work, as shown in the work histogram in Figure 6. Many frames can be processed in the range of 240-250 MHz, with a relatively small number of much larger frames. Hence the operating point for soft real-time applications can be close to the transition, but hard real-time applications must have a much higher operating frequency, when using the perfect-predictor policy.

For this reason, we compare the perfect-predictor and proven-slack policies in the transition range, shown in more detail in Figure 7. The 1-frame proven-slack policy is conservative, i.e. offers 100% quality, and uses only 0.3% more energy than the perfect predictor. Furthermore, increasing the granularity from 1 to 128 increases the energy of the proven-slack policy by only 2%. This is a positive result because it allows the power manager to run very infrequently, lowering its overhead.

Figure 8 shows that the buffer filling increases linearly with the granularity. As argued in Section III, however, storing data early does not cost any extra energy. If buffers are smaller than shown or data are always available at the input, then the application will stall, and restart when data arrive, like the race-to-idle policy. The result is a

978-1-4244-2541-9/08 $25.00 © 2008 IEEE

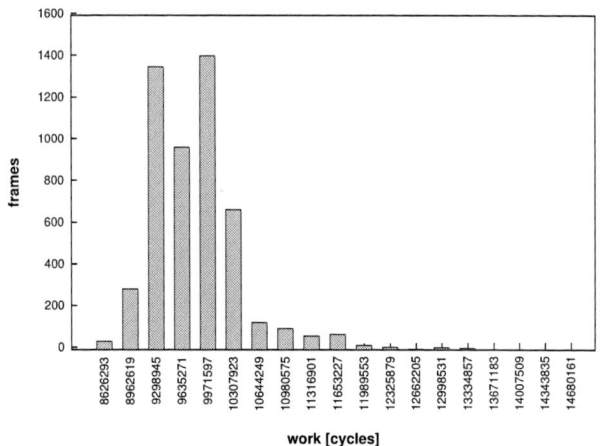

Fig. 6. Work histogram.

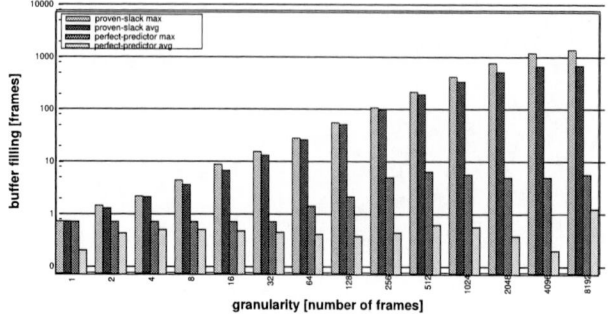

Fig. 8. Buffer fillings for different policies and granularities.

other hand, may occasionally miss a deadline, which is allowed for soft real-time applications. Both policies have been evaluated on an MPEG4 application, for a number of different *granularities*, i.e. frequency of operating point changes (power manager activations).

From the experiments we draw the following conclusions. 1) A long tail in the work distribution results in a steep quality improvement: from almost 0% to almost 100% at an additional energy cost of only 3%. This means that for soft real-time applications, there is a *clear optimum in the energy-quality trade-off*. The work distribution of many applications exhibits such a long tail, which is why hard real-time guarantees are usually considered expensive to offer. 2) The proven-slack policy offers 100% quality at only 0.3% more energy than the perfect-predictor policy, which is theoretical upper bound and hard to achieve in practice. Hence, *conservative power management for hard real-time guarantees, is only negligibly more expensive* than non-conservative power management for soft real-time guarantees. 3) The energy of the policies increases by only 2% when increasing the granularity to 128 frames. Hence, the *power manager can run very infrequently, at an insignificant energy cost*.

ACKNOWLEDGEMENTS

We would like to thank Albert Molderink of the University of Twente for the MPEG4 traces on which this work is based.

REFERENCES

[1] M. Bekooij, O. Moreira, P. Poplavko, B. Mesman, M. Pastrnak, and J. van Meerbergen, "Predictable embedded multiprocessor system design," *Lecture notes in computer science*, vol. 3199, pp. 77–91, 2004.

[2] [Online]. Available: http://www.arm.com/products/CPUs/ARM946E-S.html

[3] M. Meijer, J. Pineda de Gyvez, and R. Otten, "On-chip digital power supply control for system-on-chip applications," in *ISLPED '05: Proceedings of the 2005 international symposium on Low power electronics and design*. New York: ACM Press, 2005, pp. 311–314.

[4] A. Azevedo, I. Issenin, R. Cornea, R. Gupta, N. Dutt, A. Veidenbaum, and A. Nicolau, "Profile-based dynamic voltage scheduling using program checkpoints," *Design, Automation and Test in Europe Conference and Exhibition, 2002. Proceedings*, pp. 168–175, 2002.

[5] N. AbouGhazaleh, D. Mossé, B. Childers, and R. Melhem, "Collaborative operating system and compiler power management for real-time applications," *ACM Transactions on Embedded Computing Systems (TECS)*, vol. 5, no. 1, pp. 82–115, 2006.

[6] N. AbouGhazaleh, D. Mosse, B. Childers, and R. Melhem, "Toward the placement of power management points in real-time applications," *Compilers and operating systems for low power table of contents*, pp. 37–52, 2003.

[7] K. Choi, K. Dantu, W. Cheng, and M. Pedram, "Frame-based dynamic voltage and frequency scaling for a MPEG decoder," *Digest of technical papers- IEEE/ACM International Conference on Computer-Aided Design*, pp. 732–737, 2002.

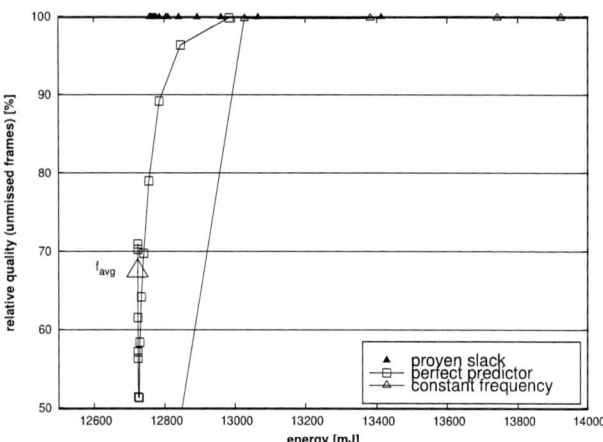

Fig. 7. Energy-quality trade-off for different policies and granularities.

conservative policy, although less energy efficient.

V. RELATED WORK

The speed of DVFS infrastructure is increasing [3], enabling power management at very fine granularity. This was the motivation for the study that we present. Azavedo [4] uses the compiler to place the checkpoints in program code at the boundaries of basic blocks, which represents fine granularity solution that uses variable granularity but in a limited range. AbouGhazaleh [5] presents the collaboration between compiler and operating system and by inserting instrumentation code into the program code to achieve to vary the granularity. The same authors propose theoretical solution for choosing the optimal granularity in [6]. Choi [7] presents DVFS technique for an MPEG decoder with sub-frame granularity by differentiating between invariable and variable parts of a decoder.

VI. CONCLUSIONS

In this paper we introduce the concepts of *work* of tokens in an application, and the difference between the worst case and actual case work (*slack*). We use slack for dynamic voltage and frequency scaling in combination with two policies: the perfect predictor and proven slack. The *proven-slack policy is conservative*, which means that it never misses deadlines (late completion of work), as required for hard real-time applications. The *perfect-predictor* policy, on the

978-1-4244-2541-9/08 $25.00 © 2008 IEEE

AUTHOR INDEX

Ahmad, Balal 155

Ahmadinia, Ali 155

Ahn, Heesun 151

Airoldi, Roberto 85

Alle, Mythri 51

Aminzadeh, Hamed...................................28

Amiri, S. Hamid 147

Arjomand, Mohammad 147

Arpinen, Tero..107

Arslan, Tughrul.......................................155

Auguin, Michel 69

Ben-Asher, Yosi......................................119

Benmohammed, Mohammed 69

Berg, Heikki..143

Bhattacharyya, S. S.....................................6

Bilavarn, Sébastien 69

Boukhechem, Sami....................................55

Bourennane, El-Bay...................................55

Boutekkouk, Fateh 69

Brox, M. 113

Brunelli, Claudio 85, 143

Carbognani, F..40

Carlsson, Mats..129

Carvalho, Ewerson 14

Choi, Jongchan 151

Dimitroulakos, Grigoris 139

Elrabaa, Muhammad E. S.........................121

Felber, N. 40

Fichtner, W. 40

Garga, Ganesh 51

Garzia, Fabio...85

Georgiopoulos, Stavros 139

Gersnoviez, A. 113

Giliberto, Carmelo......................................85

Goossens, Kees 163

Goutis, Costas E. 139

Guzma, V. 6

Hamalainen, Timo D. 44, 107, 133

Hannikainen, Marko.......................107, 133

Hedenäs, Charlotta................................129

Henzen, L. 40

Herkersdorf, Andreas 89

Hessabi, S. 65

Heusala, H. 103

Holma, Kalle 107

Huang, Zhengfeng 24

Hurskainen, Heikki 75

Jain, Jawahar 18

Jalier, Camille...61

Jamadagni, H. S. 51

Jonsson, Fredrik 129, 159

Kariniemi, Heikki 95

Kellomaki, P. 6

Kim, Chulwoo 151

Kim, Kynnyun 151

Kim, Taeyoon 151

Kinniment, David 125

Komulainen, J. 103

Kooti, H. 65

Kulmala, Ari 44

Lattard, Didier...61

Lee, Sangyong 151

Lee, Sungchul..151

Liang, Huaguo 24

Lotfi, Reza 28

Lücking, Ulf 143

Ma, Ning...10

Mafinezhad, Khalil 28

Meitinger, Michael 89

Milutinovic, Aleksandar 163

Min, Kyongwon 151

Minas, Nikolaos 125

AUTHOR INDEX

Mirza-Aghatabar, M. 65

Mohammadi, Siamak 81

Montijano, M. 113

Moraes, Fernando 14

Moreno, C. 113

Nandy, S. K. 51

Nikunen, K. 103

Nurmi, Jari 75, 85, 95

Ohlendorf, Rainer 89

Orsila, Heikki 133

Ortiz, M. 113

Pang, Zhibo 10

Park, Wonki 151

Quiles, F. 113

Raasakka, Jussi 75

Rong, Liang 129

Rotem, Nadav 119

Russell, Gordon 125

Safari, Saeed 81

Salminen, Erno 44, 107, 133

Sarbazi-Azad, Hamid 147

Sassatelli, Gilles 61

Schweiger, Kurt 32

Smit, Gerard J. M. 163

Stergiou, Stergios 18

Takala, J. 6

Tavakkol, A. 65

Tenhunen, Hannu 1, 10

Uhrmann, Heimo 36

Valinataj, Mojtaba 81

Varadarajan, Keshavan 51

Wang, Peng 1, 159

Wild, Thomas 89

Yakovlev, Alex 125

Yi, Maoxiang 24

Zheng, Li-Rong 1, 10, 129, 159

Zhou, Dian 1, 159

Zimmermann, Horst 32, 36

CURRAN ASSOCIATES INC.
proceedings
.com

9781424425419